Contents

AQA
Advanced Maths

Level 3 CERTIFICATE

Mathematical Studies

Stan Dolan
June Haighton

Powered by MyMaths.co.uk

OXFORD

OXFORD
UNIVERSITY PRESS

Great Clarendon Street, Oxford, OX2 6DP, United Kingdom

Oxford University Press is a department of the University of Oxford. It furthers the University's objective of excellence in research, scholarship, and education by publishing worldwide. Oxford is a registered trade mark of Oxford University Press in the UK and in certain other countries

© Oxford University Press 2016

The moral rights of the authors have been asserted

First published in 2016

British Library Cataloguing in Publication Data

Data available

978-0-19-836593-8

10 9 8 7 6 5 4 3

Paper used in the production of this book is a natural, recyclable product made from wood grown in sustainable forests. The manufacturing process conforms to the environmental regulations of the country of origin.

Printed in Great Britain by Bell and Bain Ltd, Glasgow

MIX
Paper from responsible sources
FSC
www.fsc.org   FSC® C007785

We have made every effort to trace and contact all copyright holders before publication, but if notified of any errors or omissions, the publisher will be happy to rectify these at the earliest opportunity.

Links to third party websites are provided by Oxford in good faith and for information only. Oxford disclaims any responsibility for the materials contained in any third party website referenced in this work.

Approval message from AQA

This textbook has been approved by AQA for use with our qualification. This means that we have checked that it broadly covers the specification and we are satisfied with the overall quality. Full details of our approval process can be found on our website.

We approve textbooks because we know how important it is for teachers and students to have the right resources to support their teaching and learning. However, the publisher is ultimately responsible for the editorial control and quality of this book.

Please note that when teaching the AQA Level 3 Certificate Mathematical Studies course, you must refer to AQA's specification as your definitive source of information. While this book has been written to match the specification, it cannot provide complete coverage of every aspect of the course.

A wide range of other useful resources can be found on the relevant subject pages of our website:www.aqa.org.uk.

Please note that the Practice Questions in this book allow students a genuine attempt at practising exam skills, but they are not intended to replicate examination papers.

Copyright Acknowledgements

AQA examination questions are reproduced by permission of AQA Education (AQA). AQA accepts no responsibility for the accuracy or method of working in the answers given.

p13, 27, 32, 33, 36, 38, 39, 40, 41, 47, 72, 146, 155, 156, 157, 158, 160, 164, 165, 166, 167, 169, 173, 238, contains public sector information licensed under the Open Government Licence v3.0.

The author and publisher would like to thank the following for permission to reproduce material:

p32, www.virginmoneylondonmarathon.com; p43, www.bfi.org.uk, www.rentrak.com; p44, Copyright © 2015, reused with the permission of the Health and Social Care Information Centre, all rights reserved; p58, www.andersred.blogspot.co.uk; p77, Nadeem Walayat – *The Market Oracle*; p77, www.spectator.co.uk; p104, *The Times*/News Syndication; p109, Extract from Corinne Moss-Racussin et al, 'Science faculty's subtle gender biases favor male students', *Proceedings of the National Academy of Sciences*, Vol 109 no 41, 21 August 2012, reprinted by permission from National Academy of Sciences; p116, Figure from J. R. Petit, J. Jouzel, D. Raynaud, N. I. Barkov, J. M. Barnola et al., 'Climate and atmospheric history of the past 420,000 years from the Vostok ice core, Antarctica', *Nature*, Vol 399 no 6735, 3 June 1999, Copyright © 2016, reproduced by permission from Macmillan Publishers Ltd; p154, 170, Contains public domain material from websites or documents of the *CIA World Factbook*; p230, © European Agency for Safety and Health at Work; p231, www.which.co.uk; p313, Policy Exchange, 'Crossing the line', 2015, www.policyexchange.org.uk; p313, www.ascl.org.uk

The publisher would like to thank the following for permission to use their photographs:

COVER: PanuRuangjan/Thinkstock; p2: mathagraphics/Shutterstock; Teddy Leung/Shutterstock; Alen Hunjet/Shutterstock; p4: Rido/Shutterstock; p6: DAVID PARKER/SCIENCE PHOTO LIBRARY; p8–9: peppi18/iStockphoto; p12: Aleksandr Markin/Shutterstock; p14: Christopher Elwell/Shutterstock; p20: sportsphotographer.eu; p27: david muscroft/Shutterstock; p32: Bikeworldtravel/Shutterstock; p46: Syldavia/iStock; p47: Stripped Pixel/Shutterstock; p48–49: gpointstudio/Shutterstock; p51: sanjagrujic/Shutterstock; p56: ShaunWilkinson/Shutterstock; p58: John David Bigl III/Shutterstock; p64: bikeriderlondon/Shutterstock; p70: Gumpanat/Shutterstock; p72: Steve Allen/Shutterstock; p78–79: Triff/Shutterstock; p80: Bonnie Fink/Shutterstock; p82: NikoNomad/Shutterstock; p85: Vladimir Grigorev/Dreamstime.com; p87: Ingram Publishing/thinkstock; p92: naran/Thinkstock; p93: magnetix/Shutterstock; BlueRingMedia/Shutterstock; p94: toysf400/Shutterstock; p99: Dmitry Kalinovsky/Shutterstock; p96: Milagli/Shutterstock; p98: Gail Johnson/Shutterstock; p100: Cloud7Days/Shutterstock; Volodymyr Goinyk/Shutterstock; p101: leonello/Shutterstock; p102–103: VLADGRIN/Shutterstock; p104: Stephen Coburn/Shutterstock; p106: Eric Isselee/Shutterstock; p107: Monkey Business Images/Shutterstock; p108: Bananaboy/Shutterstock; p111: PhotoSmart/Shutterstock; p113: northallertonman/Shutterstock; p114: mTaira/Shutterstock; p119: Syda Productions/Shutterstock; p120–121: Kim Reinick/Shutterstock; p122: Mikhail Bakunovich/Shutterstock; p125: Samot/Shutterstock; p129: Gelpi JM/Shutterstock; p134: Heritage Image Partnership Ltd/Alamy Stock Photo; Georgios Kollidas/Shutterstock; Nicku/Shutterstock; p135: WHITE RABBIT83/Shutterstock; p136–137: Rido/Shutterstock; p138: sezer66/Shutterstock; p140: photowind/Shutterstock; p142: outdoorsman/Shutterstock; p146: branislavpudar/Shutterstock; p148–149: potowizard/Shutterstock; p154: Olinchuk/Shutterstock; p156: Arne Bramsen/Shutterstock; p161: JOHN READER/SCIENCE PHOTO LIBRARY; p162: SCIENCE PHOTO LIBRARY; p163: potowizard/Shutterstock; p166: Mastofa Mosharrof Hosen/Shutterstock; p170: © Roger Cracknell 01/classic/Alamy Stock Photo; p171: Nattika/Shutterstock; p172: wavebreakmedia/Shutterstock; sippakorn/Shutterstock; p174–175: EvrenKalinbacak/Shutterstock; p176: OliverSved/Shutterstock; p181: Joseph Sohm/Shutterstock; p189: OliverSved/Shutterstock; p190: AlexanderJE/Shutterstock; p192: Paul Barnwell/Shutterstock; p200: © Lanmas/Alamy Stock Photo; p201: Poznyakov/Shutterstock; p202–203: PaulPaladin/Shutterstock; p207: Brendan Howard/Shutterstock; p212: wavebreakmedia/Shutterstock; p214: Sonulkaster/Shutterstock; p215: sakkmesterke/Shutterstock; p216: Fotokostic/Shutterstock; p218: J-L CHARMET/SCIENCE PHOTO LIBRARY; p222: Goran Bogicevic; p223: hxdbzxy/Shutterstock; p224–225: ssuaphotos/Shutterstock; p226: 2xSamara.com/Shutterstock; p230: Alex Caparros/Getty; p232: Blend Images/Shutterstock; p238: Intellistudies/Shutterstock; p239: Baloncici/Shutterstock; p240–241: szefei/Shutterstock; p244: Tom Gowanlock/Shutterstock; p255: bikeriderlondon/Shutterstock; p256: dvoevnore/Shutterstock; p260: stocker1970/Shutterstock; p262: PRILL/Shutterstock; Eric Isselee/Shutterstock; Fotos593/Shutterstock; p264–265: 3Dsculptor/Shutterstock; p270: Npeter/Shutterstock; Richie Ji/Shutterstock; sss615/Shutterstock; p273: Zygotehaasnobrain/Shutterstock; p276: Paul S. Wolf/Shutterstock; p278: Brian A Jackson/Shutterstock; Tim Large/Shutterstock; p280: 3Dsculptor/Shutterstock; p286: BsWei/Shutterstock; wavebreakmedia/Shutterstock; p288–289: AsiaTravel/Shutterstock; p290: Tinydevil/Shutterstock; p293: hakandogu/Shutterstock; p294: Andrey Pavlov/Shutterstock; p301: reptiles4all/Shutterstock; p304: Daniel Prudek/Shutterstock; p305: IrinaK/Shutterstock; p307: Crystal Eye Studio/Shutterstock; p308: Maridav/Shutterstock

There are instances where we have been unable to trace or contact the copyright holder. If notified the publisher will be pleased to rectify any errors or omissions at the earliest opportunity.

About this book

This book is about **applying** mathematics to problems and has been specially written by two experienced teachers with examining experience.

A broad variety of applications will be covered as students work through this book and hopefully all students will find much of interest. Many sections of the book should be of general interest. This is perhaps especially true of the chapters on personal finance and critical analysis. However, above all, it is a textbook designed to be used by students on the AQA Level 3 Certificate Mathematical Studies course and the chapters are conveniently numbered so that Chapter X corresponds to 3.X of the AQA specification.

Some chapters such as Chapter 1 (Data analysis) use methods students will have met before and the work will lead on smoothly from GCSE level to more challenging contexts. Other chapters such as Chapter 3 (Modelling and estimation) will introduce ways of tackling problems that students are unlikely to have used before.

 Throughout the book, four-digit MyMaths codes are provided linking directly to related lessons on the MyMaths website. This enables a student to work independently and revise any GCSE topics as necessary.

Each chapter starts with a number of questions motivating the study of the relevant area of mathematics.

Activity Each section then typically introduces a mathematical topic, using activities to ensure that reading the explanations does not become too passive.

Key point

Example Key points and worked examples should enable students to tackle the exercises with growing confidence. The emphasis throughout the book is on genuine applications of mathematics rather than abstract mathematical ideas.

At the end of each chapter there is a consolidation exercise providing mixed questions on the chapter's content. These questions are mainly of examination style and standard.

The book also includes investigations and practice questions. Additional practice questions, mark schemes and teacher notes can be found online at **www.oxfordsecondary.co.uk**.

At the end of the book there is a full set of answers to all the exercises and the practice questions. Answers to the activities and notes on the various investigations can be found online.

The authors are indebted to staff at AQA and OUP for their assistance and in particular to Katherine Bird for her careful work and suggestions. Thanks are also due to AQA Education (AQA) for kindly allowing the use of questions from past Use of Mathematics papers.

Stan Dolan and June Haighton 2016

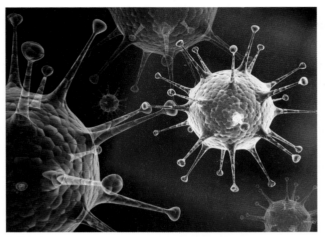

A flu virus

Simple car finance – a representative example

Loan amount £7500	Cost of credit £2491.14
Representative APR 15.95%	Loan term 48 months
Deposit £0	Monthly repayments £208.15
Rate of interest 15.95% pa	Total amount £9991.14

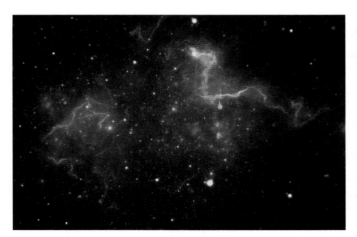

A deep space star field with nebula

Is it worth having a flu vaccine?

This year's seasonal flu vaccine is barely able to protect vulnerable people.

Only 3 out of every 100 vaccinated people are protected.

Public Health England, Feb 2015

It's crucial that these results do not discourage people in at-risk groups from having flu vaccination now, or in the future.

Public Health England, Feb 2015

$$C = \sum_{k=1}^{m} \frac{A_k}{(1 + i)^{t_k}}$$

To give consumers a way of comparing different deals, a figure called the Annual Percentage Rate (APR) is used. This formula for i, the APR expressed as a decimal, will be explained in Chapter 2.

Mathematics is the language in which God has written the universe.

Galileo Galilei

$$f = C\left(\frac{1}{b^2} - \frac{1}{a^2}\right)$$

Johann Balmer

This connects the structure of an individual atom to the analysis of light from immensely large and immensely distant stars and galaxies!

As you work through this book you will constantly be involved in mathematical modelling. It will make your studies much easier if you realise that you have already had considerable experience of mathematical modelling!

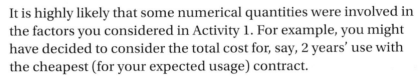

Activity 1

★ Before reading on, make a list of all the factors that you will consider when choosing your next mobile phone and tariff.

★ How will you decide?

It is highly likely that some numerical quantities were involved in the factors you considered in Activity 1. For example, you might have decided to consider the total cost for, say, 2 years' use with the cheapest (for your expected usage) contract.

Considerations other than total cost, such as what phones your friends have, may play a part in your eventual decision. However, total cost represents an important factor that most people would take into account and working out this cost is an example of an important part of mathematical modelling.

The following is a simplification of the process followed by one person when choosing a mobile phone and tariff.

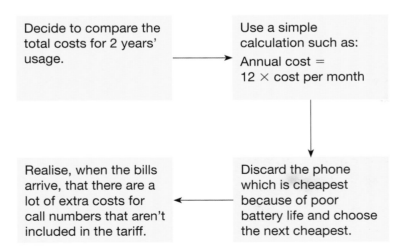

This particular example of mathematical modelling is very simple but, nevertheless, involves all the stages of the modelling cycle that will be referred to throughout this book.

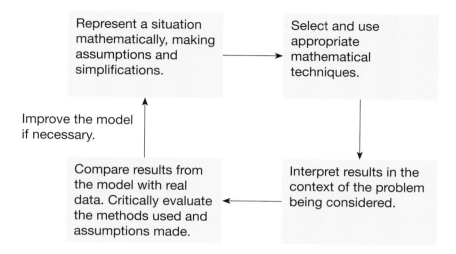

Making a good choice of mobile phone might need only elementary mathematics but it does require skill in making sensible assumptions and in interpreting your results. You already know many of the mathematical techniques you need to be able to tackle the problems in this book. As you apply these techniques in various contexts you should become

- more confident in using the techniques themselves
- more aware of what assumptions have been made in order to be able to apply these techniques
- more skilled in recognising both the implications and limitations of any results.

You should now appreciate that the stages of

- making and simplifying assumptions
- interpreting results
- critically evaluating a model

are as vital to good mathematical modelling as the stage of actually applying mathematical techniques.

Activity 2

★ Develop a good mathematical model for selecting a mobile phone (or another purchase of your own choice).

★ You could use a spreadsheet to perform any calculations.

The first stage of the modelling cycle involves making assumptions and simplifications. Since you will be relating mathematics to real applications, you will need to use data relevant to these applications. Chapter 1 of this book will consider important aspects of the collection and analysis of such data. However, in Chapter 3 you will learn how to deal with cases where not all the data are available.

Consider the following two very different contexts.

- Knowing, at least roughly, how many fish there are in the ocean is highly relevant to important questions concerning the sustainability of various types of fishing.
- Having a good idea of how many potential customers a new business would attract is essential to deciding whether or not to go ahead with the business plans.

Both of these contexts involve the estimation of quantities that seem difficult or even impossible to determine.

The Nobel Laureate physicist, Enrico Fermi, was well-known for his ability to successfully make estimates of this type and there are a number of techniques that can be used when making these, **Fermi Estimates**.

In general, when making Fermi Estimates you will carry out these important activities:

- estimating quantities
- numerical reasoning
- logical thinking
- communicating your reasoning and results.

The search for extra-terrestrial intelligence, SETI, provides some good examples of this form of mathematical modelling.

The radio-astronomer Frank Drake was one of the founders of SETI. Drake was interested in estimating the number, N, of civilisations in our galaxy transmitting radio signals that we could detect. A simplified version of the Drake equation for this number is

$$N = TShl$$

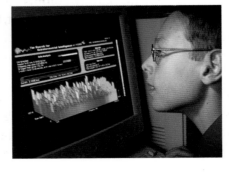

SETI@home is a computing project enabling you to use your home computer to analyse radio signals from space.

- T is the length of time in years for which a technological civilisation sends radio signals into space
- S is the number of stars formed in our galaxy each year

- h is the proportion of stars that have a planet in the habitable zone where life could occur and develop
- l is the proportion of planets in habitable zones where intelligent life develops and creates a technological civilisation.

The Drake equation is a *mathematical model* for the number of civilisations. However, Drake still had to estimate the quantities T, S, h and l.

As our technology has improved, we have increasingly good evidence for estimates of S and h. Recent estimates are $S = 7$ (Drake used $S = 1$) and $h = 0.2$.

It is the value of l which is the most highly disputed. Some would argue that the development of intelligent life requires a series of chance events of combined probability virtually zero. The contrary view that l is close to 1 is based upon the fact that life appeared to develop on Earth almost as soon as conditions became favourable. Then, once life has started, evolutionary pressure may make the development of intelligence virtually certain.

The transmission of radio signals as a commercial proposition was developed around the start of the 20th century. So a value of greater than 100 for T is not unreasonable.

If we choose $l = 0.1$ and $T = 500$, then our estimate for N would be

$$N = TShl \approx 70.$$

So, based upon these assumptions, in the entire galaxy there may be a mere 70 planets transmitting radio signals. You can see that it would therefore be very difficult to detect radio signals from extra-terrestrial civilisations, especially since our receivers would need to be tuned to the appropriate frequency.

Where are they?

Advanced civilisations may well have an average lifetime far in excess of the 500 years used in the calculation given above. However, if that were the case, then the difficulty known as the Fermi paradox would need to be explained.

Q Use the internet to find out more about the Fermi paradox.

Well before Drake had formulated his equation, Fermi had estimated how long it would take humans to colonise habitable planets throughout the galaxy. His conclusion was that, using only the forms of interstellar travel which were nearly within the capabilities of human technology, this colonisation would take some 10 million years.

In human terms this might seem a very long time but the lifetime of the Universe is measured in billions of years and Fermi had concluded that there should have been many civilisations on planets circling stars much older than ours that would already have colonised our galaxy.

1 Analysis of data

Which team is better at scoring goals?

What is the typical price for a Premiership season ticket?

What do football fans at the match think of the new facilities?

In this chapter you will learn about collecting, analysing and interpreting data. You will also use statistical diagrams such as stem-and-leaf diagrams, cumulative frequency graphs, box and whisker plots and histograms to illustrate datasets. These methods will help you to solve problems and draw conclusions.

How many people usually come to the match?

How does this vary?

Are Championship football clubs spending more than they earn?

Who is the most consistent goal scorer?

The worked examples and questions in this chapter use these skills from GCSE mathematics, but you may find other ways of tackling them.

You should know how to

- interpret and use the following terms:
  primary data, secondary data, discrete data and continuous data ⊕ 1248
- construct and interpret frequency tables, pie charts and bar charts ⊕ 1193, 1206, 1207
- find the median, mean, mode and range of ungrouped datasets ⊕ 1192, 1254
- find the median, mean, modal class and range of grouped datasets ⊕ 1201
- use measures of average and spread to compare datasets
- calculate percentages.   1963

The data handling cycle shows the main steps in carrying out investigations or solving problems in statistics.

This section focuses on collecting data. The rest of the chapter is about analysing, presenting and interpreting data.

The **population** consists of all the members in the group being considered.

When it is too costly or difficult to include the whole population, a **sample** is used instead. To be useful, this sub-group needs to give a good indication of what results would be like from the whole population. The bigger the sample, the more likely it is to do this.

A **random sample** is a sample taken in such a way that every member of the population has an equal chance of being included. Random sampling methods include:

- drawing 'names' out of a 'hat'
- numbering all members of the population, then using random numbers from a calculator, computer or tables to select the sample.

> Define the problem or aim of the investigation.
>
> Collect data.
>
> Analyse and present the data.
>
> Interpret and/or report the results.

> A population could be all the people at a football match, all the fish in a pond, all the cars made in a factory or all the houses in a town.

> A sample is **representative** if it reflects the population.
>
> When a sample is not representative, it is said to be **biased**.

Activity 1

★ List six examples of qualitative data and six examples of quantitative data. Which examples are continuous data?

💬 Find the random number generator on your calculator. Describe how you could use it to select a random sample of students from your school or college.

★ Which is likely to give more representative results: a sample of 100 students or a sample of 30?

Stratified sampling

Simple random sampling methods may not give a representative sample, so the population is sometimes stratified to make this more likely.

> **Qualitative** data are descriptive (for example, colour, flavour, gender).
>
> **Quantitative** data are numerical and may be discrete or continuous.
>
> **Discrete** data can take only exact values (for example, number of students, shoe size, ticket price).
>
> **Continuous** data can take any value in a range (for example, height, weight, time, speed, area).

Key point

To select a stratified sample:

- List characteristics that may affect the investigation (for example, age, gender, occupation, income group).
- Find out how many of the population lie in each group.
- Divide the sample in the same proportion.

> A **stratified** sample is divided into groups, or strata.

- Use a random sampling method to select individuals from each group.

Example 1 The table shows the number of students at a college.

The principal wants to select 20 students for a focus group.

How many students should she select from each group?

Student population

Gender	Part-time	Full-time
Female	465	236
Male	624	197

Total population = 465 + 236 + 624 + 197 = 1522

For a sample of 20:

Gender	Part-time	Full-time
Female	$\frac{465}{1522} \times 20 = 6$	$\frac{236}{1522} \times 20 = 3$
Male	$\frac{624}{1522} \times 20 = 8$	$\frac{197}{1522} \times 20 = 3$

$\frac{465}{1522}$ of the total population are female and study part time.

Check that you have the right number for the sample:
6 + 3 + 8 + 3 = 20

Activity 2

★ Calculate the number of students from each category for a focus group of 30 students.

★ Why might a focus group of 10 students be better than 50?

★ Why might a focus group of 50 students be better than 10?

For large populations stratified sampling can be very costly and time-consuming. Two cheaper, more convenient methods are described below.

Cluster sampling

A population may fall naturally into sub-groups, called clusters. If so, it may be much easier to collect data from a random sample of these clusters instead of the whole population.

For example, in a survey involving schools, the Local Authorities in England could be used as clusters. A sample of Local Authorities is then chosen at random and all schools in the selected Local Authorities are included.

Quota sampling

Quota sampling is used widely in opinion polling and market research. Each interviewer is given a 'quota' of types of people to interview. The sizes of the quota are in proportion to the relevant sub-groups of the population.

Activity 3

★ Suppose a market researcher is asked to interview 60 adult men, 60 adult women, 15 teenage girls and 15 teenage boys about their use of mobile phones. Do you think this will give a representative sample of all people who use mobile phones?

★ Describe three different ways of selecting a sample from the fans of a football club. Which method do you think is best and why?

Exercise 1A

1   List which of the following types of data are:

   a) qualitative   b) discrete

   c) continuous.
      i   number of texts sent
      ii   colour of flowers
      iii   weight of puppies
      iv   exam grade
      v   shoe size
      vi   length of feet
      vii   time asleep
      viii leaf area
      ix   ticket prices
      x   gender of football fans
      xi   age of football fans
      xii   nationality of players

2   You are asked to write a report on students' earnings from part-time work. Describe how you might find relevant

   a) primary data   b) secondary data

   > **Primary** data is information you collect yourself.
   >
   > **Secondary** data is information that has been collected by someone else.

3   a) A market research company wants to ask a sample of people from a town about a plan to build a new cinema complex in the town.
        What questions might the company want to ask?

   b) Comment on these methods for carrying out this survey.
      i   Interview people who use the town's bus station between 6 p.m. and 9 p.m. one Saturday evening.
      ii   Send questionnaires by post to all houses within 3 miles of the town centre.
      iii   Ask a local radio and/or TV station to include something about the cinema complex and invite people to phone or email to give their views.

I make £5 for every survey I do and it only takes a few minutes!

4   Some people do surveys online to earn money.
    Do you think the results from such surveys will be reliable?

5   Describe how you would select a representative sample and carry out a survey for

   a) a sample of 100 students from a college to answer questions about canteen facilities

   b) a sample of 1000 people from a town to answer questions about car parking.

6   Describe how you could use quota sampling to ask 500 people in a city questions about having a new international airport near the city.

7   a) Describe how you could use cluster sampling to find a large sample of people in England to answer questions about health care.

   b) Would your sample be a random sample of England's population?

8   The table gives the number of students in each year group in a school.

Year	Number of students
7	128
8	136
9	128
10	120
11	112
12	96
13	80

Simon wants to carry out a survey about computer games.

He says 'I will ask the first person from each year group to arrive at school tomorrow morning.'

Suggest ways in which Simon can improve his method and describe any problems he may encounter.

9 A dentist wants to ask a random sample of patients how satisfied they are with the service they receive from her.

Describe how she could select a random sample and what decisions and problems may be involved.

10 A botanist wants to study the distribution of plants in a random sample of 20 square metres of land from a rectangular site that is 50 metres long and 40 metres wide.

How could he use random numbers to select a sample?

11 A survey is to be carried out in a town to find out what teenagers think of the organised activities in the town.

Describe a way of using each of these methods and list the advantages and disadvantages of each method.

a) Simple random sampling
b) Stratified sampling
c) Cluster sampling
d) Quota sampling

12 The table gives the number of part-time and full-time employees of a company.

Number of employees	Male		Female	
	Full-time	Part-time	Full-time	Part-time
Factory	239	17	117	91
Warehouse	55	20	28	15
Office	29	7	42	57
Delivery	17	4	12	0

The manager wants a representative sample of 200 employees to complete a questionnaire about health and safety at work.

a) Calculate how many employees from each sub-group are needed.
b) What other characteristics of the workforce may be relevant?
c) What problems are likely to arise when gathering information by this method?

13 The table gives the number of entrants to higher education in one year.

Age	18	19	20	Other	Total
Female	94 775	42 640	11 950	33 980	183 345
Male	79 020	40 625	12 075	26 230	157 950
Total	173 795	83 265	24 025	60 210	341 295

Source: www.gov.uk (HEIPR Tables 2012–13)

a) A polling organisation wants to invite 10 000 of the students to take part in a survey about their first experiences of higher education. The sample is to be stratified by age and gender.
   i   Calculate how many students from each sub-group should be invited.
   ii List other characteristics of the population that may be relevant.
   iii Describe any problems that you think the polling organisation may have and suggest ways these might be overcome.

b) Use the internet to find out about the National Student Survey (NSS) carried out each year to collect feedback from undergraduate students about their courses. Write a short report.

14 You are asked to carry out a survey in your school or college to find out what students think about reducing the voting age to 16 and compulsory voting in national elections.

a) Describe in detail how you would use stratified sampling to select a random sample of 25% of the students.
b) Write a list of questions you could ask.
c) Describe how you would display the results.

1.2 A fair representation?

A driving test centre records the gender and age of learners who pass their practical driving test. The list gives the results for one day:

F 19   M 20   F 26   F 17   M 21   M 26   M 18   F 49   M 20

M 18   F 25   F 21   M 17   F 22   F 19   F 17   M 19

F 18   M 23   M 17   M 22   M 18   F 17   M 22   M 18

> This list shows **raw data**, recorded as the results were collected and not analysed in any way.

Activity 4

Check the averages given in the table for the females and males.

Which average do you think gives the best representative values for the data sets?

Explain your decision.

Averages	Female	Male
Mode	17 years	18 years
Mean	22.7 years	19.9 years
Median	19 years	19.5 years

You can find the mean using the in-built function on your calculator or

$$\bar{x} = \frac{\sum x}{n}$$

$$= \frac{\text{sum of values}}{\text{number of values}}$$

The best average to use depends on the data and context. The table summarises the advantages and disadvantages of each average:

Averages	Advantages	Disadvantages
Mode	Easy to find. Always one of the data values (assuming there is a mode).	There may be no mode. There may be more than one mode. Does not use all of the data.
Mean	Uses all of the data.	Affected by extreme values. Involves more calculation. Not usually one of the data values.
Median	Often (though not always) one of the data values. Not affected by extreme values.	Does not use all of the data.

Stem-and-leaf diagrams

A stem-and-leaf diagram puts the data in order and shows the 'shape' of the data set.

Key point

To draw a stem-and-leaf diagram ⬤ 1215

- List the data in order.
- Choose a stem (usually the first digit of the data values).
- List the leaves (usually the second digit) in order.
- Include a title and key.

Example 2 The list gives the marks achieved by students in a test out of 60.

39	37	56	44	32	40	26	58	35
31	42	37	51	29	38	28	42	49

a) Draw a stem-and-leaf diagram.

b) Find   **i**   the mode   **ii**   the median.

Count to check the values in your table to make sure they are all there (or cross out the original values as you enter them). For large datasets draw the diagram in two stages: list the leaves as they arise, then order them later.

a) Order the marks in a table, putting the first digit in the stem. Write a key.

Marks out of 60 in a test

```
2 | 6  8  9
3 | 1  2  5  7  7  8  9
4 | 0  2  2  4  9            Key
5 | 1  6  8                  5 | 1 means 51 marks
```

For n values the position of the median is $\frac{n+1}{2}$.

b) i There are two modes: 37 and 42.

   ii Find the middle value.

For 18 values, the middle is between the 9th and 10th.

The median is the mean of the 9th and 10th values.

$$\text{Median} = \frac{38 + 39}{2} = 38\frac{1}{2} \text{ marks.}$$

You can compare two data sets using a **back-to-back stem-and-leaf diagram.** ⊞ 1215

Here is the driving test data from page 14.

Age (in years) on passing driving test

```
   Females |   | Males
         9 | 4 |
           | 3 |
 6  5  2  1| 2 | 0  0  1  2  2  3  6
 9  9  8  7  7  7| 1 | 7  7  8  8  8  8  9
```

Key
5|2|6 means
Female 25, Male 26 years old

In this case the female ages are more widely spread than the male ages.

Activity 5

★ This list gives the ages of a group of holidaymakers.

32	18	23	19
43	27	21	35
26	19	23	31
24	37	29	23

a) Draw a stem-and-leaf diagram.

b) Find
   i   the mode
   ii   the median.

Activity 6

An office worker records the number of emails he receives and sends each day:

Received	36	29	19	44	26	18	32	24	25	42	20	26	31	29	38
Sent	15	21	32	27	19	17	39	28	14	35	9	27	19	35	27

a) Draw a back-to-back stem-and-leaf diagram.

b) Compare the modes, medians and spread of the Received and Sent datasets.

The marks out of 100 achieved by two classes in a maths test are listed:

Class A   8 54 55 56 58 59 59 59 62 62 63 63 63 64 65
Class B   25 28 33 37 41 47 59 59 63 69 70 73 76 80

The range, like the mean, is affected by extreme values. To avoid this, you can use the range of the middle 50% of the distribution, called the **interquartile range**, instead of the full range.

> ### Key point
> To find the interquartile range (IQR)
> - List the data in order.
> - Find the lower quartile (LQ) – the value that is $\frac{1}{4}$ of the way through the data set.
> - Find the upper quartile (UQ) – the value that is $\frac{3}{4}$ of the way through the data set.
> - Subtract   IQR = UQ − LQ

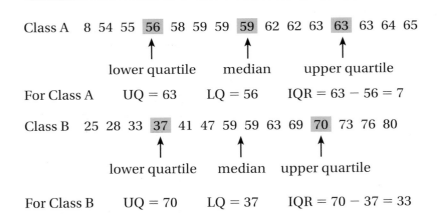

Class A   8 54 55 **56** 58 59 59 **59** 62 62 63 **63** 63 64 65

         lower quartile   median   upper quartile

For Class A   UQ = 63   LQ = 56   IQR = 63 − 56 = 7

Class B   25 28 33 **37** 41 47 59 59 63 69 **70** 73 76 80

         lower quartile   median   upper quartile

For Class B   UQ = 70   LQ = 37   IQR = 70 − 37 = 33

The IQR for Class B is larger than that for Class A, showing that the marks from Class B vary more.

Activity 7
Find the range of marks for each class. Does the range clearly show the variability or consistency of the marks in these classes?

When $n \leqslant 20$, where n is the number of values, treat the values at either side of the median as a separate data set. When $n > 20$, use $\frac{n}{2}$ for the position of the median and $\frac{n}{4}, \frac{3n}{4}$ for the quartiles.

Extreme values like 8 in Class A are sometimes called **outliers**. They affect the mean and range, but not the median and interquartile range.

Activity 8
★ Find and compare the IQRs of the times taken by these two groups of apprentices to complete a task.

Group P (min)	32	18	23	19	43	27	21	35	26	19	23
Group Q (min)	31	24	37	29	20	35	69	21	25	33	

★ Why is the IQR a better measure of spread than the range in this case?

Standard deviation

The **standard deviation** is used more widely in statistical analysis than the range or interquartile range. *All* of the values in a data set are used in one of these formulae:

$$\sigma_n = \sqrt{\frac{\sum(x - \bar{x})^2}{n}} \text{ or } \sigma_{n-1} = \sqrt{\frac{\sum(x - \bar{x})^2}{n - 1}}$$

Scientific calculators and spreadsheets have in-built functions to work them out.

σ_n and σ_{n-1} are usually very similar. For this course always use σ_{n-1}.

Key point

To find the mean and standard deviation

- Enter the data into a calculator or spreadsheet.
- Use the in-built functions to find $\bar{x}$ and σ.

For the standard deviation:
- use σ_{n-1} on a calculator
- use the formula STDEV on a spreadsheet.

Activity 9

Find out how to use *your* calculator to find averages and measures of spread.

Use the data for Class A, Class B, Group P and Group Q from page 16.

Use a spreadsheet to check your results and compare your answers with other students.

If any of your results disagree, try to find out why.

Variance = (standard deviation)^2 is also used in some statistical work.

The table summarises the advantages and disadvantages of each measure of spread.

Measure of spread	Advantages	Disadvantages
Range	Easy to find.	Affected by extreme values. Does not use all of the data.
IQR	Not affected by extreme values.	Does not use all of the data. May not give a fair comparison if one dataset has more values than the other.
Standard deviation	Uses all of the data. Gives a fair comparison when one dataset has more values than the other.	More difficult to calculate. (But on scientific calculators and spreadsheets you only need to enter the data and use the correct key or formula.)

For standard deviation you just need to remember that higher values mean the data are more spread out and lower values mean the data are closer together. The units are always the same as those of the data.

Activity 10

Use the internet to find out which of the averages and measures of spread are used to summarise data or describe distributions.

List the examples that you find in government information and in statistical research.

To describe or represent a population or sample, give one average and one measure of spread. It is usual to give the mean and standard deviation or the median and interquartile range, but occasionally the mode and range may be more useful.

Exercise 1B

1   For each dataset find
    i   the mode   ii   the median
    iii   the mean   iv   the range.

   a)   Price of the cheapest season ticket
        at each Premiership football club in
        2013–14.

£1014	£335	£499	£595
£550	£544	£501	£365
£710	£299	£532	£383
£499	£608	£459	£400
£449	£795	£349	£640

   b)   Estimated annual earnings, in millions
        of pounds, of players in England's
        national team.

£1.4m	£1.0m	£1.6m	£4.7m
£5.3m	£3.1m	£3.4m	£2.6m
£2.6m	£0.8m	£3.6m	£15.5m
£2.9m	£4.2m	£5.0m	£3.9m
£6.2m	£6.0m	£5.2m	

2   Stacy is writing a report about the rates of
    pay of workers at a warehouse.

    She finds this table which gives the mean
    rate of pay for packers and fork-lift truck
    drivers.

	Number of workers	Mean wage (£ per hour)
Packers	42	£9.50
Fork-lift truck drivers	14	£13.50

    Stacy says that the overall mean must be
    £11.50 per hour.

   a)   Explain why Stacy is wrong.
   b)   Find the overall mean wage.

    > Start by finding the total wages for the 56
    > workers.

3   Jack's mean score from 8 matches on the
    cricket team is 45. The coach says he must
    increase his mean score to 50 in the next
    match to keep his place on the team.
    How many runs must Jack score in the
    next match?

4   The bar chart shows the number of students
    studying A level sciences at a school.

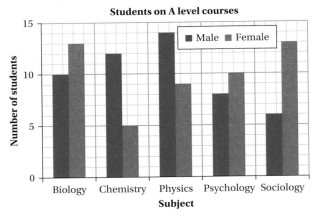

Sam says the average class size is 23, Tao
says it is 20 and Lily says it is 19.

   a)   Explain how they can all be correct.
   b)   Which average do you think best
        represents the data? Give reasons.

5   a)   For each dataset find the median and
        interquartile range.
        i   Number of minutes that the
            school bus is late each morning
            in 4 weeks:
            2 0 1 0 1 0 19 3 0 1
            0 0 2 0 0 1 4 1 2 0
        ii   Number of plants that germinate
            from 10 packets of seeds:
            42   39   37   40   43   13   45
            38   41   44
        iii   Amount spent on lunch by each
            student on a college course:
            £2.55   £4.36   £2.97   £9.10
            £2.87   £3.45   £1.99   £2.75
            £3.52

   b)   Explain how you can tell that the
        median and interquartile range will
        give more useful values than the mean
        and range for these datasets.

6   Find the mean and standard deviation of
    each dataset.

   a)   Marks in a test out of 50:
        46   34   45   49   32   44   47   35
        29   38   46   29

b) Temperatures at midday:

28 °C   21 °C   23 °C   22 °C   23 °C
24 °C   27 °C

c) Journey time to college (in minutes):

36   24   32   29   35   45   27   36
40   28

d) Number of runs scored in a game of cricket:

34   98   46   0   29   13   84   72
51   46   0   18

7 Which of the datasets in question **6** contain

a) discrete data   **b)** continuous data?

8 The stem-and-leaf diagram shows the heights of the students in a class.

Heights of students

18	1
17	0 4 4 6 7 8
16	0 1 3 6 7 8
15	3 7 7 8
14	9
13	4

Key 13 | 4 means 134 cm

a) Find three different averages and three different measures of spread.

b) Which average and which measure of spread would you use to represent this dataset? Give reasons.

9 A car park attendant records the time for which each car is left in a short stay car park. The results in minutes are listed below.

45 59 42 15 54 59 18   9 44
30 58 35 28 33 58 59 24 45
58 47 15 59 10 55 59 48 36
46 56 21 59 36 51 43 47 58

a) Draw a stem-and-leaf diagram.

b) Find

   i   the mode
   ii the range
   iii the median
   iv the interquartile range.

c) Use your answers to write a short paragraph about the results.

10 Rosie can drive to college by two different routes. She keeps a record of how long it takes her to get to college by each route for 3 weeks. The back-to-back stem-and-leaf diagram shows her results.

Time taken for journey to college

Route A		Route B
9 9 8 7	1	9
7 6 5 3	2	0 2 3 4 6 8 9
9 8 4 2	3	0 1 1 2 3 4
3 1 0	4	0

Key 7 | 1 | 9 means
Route A takes 17 minutes
Route B takes 19 minutes

a) For each route, find

   i   the mode   **ii** the range
   iii the median **iv** the interquartile range.

b) **i** Use the stem-and-leaf diagram and your answers to part **a** to write a paragraph comparing the times taken using the two routes.

   ii Which route would you advise Rosie to take when she sets off from home at 08:25 and needs to get to college by 09:00?

11 The table gives the gender and ages of the people who marry at a registry office during one week.

Mon	Tues	Wed
F21 & M28	F16 & M17	F26 & M32
F35 & M35	F27 & M33	F30 & M38
F19 & M19		M35 & M28

Thurs	Fri	Sat
F29 & M42	F26 & M25	F38 & M34
F21 & M24	F32 & F28	F18 & M23
	F28 & M29	F23 & M32
	F17 & M21	F40 & M45
		F26 & M30
		F34 & M35

Write a short report comparing the ages of the men and women. Include a back-to-back stem-and-leaf diagram, an average and a measure of spread.

1.4 Box and whisker plots

A **box and whisker plot** (often called a **boxplot**) illustrates a dataset by showing the median and quartiles as well as the highest and lowest values.

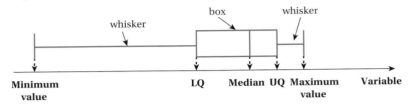

> Compare the lengths of the four parts of this box and whisker plot. The bottom half of the distribution is more widely spread than the top half.

The boxplots below illustrate the test results for the two classes on page 16.

Marks in a maths test out of 100

	Class A	Class B
Minimum	8	25
Lower Quartile	56	37
Median	59	59
Upper Quartile	63	70
Maximum	65	80

Activity 11

★ Compare the box and whisker plots for Class A and Class B. Write a paragraph to describe how the marks are distributed.

★ Draw box and whisker plots for the times taken by the two groups of apprentices (P and Q) using the data on page 16.

Example 3 The frequency table shows the number of goals scored by Manchester United in matches in the 2013–14 season.

No. of goals	No. of matches
0	9
1	9
2	9
3	7
4	4

a) Find
  i the median
  ii the interquartile range.

b) Draw a box and whisker plot to illustrate the data.

c) Calculate   i the mean   ii the standard deviation.

---

a) When $n > 20$, use $\dfrac{n}{2}$, $\dfrac{n}{4}$ and $\dfrac{3n}{4}$ for the positions of the median and quartiles.

i Total number of matches $n = 38$

Median = 19th value = 2 goals

ii Position of LQ $= \dfrac{38}{4} = 9.5$th value

LQ $= \dfrac{0 + 1}{2} = 0.5$ goals

Position of UQ $= \dfrac{3 \times 38}{4} = 28.5$th value   UQ = 3 goals

IQR $= 3 - 0.5 = 2.5$ goals

b) Manchester United goals in matches in the 2013–14 season

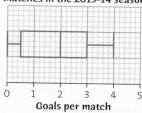

0   1   2   3   4   5
Goals per match

c) Enter the values from the original table into your calculator.

Mean $\bar{x} = 1.68$ (to 3 sf)

Standard deviation $\sigma = 1.32$ (to 3 sf)

You may find a **cumulative frequency** table useful.

x	f	CF
0	9	9
1	9	9 + 9 = 18
2	9	18 + 9 = 27
3	7	34
4	4	38

The 19th value is in the 2 goals group, so the median is 2.

The 9th value is 0 goals and the 10th value is 1 goal, so the LQ is 0.5.

The 28th and 29th values are both in the 3 goals group, so the UQ is 3.

Activity 12

★ Find out how to enter frequencies into *your* calculator.
Check the values given for $\bar{x}$ and σ in part **c** of the example.

★ The table gives the goal distribution for Manchester City in 2013–14.

No. of goals	0	1	2	3	4	5	6	7
No. of matches	4	5	10	8	6	2	2	1

Find $\bar{x}$ and σ for Manchester City.
Compare these measures with those for Manchester United.
Use a spreadsheet to check all your results for $\bar{x}$ and σ.

★ On the same graph, draw two box and whisker plots—one for Manchester United and another for Manchester City.
Describe and compare the two distributions.

★ Which of the following pairs of measures give a good comparison between the two distributions?

• mean and standard deviation

• median and interquartile range

• mode and range

Explain your answer.

The formula for the mean from a frequency table is

$$\bar{x} = \frac{\sum fx}{\sum f}$$ 1254

You can use this to check the mean from the calculator:

Total number of goals $=$
$\sum fx = 9 \times 0 + 9 \times 1$
$+ 9 \times 2 + 7 \times 3 + 4 \times 4 = 64$

Total number of games $=$
$\sum f = 9 + 9 + 9 + 7 + 4 = 38$

Mean $= \dfrac{64}{38} = 1.68$ goals per match.

Exercise 1C

1 The table shows the ages of female and male apprentices attending a course.

Age (years)	17	18	19	20	21
Number of females	6	9	11	7	0
Number of males	13	11	8	5	9

For the female and male apprentices compare

a) the modal age
b) the range of ages
c) the median age
d) the mean age.

2 The runs scored by Ben and Sanjay in this season's cricket matches are listed below.

Ben: 75 43 12 6 0 20 34 47 29 5 91

Sanjay: 28 49 27 61 29 14 58 37 25 18 46

a) Draw two box and whisker plots to compare their scores.
b) Compare Ben and Sanjay's mean score and standard deviation.
c) There is only one place left on the team for the next match. Who would you pick and why?

3 The table gives the annual wages of people who work in a supermarket.

Annual wage	£21 000	£24 000	£70 000
Number of workers	20	7	1

a) Find the median and interquartile range.
b) An outlier is defined as a value that is more than 1.5 times the interquartile range below the lower quartile or above the upper quartile. Show that £70 000 is an outlier.
c) The supermarket manager says that the average wage is £23 500. Comment on the use of this value.

4 The table gives the number of burglaries in two towns last year.

Month	Jan	Feb	Mar	Apr	May	Jun
Badley	35	21	18	15	16	12
Lootham	29	27	20	13	11	16

Month	Jul	Aug	Sep	Oct	Nov	Dec
Badley	14	7	20	29	38	40
Lootham	9	11	21	24	42	33

a) For each of the towns find
   i the mean
   ii the standard deviation
   iii the median
   iv the interquartile range.
b) Draw box and whisker plots.
c) i Kylie says that Badley is safer from burglaries than Lootham. Which of the statistics supports her comment?
   ii Winston says that this is wrong and Lootham is safer than Badley. Which of the statistics supports his point of view?

5 A school library has 15 computers. The librarian records the number of computers used each lunch hour. The table gives the results for 8 weeks.

Computers in use	15	16	17	18	19	20
No. of lunch hours	1	17	11	8	2	1

a) Write down
   i the mode   ii the range.
b) Find
   i the median
   ii the interquartile range.
   iii Draw box and whisker plots to illustrate the data.
c) Find
   i the mean
   ii the standard deviation.

6 The labels on packets of seeds say 'Average contents: 40 seeds.'

After receiving complaints from customers, a garden centre checks the first 100 packets in their stock. The table gives the results.

Number of seeds	37	38	39	40	41	42	43	44
Number of packets	11	13	24	28	23	0	0	1

a) **i** Use the data to find three averages.

ii Write a sentence that the garden centre could use in a letter defending the claim on the packets.

b) An outlier is a data value that is more than 1.5 times the interquartile range below the lower quartile or above the upper quartile. Show that 44 is an outlier.

c) Explain two ways in which the sampling method used by the garden centre could be improved.

7 Nick draws this box and whisker plot to show the turnover and wages bill of Championship football clubs in the 2012–13 football season.

> Turnover is the total amount of money that the club earns during the year.

Championship finances

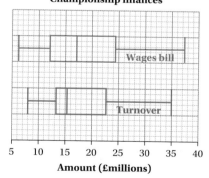

Amount (£millions)

Compare the main features of these two distributions.

Write a paragraph to describe what the boxplots tell you about the finances of Championship football clubs.

8 The table gives the ages of students who go on an exchange trip to Germany.

Age (years)	12	13	14	15	16	17	18
Number of students	10	13	11	7	14	5	0

a) Gemma says the modal age is 14 years and the range is 6 years.

i What errors has Gemma made?

ii Find the correct mode and range.

b) Mark uses his calculator and says that the mean age is 8.6 years and the standard deviation is 4.9 years.

i What error has Mark made?

ii Find the correct mean and standard deviation.

c) Sunita says the median age is 15 years and the interquartile range is 4 years.

i What error has Sunita made?

ii Find the correct median and interquartile range.

iii Draw a box and whisker plot and describe how the ages are distributed.

9 The table gives the number of errors made by workers before and after training.

Number of errors	Number of workers	
	Before training	After training
3	0	2
4	0	8
5	3	10
6	10	11
7	13	5
8	8	0
9	1	0
10	1	0

a) Draw box and whisker plots to compare the results before and after training.

b) Describe what the box and whisker plots show about the effect of the training on the workers' performance.

10 A psychology student carries out a memory test. He shows 10 objects to a group of volunteers. After removing the objects he asks the volunteers to list the ones they remember. A day later he asks them to write another list of the objects they remember. The table gives the results.

Number of items remembered	Number of volunteers	
	On day of test	After 1 day
1	0	1
2	0	1
3	0	3
4	0	3
5	2	4
6	1	5
7	3	3
8	4	0
9	5	0
10	5	0

a) Draw box and whisker plots to illustrate these results.

b) Find the mean and the standard deviation for both datasets.

c) Use your answers to write a paragraph describing the results of the experiment.

11 The chart shows the results of a survey on the number of dogs kept by people who live in a street.

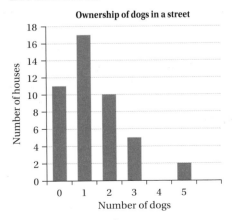

a) i Find the mode, median and mean number of dogs per house.

ii Kate says the mean is the best average to use. Give a reason why Kate may say this.

iii Mel says the mode and median are both better than the mean. Give a reason why Mel may say this.

b) i What is shown on the bar chart that would not be shown on a box and whisker plot?

ii What would be shown on a box and whisker plot that is not shown on the bar chart?

12 Gemma and Sonja serve customers in a clothes shop. Their tally chart shows the number of items customers buy in a sale on one day.

Number of items	Number of customers
0	⊞⊞ ⊞⊞ ⊞⊞ |||
1	⊞⊞ ⊞⊞ ⊞⊞ ⊞⊞
2	⊞⊞ ⊞⊞ ⊞⊞ ⊞⊞ ⊞⊞ ||
3	⊞⊞ ⊞⊞	
4			
5	
6		

The shop manager asks them to draw a chart to display these results.

a) i Gemma draws this box and whisker plot.

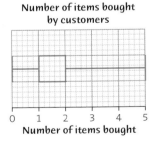

Number of items bought by customers

Sonja says there should be a vertical line across the middle of the box. Is Sonja correct? Show how you decide.

ii Describe a different error Gemma has made.

b) Sonja says this bar chart shows the data better than the box and whisker plot.

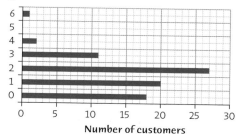

Number of customers

i Comment on this diagram and describe any improvements that could be made.

ii Which of the two diagrams do you think would be most useful to the shop manager? Give a reason.

13 Alec and Kris go swimming each week. They always end the session by having a race. The table gives their times for the last 40 weeks.

Time taken to swim 50 m (seconds)	Number of weeks	
	Alec	Kris
28	6	0
29	8	7
30	9	15
31	12	11
32	4	7
33	1	0

Alec wants to use a chart to illustrate these results.

a) Use the data to draw
   i two box and whisker plots
   ii a bar chart.
   iii Which type of chart would you advise Alec to use? Give reasons.

b) Kris wants to summarise the data by giving an average and a measure of spread.

Find suitable values for him to use and explain your choice.

14 Many clothes shops in the UK use 'vanity sizing'. This means that the measurements of their clothes are more generous than those suggested by British Standards. The table gives the actual measurements in inches of dresses sold as size 12 and size 14 at 8 high-street shops. The British Standard (BS) sizes are also given in centimetres. (1 inch = 2.54 cm)

Shop	Dress size 12		Dress size 14	
	Bust	Hips	Bust	Hips
A	35.4	39.0	37.4	42.9
B	35.9	39.0	37.8	41.0
C	35.4	37.6	37.0	40.0
D	36.2	38.0	38.2	40.0
E	36.2	38.0	38.2	40.0
F	37.0	40.2	39.0	42.3
G	37.0	40.2	38.6	41.7
H	37.8	40.2	40.2	42.3
BS	88	93	92	97

Write a short report about vanity sizing.

Use statistical measures to support your findings. Include diagrams if you wish.

15 a) Ask all members of your class to estimate one or more of the following:
   i the number of texts they send in a month
   ii the time they spend on computers in a week
   iii the distance they walk in a year
   iv the amount they spend on snacks in a year.

b) Use averages, measures of spread and/or statistical diagrams to summarise the results. Compare your estimates with those from the whole class.

Students in year 12 who have part-time jobs are asked how long they work each week. The table gives the results.

Time x (h)	No of students
$0 < x \leqslant 4$	3
$4 < x \leqslant 8$	29
$8 < x \leqslant 12$	43
$12 < x \leqslant 16$	10
$16 < x \leqslant 20$	6
$20 < x \leqslant 24$	1

The class $4 < x \leqslant 8$ includes times between 4 hours and 8 hours. It also includes 8 hours, but not 4 hours.

You can find the median and quartiles of these data from a **cumulative frequency graph**. First draw a cumulative frequency table, then the graph. ⚄ 1195

Time (h)	CF
0	0
4	3
8	32
12	75
16	85
20	91
24	92

$3 + 29 = 32$ students worked up to 8 hours.

$32 + 43 = 75$ students worked up to 12 hours.

The **cumulative frequency** tells you how many students worked less than or equal to the corresponding number of hours.

Cumulative frequency is always plotted against the *upper* class boundary.

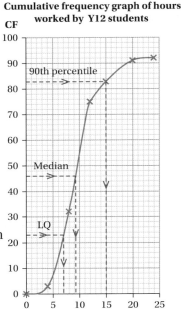

Cumulative frequency graph of hours worked by Y12 students

The total number of students, $n = 92$

Position of median $\approx 92 \div 2 = 46$th
Median ≈ 9.3 hours

Position of LQ $\approx 92 \div 4 = 23$rd
LQ ≈ 7 hours

Always remember to include a title and label the axes on graphs and charts.

When n is large, use $\dfrac{n}{2}$ to find the position of the median.

You can find other percentiles from the graph.

For the 90th percentile: 90% of 92 $= 0.9 \times 92 = 82.8$

The graph gives the value of 15 hours, meaning that 90% of the students worked less than or equal to 15 hours.

The lower quartile is sometimes called the 25th **percentile** because it is 25% of the way through the distribution.

Activities 13

★ Which percentile is the median?

★ Find the upper quartile. Which percentile is this? Calculate the interquartile range.

★ Find the 10th percentile, the 20th percentile and the 80th percentile.

★ A school report says fewer than half of year 12 students work more than 10 hours per week. Does the graph confirm this?

You can draw the same type of graph or box and whisker plot using percentages instead of frequencies.

Example 4 The table shows speed data collected on motorways by the Department of Transport. Speeds are given in miles per hour (mph), rounded to the nearest mile per hour.

Speed (mph)	% cars	% buses
Under 50	5	6
50–59	14	41
60–64	14	24
65–69	20	12
70–74	21	10
75–79	14	5
80–89	10	2
90 and over	2	0

Source: www.gov.uk (Free flow speed statistics, 2013)

a) i Find the median and interquartile range for each dataset.

ii Draw a box and whisker plot for each dataset.

iii Describe what the medians and interquartile ranges tell you about the speeds of the cars and buses.

b) A report says that on average cars travel $7\frac{1}{2}$ mph faster than buses and the bus speeds are less variable. Do you think the report is correct? Show how you decide.

a) i The table does not give the lowest and highest speeds. You can assume widths for the first and last groups, but you must state your assumptions.

> ▶ Assume that the first group is 40–49 mph and the last group is 90–99 mph.

These assumptions use the same class widths as the adjacent groups in the table. You could use a class width of 5 mph (as in the other groups) or consider what might be reasonable lowest and highest speeds on a motorway.

Think carefully about the upper boundaries of each group. As the speeds were measured to the nearest mph, the lowest group starts at 39.5 and the upper class boundaries are 49.5, 59.5, 64.5, …

Speed (mph)	CP cars	CP buses
39.5	0	0
49.5	5	6
59.5	19	47
64.5	33	71
69.5	53	83
74.5	74	93
79.5	88	98
89.5	98	100
99.5	100	100

39.5 rounds up to 40

You can save time by drawing both cumulative percentage graphs on the same axes. If you draw separate cumulative percentage graphs, use the same scales so that it is easy to compare them.

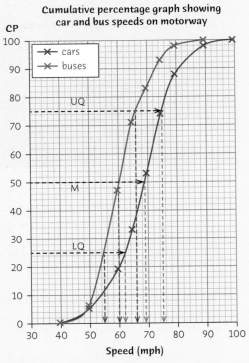

Cumulative percentage graph showing car and bus speeds on motorway

Cars
Median = 69 mph
LQ = 62 mph
UQ = 75 mph
IQR = 75 − 62 = 13 mph

Buses
Median = 60 mph
LQ = 55 mph
UQ = 66 mph
IQR = 66 − 55 = 11 mph

ii

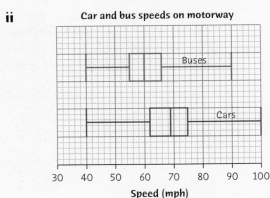

Car and bus speeds on motorway

iii The cars are on average travelling faster than the buses and their speeds are more variable.

b) The interquartile range for the cars is greater than that for the buses, so the report is correct in saying that the bus speeds are less variable.

'On average' does not make it clear which average the report is using, so consider all the averages.

The difference between the medians is 9 mph which does not agree with the $7\frac{1}{2}$ mph given in the report.

The modal class for car speeds is 70–74 mph and the modal class for bus speeds is 60–64 mph, but these do not give a difference of $7\frac{1}{2}$ mph.

To estimate the mean and standard deviation from a grouped frequency table, take x at the mid-value of each group. ⊕ 1201

Using the mid-value of each group is equivalent to assuming that all the data values are at the centre of the group, so this gives estimates rather than accurate values for the mean and standard deviation.

Speed x (mph)	% cars	CP buses
44.5	5	6
54.5	14	41
62	14	24
67	20	12
72	21	10
77	14	5
84.5	10	2
94.5	2	0

Using a calculator gives:

Cars $\bar{x} = 68.2$ mph (to 1 dp) $\sigma = 10.7$ mph (to 1 dp)

Buses $\bar{x} = 60.7$ mph (to 1 dp) $\sigma = 8.6$ mph (to 1 dp)

The difference between the means $= 68.2 - 60.7 = 7.5$ mph, so the report is correct.

Also the standard deviation of the bus speeds is less than that for the car speeds, supporting the report that the bus speeds are less variable.

Activity 14

★ Check the values given for $\bar{x}$ and σ in part **b** of the example.

★ The table gives the speed data for motorcycles on motorways.

Speed (mph)	% motorcycles
Under 50	6
50–59	27
60–64	9
65–69	13
70–74	15
75–79	13
80–89	13
90 and over	4

Find $\bar{x}$ and σ for the motorcycle speeds.

Compare these measures with those for the cars and buses.

★ Use a spreadsheet to check all your results for $\bar{x}$ and σ.

★ Draw a cumulative percentage diagram and box and whisker plot for the motorcycle speeds. Compare the distributions of speed for the three different types of vehicle.

Use the same scales as for the car and bus speeds to make comparison easier.

Exercise 1D

1   A fruit farmer records the mass of strawberries picked by his workers on each day during a summer season. The table gives the results.

> The mass of 100 kg is included in the first group, not the second.

Mass m (kg)	Number of days
$0 < m \leqslant 100$	1
$100 < m \leqslant 150$	3
$150 < m \leqslant 200$	16
$200 < m \leqslant 250$	28
$250 < m \leqslant 300$	19
$300 < m \leqslant 400$	5

a) Write down the modal class.
b) Draw a cumulative frequency graph.
c) Use your graph to find
   i   the median
   ii the lower quartile
   iii the upper quartile
   iv the interquartile range
   v   the 40th percentile
   vi the 80th percentile.
d) Find an estimate for
   i   the mean
   ii the standard deviation.

2   A football club records the number of people who attend each league match. The table gives the results for one season.

Attendance	No of matches
Under 10 000	1
10 000–20 000	2
20 000–30 000	10
30 000–40 000	19
40 000–60 000	5
60 000–80 000	1
over 80 000	0

a) Write down the modal class.
b) i   Draw a cumulative frequency graph.
   ii Use your graph to find the median and interquartile range.
   iii The manager says that over 5% of the matches have had an attendance of over 50 000. Does the graph confirm this? Show how you decide.
c) Find an estimate for
   i   the mean
   ii the standard deviation.
d) The manager says that last year the mean was 38 000 and the standard deviation was 16 000. How have attendances changed since last year?

3   After apprentices were timed on diagnosing a machine fault, they were given extra training, then timed again. The cumulative frequency graph shows the results.

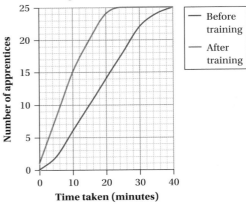

Cumulative frequency graph showing time taken by apprentices to diagnose a machine fault

a) Find the median and interquartile range of the times before and after the training.
b) Write a paragraph to describe the effect the training has had on the apprentices' performance.

4 Students on a childcare course weighed the 3-year-old children at a nursery. The table shows their results with masses given to the nearest 0.1 kg.

Mass (kg)	No of children
11.0–11.9	2
12.0–12.9	4
13.0–13.9	10
14.0–14.9	12
15.0–15.9	6
16.0–16.9	4
17.0–17.9	2

a) Find estimates for the mean mass and standard deviation for this group.

b) i Draw a cumulative frequency diagram.

ii A childcare book gives the following percentile masses for 3-year-old children.

Percentile	5	10	25	50	75	90	95
Mass (kg)	11.8	12.3	13.1	14.1	15.3	16.5	17.3

Compare values from your graph with those given in this table.

5 A survey asks students how long they spent online and how long they spent watching TV on one day.

| Time (min) | Number of students | |
	Online	Watching TV
0–29	9	0
30–59	26	17
60–89	18	29
90–119	7	8
120–149	0	5
150–179	0	1

a) Write down the modal class for each dataset.

b) i Compare estimates for the mean and standard deviation of the datasets.

ii Describe what these measures tell you about the way the students spend their time.

c) i Use cumulative frequency graphs to find the median and interquartile range for each dataset.

ii Do these measures support your comments in b) ii?

6 These diagrams both show the results of a survey about the ages of people who use Twitter

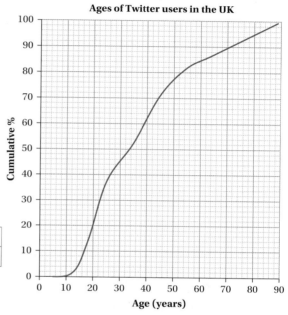

Ages of Twitter users in the UK

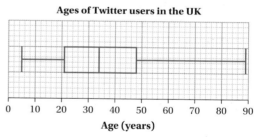

Ages of Twitter users in the UK

a) Describe the main features of this distribution.

b) i Alex says that a cumulative frequency graph shows the results better than a box and whisker plot. Give a reason why Alex might say this.

ii Gemma disagrees. She says that a box and whisker plot illustrates the data in a better way. Give a reason why Gemma might say this.

c) What information about the survey is not given on either of the diagrams?

7 Each year thousands of people raise money for charity by running in the London Marathon. The table gives the times taken by women and men in 2015.

Time taken t (hours)	Frequency	
	Women	Men
$2 < t \leqslant 2\frac{1}{2}$	15	77
$2\frac{1}{2} < t \leqslant 3$	92	1743
$3 < t \leqslant 3\frac{1}{2}$	834	3526
$3\frac{1}{2} < t \leqslant 4$	2205	5559
$4 < t \leqslant 4\frac{1}{2}$	2988	4799
$4\frac{1}{2} < t \leqslant 5$	3210	3711
$5 < t \leqslant 5\frac{1}{2}$	2233	1985
$5\frac{1}{2} < t \leqslant 6$	1412	1002
$6 < t \leqslant 7$	1157	679
$7 < t \leqslant 8$	225	128
$8 < t \leqslant 9$	1	2

Source: www.virginmoneylondonmarathon.com

a) i Draw a cumulative frequency graph to illustrate these results.

ii Use your graph to compare the performance of the women and the men.

b) The box and whisker plots show the results from the 2014 London Marathon.

2014 London Marathon results

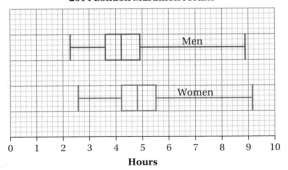

i Use these box and whisker plots to compare the performance of the women and the men in 2014.

ii Compare the performance of the women and men in 2014 and 2015.

8 In an investigation about part-time work Tracy and Josh find this information on the internet.

Percentile	Number of hours worked		Rate of pay (£ per hour)	
	Male	Female	Male	Female
10	5.0	6.4	6.31	6.31
20	8.4	10.3	6.43	6.62
25	9.6	12.5	6.57	6.87
30	11.2	14.5	6.85	7.11
40	14.9	16.5	7.33	7.70
50	17.8	19.3	8.01	8.46
60	20.1	21.2	9.13	9.50
70	23.1	23.5	11.00	11.00
75	24.1	24.6	12.50	12.28
80	25.3	25.6	15.00	14.05
90	29.2	28.8	22.64	19.37
Mean	17.4	18.4	11.62	11.11
No of jobs (thousands)	1819	5347	1819	5347

Source: ONS (ASHE, 2014 Provisional Results)

a) Tracy says 'On average, the men get over 50 pence more per hour than the women.'

Josh says 'No – on average, the women get 45 pence more per hour than the men.'

i Explain each person's statement.

ii Give a reason why this difference may occur.

b) Compare the interquartile ranges of rates of pay.

c) Write a paragraph to compare the hours worked by men and women who have part-time jobs.

d) Why is it difficult to draw box and whisker plots for this data?

9 The table gives the ages on first marriage for people in the UK.

Age (years)	Number of males		Number of females	
	1981	2011	1981	2011
16–19	18537	850	63283	2921
20–24	133652	19244	141820	33809
25–29	70978	61391	39763	72095
30–39	29501	82734	14477	66788
40–49	4100	19013	2268	12227
50–59	1619	3407	1096	2292
60–69	592	879	502	402
70 and over	127	193	159	100

Source: ONS (Marriages in England and Wales, 2012)

a) Calculate an estimate for the mean and the standard deviation for each dataset. State any assumptions you make.

b) Find the median and interquartile range for each dataset.

c) Describe what your results tell you about changes between the ages at which people married in 1981 and 2011.

d) A sociologist wants to carry out a survey to find out why people decide to marry at particular ages. Suggest a sampling technique that could be used to find a representative sample of newlyweds and describe any problems that may arise in carrying out this survey.

10 Use the internet to find population data for the UK. Write a short report comparing the age distribution of males and females, or the age distribution for different years. Include statistical measures and diagrams to support your comments.

11 A holiday company has downloaded 40 years of weather data for two seaside resorts in August from the internet.

Hours of sunshine per month

Hours of sunshine (h)	Number of months	
	Eastbourne	Whitby
$80 < h \leqslant 120$	0	1
$120 < h \leqslant 160$	1	10
$160 < h \leqslant 180$	2	8
$180 < h \leqslant 200$	3	7
$200 < h \leqslant 220$	7	8
$220 < h \leqslant 240$	12	5
$240 < h \leqslant 280$	8	1
$280 < h \leqslant 320$	6	0
$320 < h \leqslant 360$	1	0

Rainfall per month

Rainfall (mm)	Number of months	
	Eastbourne	Whitby
$0 < x \leqslant 10$	3	1
$10 < x \leqslant 20$	3	2
$20 < x \leqslant 30$	6	3
$30 < x \leqslant 40$	6	8
$40 < x \leqslant 50$	5	6
$50 < x \leqslant 60$	0	4
$60 < x \leqslant 70$	1	4
$70 < x \leqslant 80$	4	1
$80 < x \leqslant 100$	7	5
$100 < x \leqslant 120$	4	2
$120 < x \leqslant 140$	1	4

a) Write a paragraph comparing the weather at these two resorts in August. Use statistical diagrams and measures to support your comments.

b) Find weather data for a resort of your choice and compare the results with those for Eastbourne and Whitby.

1.6 Histograms

A headteacher is worried about the speed of vehicles passing her school in the lunch hour. Two students carry out a traffic survey. They measure the speeds together, but record and illustrate the data in different ways as shown below.

Sally's results

Sally collects the results in a tally chart using equal group intervals. She draws up a frequency table then a diagram to illustrate her results.

Speed (mph)	No. of vehicles
0–10	1
10–20	8
20–30	36
30–40	20
Total	**65**

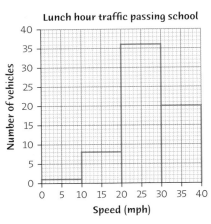

Tom's results

Tom lists the speed of each vehicle, then draws up a different table. He says the headteacher will be most interested in the data around and above the road's speed limit of 30 mph.

Speed (mph)	Frequency
0–26	17
26–30	28
30–32	14
32–34	3
34–36	0
36–38	2
38–40	1
Total	**65**

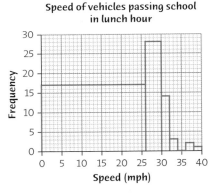

The 26–30 mph bar representing 28 vehicles is more than twice as large as the 30–32 mph bar representing 14 vehicles.

The 0–26 mph bar on Tom's diagram suggests a lot of vehicles travelled at low speeds, but only about a quarter of them were in this group.

Sally's methods are easy and straightforward, but her table and chart do not give a lot of detail about the majority of the cars passing the school, with speeds between 20 and 40 mph.

Tom's idea of using different class width to focus on the most relevant speeds is good, but his diagram is very misleading. The problem with Tom's diagram is that he has used height to represent frequency.

Key point

For a **histogram** with unequal class widths, you must use *area* to represent frequency. ⊕ 1197

To draw a histogram:

- Subtract the class boundaries to find the class widths.
- When the class widths are not equal, divide each frequency by the corresponding class width to give the **frequency density**.
- Draw the histogram using frequency density on the vertical axis.

You can only use frequency on the vertical axis when the groups have *equal* widths.

Using the notation $0 < x \leqslant 10$ makes it clear that a vehicle travelling at 10 mph would be in the first group rather than the second.

The table shows Sally and Tom's data in different groups and the corresponding frequency densities.

Speed (mph)	Frequency	Class width	Frequency density
$0 < x \leqslant 10$	1	10	$1 \div 10 = 0.1$
$10 < x \leqslant 20$	8	10	$8 \div 10 = 0.8$
$20 < x \leqslant 26$	8	6	$8 \div 6 = 1.\dot{3}$
$26 < x \leqslant 30$	28	4	$28 \div 4 = 7$
$30 < x \leqslant 32$	14	2	$14 \div 2 = 7$
$32 < x \leqslant 34$	3	2	$3 \div 2 = 1.5$
$34 < x \leqslant 36$	0	2	$0 \div 2 = 0$
$36 < x \leqslant 38$	2	2	$2 \div 2 = 1$
$38 < x \leqslant 40$	1	2	$1 \div 2 = 0.5$
Total	**65**		

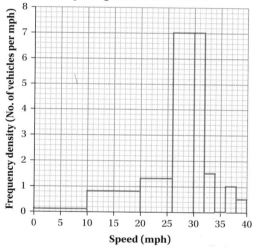

Histogram showing speed of vehicles passing school in lunch hour

The histogram shows more detail than Sally's diagram, but is not misleading like Tom's.

Activity 16

★ Check that the area of each bar gives the frequency of the group.

Estimate the number of vehicles that were travelling at less than 15 mph.

Why is this an estimate?

Write a paragraph to report the results of the traffic survey.

★ The students also measured the speeds of passing vehicles in the hour after school.

The table gives their results.

Draw a histogram and compare it with that for the lunch hour.

Speed (mph)	Frequency
$0 < x \leqslant 10$	4
$10 < x \leqslant 20$	7
$20 < x \leqslant 26$	15
$26 < x \leqslant 30$	24
$30 < x \leqslant 32$	9
$32 < x \leqslant 34$	5
$34 < x \leqslant 36$	1
$36 < x \leqslant 38$	2

Exercise 1E

1 The table gives the amounts spent by customers at a supermarket on one day.

The last class has been assumed to have the same width as the previous class.

Amount (£P)	Number of customers	Lower class boundary	Upper class boundary	Class width	Frequency density (customers per £)
$P < 25$	75	0	25	25	$75 \div 25 =$
$25 \leqslant P < 50$	97	25	50	25	$97 \div 25 =$
$50 \leqslant P < 100$	165	50	100	50	$165 \div 50 =$
$100 \leqslant P < 150$	86	100	150		
$150 \leqslant P < 250$	23	150	250		
$P \geqslant 250$	8	250	350		

a) Copy and complete the table. State any assumptions you make.

b) Draw a histogram to illustrate the data.

2 The table gives estimates for the age distribution of the UK population in June 2013.

Age (years)	No. of people (thousands)	Lower class boundary	Upper class boundary	Class width	Frequency density (thousands per year)
0–4	4014	0	5	5	$4014 \div 5 = 803$
5–15	11 179	5	16	11	$11 179 \div 11 = 1016$
16–44	21 453	16	45	29	
45–64	16 328				
65–74	6031				
75–89	4574				
90+	527				

Source: ONS (Annual Mid-Year Population Estimates for the UK, 2014)

a) Copy and complete the table. State any assumptions you make.

b) Draw a histogram to illustrate the age distribution.

3 The Intelligence Quotient (IQ) is a measure of intelligence.

All students are tested when they join Gainsby College. The table gives this year's results.

IQ	Number of students
$85 < x \leqslant 95$	21
$95 < x \leqslant 100$	76
$100 < x \leqslant 105$	137
$105 < x \leqslant 110$	129
$110 < x \leqslant 115$	108
$115 < x \leqslant 125$	83
$125 < x \leqslant 135$	16

a) i   Calculate an estimate for the mean IQ and the standard deviation for this year's students.

ii Last year's students had a mean IQ of 106 and the standard deviation was 7.4.

Write two statements comparing the results for this year's students with last year's.

b) i   Draw a histogram for the data for this year's students.

ii Use your histogram to estimate the percentage of this year's students that have an IQ of more than 106.

4 Will finds this data for flight delays at Gatwick in one month.

Number of minutes late	Number of flights
Early–15	12 618
16–30	2460
31–60	1589
61–180	1029
181–360	216
> 360	30

Data from Civil Aviation Authority

Will draws this histogram to illustrate the data.

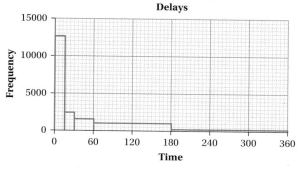

a) i Give reasons why this histogram is incorrect.

ii Use the data to draw a correct histogram. Describe any problems and assumptions you make.

b) i Use the data to estimate the mean and standard deviation.

ii Explain why the values you have found are estimates.

5 A local authority records the widths of the external doors on all of its public buildings. The table gives the data with widths rounded to the nearest 10 mm.

Door width (mm)	Number of doors
Less than 750	0
750–790	24
800–890	146
900–990	124
1000–1090	68
1100 and over	21

Official guidelines say that external door widths must be at least 775 mm.

The local authority wants to increase all external door widths to at least 850 mm.

Use a histogram to estimate:

a) the number of doors that do not comply with the official guidelines

b) the percentage of the doors that the local authority wants to widen.

> Take care with class boundaries here. The 750–790 group starts at 750, but ends at 795. Can you see why?

6 The histogram shows the times that patients spent at a hospital's Accident and Emergency department during one day.

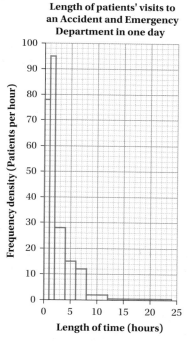

The target for hospitals is for 95% of patients to be seen in under 4 hours.

The average time for hospitals in England is 135 minutes.

Use data from the histogram to compare this hospital's performance with these statistics.

> You could draw a grouped frequency table for the data.

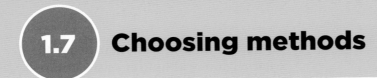

The Hadley Centre website gives weather statistics dating back to 1766.

> A time series graph shows how a quantity changes over time.

The total rainfall data for 1770–1779 and 2000–2009 in the tables below are shown by the two different lines on the **time series** graph.

Year	1770	1771	1772	1773	1774	1775	1776	1777	1778	1779
Total rainfall (mm)	1079.4	792.9	1031.8	1033.8	994.7	1012	847.4	860.3	887.9	900.5

Year	2000	2001	2002	2003	2004	2005	2006	2007	2008	2009
Total rainfall (mm)	1232.5	970	1117.8	761.4	973.6	825.1	904.8	1022.7	1089.6	977.1

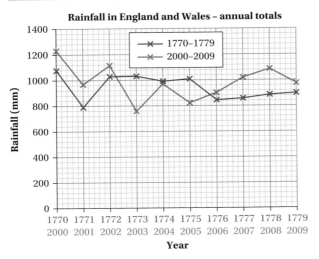

Activity 17

★ Use the graph to compare the rainfall for 1770–1779 with the rainfall for 2000–2009. Describe any similarities or differences.

★ Use the data in the table to compare

- the mean annual rainfall and the standard deviation

- the median and interquartile ranges for the 1770 and 2000 decades.

When you are solving a problem or summarising your findings from an investigation, the methods you choose will depend on the data you have collected and what you are trying to show. Here is a list of some of the methods you could use:

Averages & measures of spread	Notes
mean and standard deviation	Uses all of the data, but affected by extreme values.
median and interquartile range	Does not use all of the data, but not affected by extreme values.
mode and range	Both are very easy to find. There may not be a mode or there may be more than one. The range is affected by extreme values.
Charts and diagrams	**Notes**
pictograms	Often used for qualitative data. Visually attractive, but time-consuming to produce.
pie charts	Can be used for qualitative as well as quantitative data. Shows the proportion of the total that is in each category, but not the frequency.

Paper 1

Charts and diagrams	Notes
bar charts	Used for qualitative and ungrouped discrete data.
	Relatively easy to draw and can be used to compare two or more datasets.
vertical line graphs	Used for qualitative and ungrouped discrete data.
	Easy to draw and shows the frequency clearly.
stem-and-leaf diagrams	A useful way of visualising quantitative data.
	Organises raw data and enables the mode, median, quartiles and range to be found.
box and whisker plots	Used for quantitative data.
	Shows the minimum, maximum, median and quartiles clearly, but not the frequency.
cumulative frequency diagrams	Used for continuous and grouped discrete data.
	Allows you to find the median and quartiles (as well as other information such as percentiles).
	You must remember to plot points at the upper class boundaries.
histograms	A useful way of visualising continuous and grouped discrete data.
	Area represents frequency, so when the class widths are unequal you must remember to use frequency density rather than frequency on the vertical axis.

Whatever methods you use, you must always think carefully about whether they are appropriate and help to get your message across.

Activity 18

A group of students looked for patterns or changes in rainfall since the Hadley records began. The following pages show some of the statistical methods the students used.

Consider each student's work.

- Was the method they used appropriate?
- Describe any patterns or changes that they found.
- Describe any ways their work could be improved.
- What other methods could they have used?

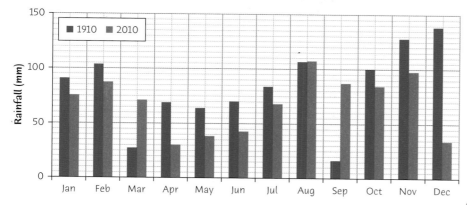

Ahmed says 'I drew this bar chart to compare the monthly rainfall for two years, a century apart.'

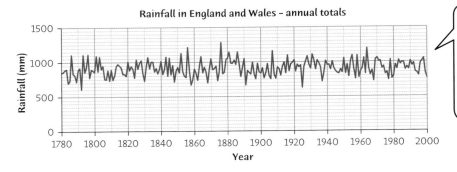

Rainfall in England and Wales – annual totals

Kayleigh says 'I wanted to see whether the annual rainfall had changed over a long period, so I used my computer to draw this graph.'

Cilla says 'I decided to see how the mean and standard deviation of the monthly rainfall amounts varied. I didn't have time to work them out for all of the years, so I just did it for some early years and some later years. Here are my results.'

Year	1766	1767	1768	1769	1770	2010	2011	2012	2013	2014
Mean rainfall (mm/month)	63.0	78.4	103.9	79.2	90.0	68.5	65.6	103.7	76.4	92.1
Standard deviation	33.8	32.0	38.7	27.9	41.0	25.1	28.5	46.9	35.4	45.2

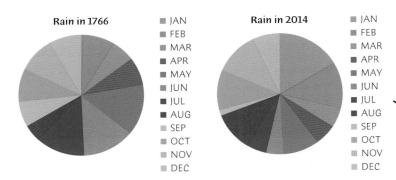

Rain in 1766

Rain in 2014

Kirsty says 'I used my computer to draw these pie charts for the first and last years in the table.'

Years	1771–1800	1801–1830	1831–1860	1861–1890	1891–1920	1921–1950	1951–1980	1981–2010
Median annual rainfall (mm)	888.35	903.55	873.95	912.45	907.45	925.4	904.5	971.8
IQR (mm)	199.6	157.35	169.3	183.95	155.45	165.65	206.65	170.25

Tim says 'I split the years into groups of 30, then worked out the median and interquartile range for each.'

Activity 19

Q Use the internet to find other weather data and use statistical methods to investigate any changes over time.

★ Research how modelling and statistics are used to investigate climate change.

★ Find out what 'Big Data' means, and how it is used.

Exercise 1F

1 A tutor on a childcare course gives her students this table which shows the weights of newborn babies born in one month at a local hospital.

The tutor asks for a statistical diagram to show this information.

Weight (kg)	No of babies
$1 < x \leqslant 2$	0
$2 < x \leqslant 3$	12
$3 < x \leqslant 3.5$	29
$3.5 < x \leqslant 4$	31
$4 < x \leqslant 4.5$	8
$4.5 < x \leqslant 5$	3

Helen draws this cumulative frequency graph.

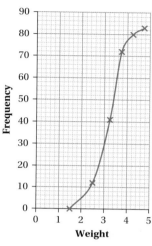

Liam draws this histogram.

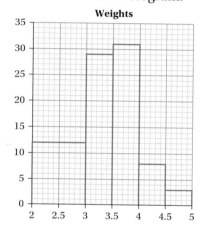

a) List the errors each student has made.

b) Draw a better cumulative frequency graph and histogram.

2 The tables give the size of households and age distribution of the population for three adjacent postcodes in North Yorkshire.

Number in household	Number of dwellings		
	YO11	YO12	YO13
1	5472	6175	1366
2	4848	6828	2211
3	1747	2873	652
4	1087	2204	507
5	374	697	153
6	148	280	58
7	35	46	10
8+	17	33	8
Total	13 728	19 136	4965

Source: ONS

Age group	Number of people		
	YO11	YO12	YO13
0–4	1531	2536	455
5–9	1245	2340	483
10–15	1697	3165	675
16–19	1618	2180	443
20–29	4097	4827	837
30–44	4821	7795	1604
45–59	5761	8979	2528
60–74	5245	7711	2722
75–89	2667	3764	1267
90+	372	452	106
Total	29 054	43 749	11 120

Use statistical diagrams and measures to display and analyse this information.

Write a report on your findings. Include a comparison of the distribution in ages and household sizes in the three areas.

Consolidation exercise 1

1 a) Describe what is meant by
   i   a random sample
   ii a representative sample.
   b) Give one example of the use of
   i   cluster sampling
   ii quota sampling.
   c) The table gives the number of students in each year of an engineering course.

	Year 1	Year 2	Year 3
Male	17	19	23
Female	12	11	8

   The course tutor wants to select a representative sample of 20 students.
   i   How many students from each group should he select?
   ii Suggest two ways he could select the students he needs from each group.

2 A biology student measures the heights of plants grown outdoors and indoors.

   The results in centimetres are in the table.
   a) i   Draw a back-to-back stem-and-leaf diagram.
      ii Write down the modes for each group of plants.
      iii Work out the range of heights for each group.
   b) i   Draw a box and whisker plot for each distribution.
      ii Describe and compare the distributions.

Outdoors	15	22	11	26	31
	27	16	18	25	9
	32	10	13	23	19
	11	17	21	20	17
Indoors	18	29	29	15	22
	17	23	35	27	32
	22	27	18	14	10
	27	28	30	34	12

3 The table gives college students' marks in an A level French test.
   a) i   Draw a cumulative frequency graph.
      ii Find the 40th percentile. What does this tell you?
      iii The pass mark for the test is 40%. Estimate the number of students who passed.
   b) Draw a box and whisker plot to illustrate the distribution of marks.

Marks	Number of students
1–10	0
11–20	8
21–30	11
31–40	23
41–50	17
51–60	5

4 In a survey Julie asks students how much they spent on clothes last month.

   She groups her results in this table.
   a) Julie says 'My data is continuous, secondary data.'
       Is Julie correct? Explain your answer.
   b) Draw a histogram for Julie's data.
   c) Find an estimate for the mean amount students spent on clothes and the standard deviation.

Amount (£P)	Number of students
$P < 20$	8
$20 \leqslant P < 30$	19
$30 \leqslant P < 40$	27
$40 \leqslant P < 60$	18
$60 \leqslant P < 100$	14
$P \geqslant 100$	2

5   Tanya and Oliver have drawn these diagrams to show the consumption of petrol and diesel
    in the UK between 2000 and 2013.

Tanya's graph

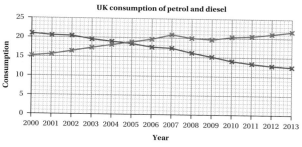

Oliver's chart

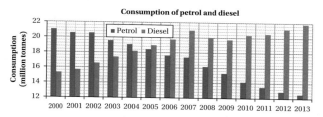

Data from www.gov.uk (Department for Transport)

a)   List the ways in which each diagram can be improved.

b)   Describe how the consumption has changed over this time period.

6   The Local Authority in Blackpool has 39 primary schools, 17 secondary schools and 7
    colleges for 16–18 year olds. The Local Authority wants to find out the opinions of teachers
    in the area. One suggestion is for the Local Authority to choose one school or college from
    each group at random, then interview two teachers from each.

a)   Give two reasons why this is not a good sample.

b)   Describe a better sampling method the Local Authority could use.

7   The table shows box office takings for films.

a)   Write down the modal class.

b)   Give an estimate for **i** the mean **ii**
     the standard deviation.

c)   Estimate the median and interquartile
     range. Describe any problems that arise.

d)   Write three sentences explaining what
     the data shows.

Box office takings (£T million)	Number of films
$0 < T < 0.1$	433
$0.1 \leqslant T < 1$	133
$1 \leqslant T < 5$	70
$5 \leqslant T < 10$	28
$10 \leqslant T < 20$	20
$20 \leqslant T < 30$	6
$30 \leqslant T < 50$	8

Source: BFI Statistical Yearbook 2014 (Rentrak)

8   For non-emergency treatment, the waiting
    time to see a consultant should be no more
    than 18 weeks from referral. The table gives
    the waiting time results for one hospital.

    Comment on the hospital's performance.

    Use statistical measures and/or diagrams to
    support your comments.

Waiting time	% of patients
Less than 5 weeks	2
5–9 weeks	17
10–15 weeks	26
16–17 weeks	38
18 weeks	12
19 weeks	4
20 weeks	1
More than 20 weeks	0

UK fails to bridge the gender gap (October 2014)

The UK has fallen to 26th in the ranking of countries in the World Economic Forum's recent Global Gender Gap Report. This report ranks countries according to male/female equality in four categories: Economy, Education, Health and Politics. The UK's overall ranking has steadily declined from the forum's first league table in 2006 when the UK was 9th.

One area considered in the report is Health.

The table below gives information about people who died prematurely in the UK in 2013.

Age	Years of life lost		Population		Number of deaths	
	Female	Male	Female	Male	Female	Male
20 to 24	5544.0	4582.0	1 774 400	1 829 400	90	79
25 to 29	9128.7	9256.8	1 844 700	1 840 600	161	174
30 to 34	12 873.3	13 620.6	1 850 300	1 831 800	249	282
35 to 39	18 806.9	19 445.6	1 687 700	1 681 400	401	446
40 to 44	33 890.5	34 221.6	1 912 300	1 877 600	805	882
45 to 49	46 401.2	51 744.6	1 986 100	1 939 700	1244	1513
50 to 54	55 094.0	65 072.7	1 824 700	1 792 800	1690	2191
55 to 59	61 229.9	79 138.4	1 574 600	1 539 600	2179	3128
60 to 64	71 076.3	95 329.8	1 499 800	1 440 000	2999	4518
65 to 69	86 775.0	114 827.2	1 498 100	1 419 700	4450	6676
70 to 74	89 900.0	111 577.5	1 106 900	998 900	5800	8265
75 to 79	761.6	877.2	944 500	798 900	64	86
80 to 84	556.8	569.8	751 100	549 100	64	77
85 to 89	323.3	291.2	495 500	291 900	53	56
90+	170.0	93.0	323 100	127 300	50	31

Premature death means it could have been prevented if the people had been given timely and effective healthcare.

Source: Health & Social Care Information Centre (www.hscic.gov.uk)

Here is a question you could ask based on this information:

How does the mean number of years lost by the females in each age group compare with the mean number of years lost by the males?

Write down two more questions that you could ask.

10 One of the other areas in the Global Gender Gap Report is Economy.

Q Use the internet to find out more about gender inequality at work.

Write a report or prepare a PowerPoint presentation on gender inequality including statistical measures and diagrams that support your findings.

Paper 1

11 Q The Civil Aviation Authority collects data about passengers using UK airports.
The tables give some of the data for people arriving at Gatwick and Heathrow in 2013.

Travelling group sizes

No of people in group	% of groups	
	Gatwick	Heathrow
1	36.8	63.3
2	36.1	23.9
3	4.9	4.5
4	12.8	4.1
5	3.7	1.2
6 or more	5.8	3.1
Thousands of passengers	32 402	45 744

Age distribution

Age (years)	% of passengers	
	Gatwick	Heathrow
10 or less	3.2	0.9
11–15	3.4	1.6
16–19	4.9	4.2
20–24	10.5	9.8
25–34	18.5	24.0
35–44	16.0	20.3
45–54	17.9	18.5
55–59	7.6	7.3
60–64	8.3	6.1
65–74	8.0	5.9
Over 74	1.7	1.4
Thousands of passengers	32 402	45 744

Source: CAA Passenger Survey report

Trip length

No of days x	% of passengers	
	Gatwick	Heathrow
$x < 1$	2.7	4.7
$1 \leqslant x < 3$	11.4	13.1
$3 \leqslant x < 6$	30.4	25.1
$6 \leqslant x < 8$	24.3	12.7
$8 \leqslant x < 15$	23.4	21.8
$x \geqslant 15$	7.8	22.7
Thousands of passengers	32 402	45 744

a) Use the data to compare the use of Gatwick and Heathrow by passengers who travel by air.

Q b) Use the internet to find and compare data for other airports.

Review

After working through this chapter you should:

- know the meaning of these terms applied to data: qualitative, quantitative, discrete, continuous, primary, secondary
- appreciate the strengths and limitations of random, cluster, stratified and quota sampling methods and use this understanding to design and use sampling strategies
- understand that increasing the size of a sample can improve accuracy, but may cost more in both time and money.

You should now be able to

- represent data numerically by finding the mean, median, mode, quartiles, interquartile range and standard deviation (from raw data, a frequency table or statistical diagram)
- represent data appropriately using histograms, cumulative frequency graphs, stem-and-leaf diagrams and box and whisker plots
- interpret statistical measures and diagrams and reach conclusions based on them.

Investigation

The housing crisis

- Home ownership now out of reach for young people
- House prices now almost 7 times income
- More than 2000 people are sleeping rough
- Young adults cannot escape the family nest
- $\frac{1}{3}$ of privately rented homes are not fit for purpose
- 250 000 new homes needed each year

Many of the problems are blamed on the lack of housing and it is estimated that 250 000 new homes are needed per year to solve the problems.

UK house prices

	All dwellings (%)		1st time buyers (%)	
	2006	2014	2006	2014
Under £80 000	6.9	4.4	14.5	8.2
£80 000 to under £100 000	8.0	4.7	14.2	7.9
£100 000 to under £125 000	11.9	8.5	18.2	13.2
£125 000 to under £150 000	13.1	9.9	15.9	13.5
£150 000 to under £200 000	23.0	19.8	20.4	21.6
£200 000 to under £250 000	15.0	15.5	9.4	13.3
£250 000 to under £300 000	7.4	10.0	3.5	7.5
£300 000 to under £400 000	7.6	12.0	2.4	7.7
£400 000 to under £500 000	3.4	6.4	0.8	3.6
£500 000 to under £1 000 000	3.3	7.4	0.6	3.0
£1 000 000 and over	0.5	1.5	0.1	0.5

Data from the Regulated Mortgage Survey

Borrowers' incomes (UK)

	Under £20 000	£20 000 to £24 999	£25 000 to £29 999	£30 000 to £39 999	£40 000 to £49 999	£50 000 and over
1996	47.1%	16.6%	11.4%	11.9%	5.5%	6.5%
2014	5.0%	6.1%	7.8%	18.1%	16.4%	46.5%

Data from the Regulated Mortgage Survey

Age of borrower: years (UK)

	Under 25 (%)	25 to 34 (%)	35 to 44 (%)	45 to 54 (%)	55 or over (%)
All dwellings					
1996	12.0	44.7	24.8	12.7	5.7
2014	6.1	41.4	30.4	16.8	5.3
First time buyers					
1996	22.1	52.7	16.0	6.3	2.8
2014	13.7	60.2	18.8	6.0	1.3

Data from the Regulated Mortgage Survey

UK Households by size

No of people in household	No. of households (thousands)	
	1996	2014
1	6 608	7586
2	8 094	9342
3	3 918	4311
4	3 469	3708
5	1 186	1208
6 or more	463	548
Total	**23 738**	**26 703**

Source: ONS (Labour Force Survey)

The charity Shelter estimates that there are 29 000 privately owned houses standing empty in London.

Use data given in the tables and/or from the internet to investigate the reported problems.

What effect do you think other aspects such as unemployment, relationship break-up, migration and lending by banks and building societies have had on the housing situation?

Could enforcing the use of empty buildings and building on brown-field sites help?

Use the internet to find data to test your ideas.

Write a report or prepare a presentation on your findings.

Include:

- calculation and interpretation of statistical measures
- statistical diagrams
- comments on how you might have done things differently.

2 Personal finance

Before borrowing money from a friend, decide which you need most.

There are three kinds of people, the haves, the have-nots and the have-not-paid-for what-they-haves.

Earl Wilson

In this chapter, you will study things that affect how you can make the most of your income. These include taxation, interest on investments, the cost of credit and the effect of inflation.

Invest in yourself. Your career is the engine of your wealth.

Paul Clitheroe

Many work hard for money, a few make money work hard for them.

You should know how to:

- use spreadsheet formulae such as A1+A2+A3   2*B3   SUM(A1:A10)
- substitute numerical values into formulae and expressions 1186, 1187
- interpret fractions and percentages as operators 1962
- work out one value as a percentage of another 1302
- solve percentage problems using a multiplier. 1060, 1073

Paper 1

Annual income £10 000, annual expenditure £9999, result happiness.

Annual income £10 000, annual expenditure £10 001, result misery.

(after Wilkins Micawber, David Copperfield, Dickens)

This chapter is about understanding and managing the flows of money illustrated in this diagram.

The individual parts of the diagram are covered in the indicated sections of this chapter.

This section gives a general overview of the process.

Interest on investments [2.7]

Earned income

Interest on loans [2.4, 2.5, 2.6]

Tax [2.2, 2.3]

Other expenditure [2.8]

Many people keep track of the flow of their money by reading their current account statements, which give details of all the transactions that have taken place.

A current account is your main day-to-day bank account where you can pay in and withdraw money and set up direct debits for regular payments.

HIGH STREET BANK

Mr Name

His address

31 Aug – 30 Sept 2015

Sort code 00-14-25

Account number 12345678

Date	Payment details	Paid out	Paid in	Balance
31 Aug 15	**Brought fwd.**			**87.15**
1 Sep 15	DD Credit card	30.00		57.15
1 Sep 15	ATM Cash	100.00		OD 42.85
1 Sep 15	Unauthorised OD	25.00		OD 67.85
3 Sep 15	My employer		1100.00	1032.15
4 Sep 15	VIS Megastore	95.00		937.15
8 Sep 15	DD Landlord	500.00		437.15
10 Sep 15	DD TV Licence	12.12		425.03
12 Sep 15	VIS Hyperstore	90.78		334.25
16 Sep 15	CHQ 123456	100.00		234.25
18 Sep 15	ATM Cash	100.00		134.25
24 Sep 15	DD Visiphone	20.00		114.25
30 Sep 15	DD Gaslec	53.60		60.65
30 Sep 15	OD Interest	0.45		60.20
	Carried fwd.			**60.20**

DD Direct debit
ATM Cash machine (automated teller machine)
OD Overdrawn (when funds do not cover a payment)
VIS A debit card charged directly to the account
CHQ Cheque

Activity 1 💬

★ Carefully study the bank statement and check that you understand each entry. Use the summary of abbreviations to help you.

★ Think about how Mr Name has organised his finances. Make a list of things he has done which seem good.

★ What aspects of the management of Mr Name's finances are not very good?

★ What are Mr Name's total outgoings (expenditure) for this month? Compare this with his income.

When you are organising your finances some things, such as income and housing costs, are often fixed. However, you can often make some small changes which help you save money and avoid paying excess charges.

Mr Name appears to be spending almost exactly what he earns. This makes it even more important not to lose £25.45 in overdraft charges through poor management.

Activity 2

How could the £25.45 have been saved by planning ahead?

There is one particular item on this bank statement which might be of concern and that is the Direct Debit of £30 to the credit card company. It looks like this could be a small fixed repayment rather than a payment for the full credit card bill. If that is the case, then this is an example of poor management of money.

Some people wait for their current account statement to arrive each month and are then a bit surprised, as it may be better or worse than they expected. An improvement on this is to use internet banking so you can look at your statement at any time and keep an eye on your finances. However the only way to budget carefully is to plan ahead. This can easily be done by setting up a spreadsheet for yourself just as you would for a small business.

Activity 3

★ Set up a spreadsheet that you can use to replicate Mr Name's statement.

Your spreadsheet should use his income and direct debits as fixed amounts. It also needs to be able to deal with the entry of cheques, ATM withdrawals and debit card payments as you go through the month.

★ Now suppose that you are living away from home, either at university or working on your first job.

Making any necessary assumptions about costs and income, modify your spreadsheet to allow you to budget for this period.

If you withdraw money or pay for something, but there is not enough money in your account to cover the transaction, you may end up being overdrawn. The bank may still carry out the transaction but will charge you for this service. If you authorise an overdraft with your bank the charge will be considerably smaller.

When you use a credit card to buy something, the credit card company lends you the money. They usually charge high interest rates so if you don't pay off your bill in full you can end up paying back much more than you originally borrowed.

You will learn more about managing debt in Section 2.4.

2.2 Income tax

Income tax is a charge on personal incomes. Each year, the personal tax allowances, income tax rates and tax bands are set by the government.

Most people have a personal allowance. This is the amount of tax-free income you are allowed each year. The personal allowance varies each year. For 2015–16 it was £10 600.

The amount of income tax you pay depends on how much taxable income you have above your personal allowance. There are different rates of income tax depending on what your income is.

This table shows the income tax rates and taxable bands for 2015–16.

The government uses income tax to raise money to pay for things such as defence, health care, education, benefits, law and order, transport and interest on the national debt.

Rate	Taxable income
Basic: 20%	£0 – 31 785
Higher: 40%	£31 786 – £150 000
Additional: 45%	Over £150 000

Key point

To calculate your income tax (for all but the highest earners):

Find your taxable income by subtracting your personal tax allowance from your annual income.

You pay income tax at 20% on the first £31 785 of your taxable income.

You pay income tax at 40% on your taxable income over £31 785.

Example 1 During the year 2015–16, Katherine earned £85 000. She had a personal tax-free allowance of £10 600.

a) Calculate the amount of income tax which Katherine paid in the year.

b) What percentage of her salary did she pay in income tax?

a) Income   £85 000   First calculate her taxable income.

Allowance   £10 600

Taxable income   £74 400

Then split the taxable income into different bands.

Work out how much income is in the basic tax band and how much is in the higher tax band.

Basic   £31 785

Higher   £74 400 − £31 785 = £42 615

Then calculate the actual amount of tax to be paid.

Basic   20% of £31 785 = £31 785 × 0.2 = £6357

Higher   40% of £42 615 = £42 615 × 0.4 = £17 046

The total amount of tax Katherine pays = £6357 + £17 046 = £23 403

$20\% = \frac{20}{100} = 0.2$
A quick way to calculate 20% of a quantity is to multiply by 0.2   1962

b) To work out the amount of tax as a percentage of her salary calculate

$\dfrac{\text{total tax}}{\text{total income}} \times 100\%.$ ⊕ 1961

$\dfrac{23\,403}{85\,000} \times 100\% \approx 27.5\%$

Example 2
During the year 2015–16, Hussein earns £4035 per month and has a tax-free allowance of £10 600.

a) How much income tax is taken from his salary each month?

b) What percentage of his salary is taken in income tax?

c) Comment on your answers to Example 1b and Example 2b.

---

a) Annual income   £4035 × 12 = £48 420

First calculate his taxable income.

Allowance	£10 600
Taxable income	£37 820

Then split the taxable income into different bands.

Basic	£31 785
Higher	£6035

Then calculate the tax.

£31 785 × 0.2 = £6357

£6035 × 0.4 = £2414

Total   £8771

Convert this yearly tax to monthly tax.

Monthly $\dfrac{£8771}{12} = £730.92$

Hussein pays £730.92 per month in tax.

> For 2015–16, a person paying any amount of higher rate tax will always pay tax of £6357 at the basic rate.

> Tax systems (such as in the UK) where people with a higher income pay a higher overall percentage of their income in tax are said to have a 'progressive' tax system.

b) Find £730.92 as a percentage of £4035.

$\dfrac{730.92}{4035} \times 100\% \approx 18.1\%$

c) The person with the higher income pays an overall higher rate of tax.

Activity 4

★ Set up a spreadsheet to calculate how much income tax a person with an annual income of £100 000 or less would have to pay in 2015–16.

🔍 Research what special rule applies to a person whose income is over £100 000.

Exercise 2A

1   During the year 2015–16, John is employed on an annual salary of £21 588. He has a personal tax-free allowance of £10 600. What income tax does he pay that year?

2   During the year 2015–16, Salma is employed on an annual salary of £51 588. She has a personal tax-free allowance of £10 600. What income tax does she pay that year?

2.3 Your payslip

When you start work you will receive a payslip each time you are paid. This is usually once a month.

Most of the entries on the payslip are calculated automatically from just two numbers, your salary and your personal tax allowance. Errors are therefore not that common, but you should certainly check that you are on the correct salary scale and (perhaps the most likely source of error) that the correct tax allowance is being used.

> Almost every adult resident in the UK has a unique NI number. It is used for social security, taxation and identity purposes.

Employee No.	Employee	Pay Date		National Insurance No.	
01234	Mrs Name	31/09/2015		AB123456C	

Payments	Units	Rate	Amount	Deductions	Amount
Salary	1	1750.00	1750.00	PAYE Tax	173.33
				PAYE NI	129.36
				Personal pension	50.00
				Student loan	27.48
				Total deductions	380.17

Tax period 6	Total gross pay 1750.00			Totals year to date	
				Total gross pay TD	10500.00
				Tax paid TD	1039.98
				NI TD	776.16
				Pension TD	300.00
				Student loan TD	164.88
Tax code 1060L	Payment method: BACS			Net Pay	1369.83

> Gross pay – your pay before any deductions such as tax or National Insurance.

> April is the first month in the tax year and is called tax period 1. September is therefore tax period 6.

> BACS – Bank Automated Clearing System. The salary is paid direct to a bank account.

> This is your personal allowance divided by 10.

> PAYE – Pay As You Earn. This is the system of paying tax and National Insurance contributions automatically from your wages.

> This is put into an account to supplement your state pension.

> Your pay slip is likely to contain running totals for the tax year. In this example the deductions have been the same each month and so the totals are just 6 times the monthly amounts.

> Net pay (or 'take home pay') – your pay after all deductions.

Activity 5 💬

★ Carefully read through the payslip and check that you understand the various entries.

★ Use your knowledge of taxation to show that the figure of £173.33 for the PAYE tax is correct.

★ What percentage of her gross pay does Mrs Name actually receive?

★ You might well be surprised by how much of Mrs Name's salary is taken in deductions. Many people are surprised by this when they receive their own first payslip!

National Insurance

National Insurance (NI) is another charge on a person's salary.

This table shows National Insurance payments for 2015–16.

Percentage National Insurance due	Minimum weekly income	Maximum weekly income	Minimum monthly income	Maximum monthly income
Non-contracted out				
Nil		below £112		below £486
0%	£112	£155	£486	£672
12%	£155.01	£815	£672.01	£3532
2%	above £815		above £3532	
Contracted out				
10.6%	£155.01	£770	£672.01	£3337

If you have a weekly income of £250, you will pay 12% on the amount of your income above £155.

If you have a weekly income of £850, you will pay 12% on the amount of your income between £155.01 and £815 plus 2% of the amount above £815.

> **Example 3** During the year 2015–16, Karl earned £23 484. He paid NI at the non-contracted out rate. Calculate the amount of NI he paid each month.
>
> ---
>
> First calculate Karl's monthly salary.
>
> Monthly salary $= \dfrac{£23\,484}{12} = £1957$
>
> Karl pays 0% on the first £672.   £672 × 0 = 0
>
> Karl pays 12% on the amount above £672.
>
> Income above £672 is £1957 − £672 = £1285
>
> 12% of £1285 is £1285 × 0.12 = £154.20
>
> Karl paid £154.20 NI each month.

Student loans

Once you are earning above a certain amount, you have to start repaying your student loan.

> **Key point**
> During the year 2015–16, student loans were paid back at a rate of 9% of gross earnings in excess of £17 335.

Both employees and employers make NI contributions. The money pays for state pensions and other workers' benefits.

For the purpose of calculating your NI contributions you can treat the nil band and the 0% band as being just one band.

A person is 'contracted out' if they are a member of a contracted out occupational pension scheme or personal/stakeholder pension. They pay a slightly lower rate of National Insurance as the state does not have to pay them as much pension. Contracting out ends in April 2016.

Karl's monthly income is less than £3542 so you only need to look at the information for the 0% and 12% bands.

$12\% = \dfrac{12}{100} = 0.12$ so you can calculate 12% by multiplying by 0.12
⊕ 1963

The size of your loan doesn't affect the repayments. The rate you pay back each year depends only on your income and not on how much you borrowed.

Example 4 Sean has a gross income of £18 000 p.a. What amount of his student loan does he have to repay every month?

p.a. stands for per annum so Sean's annual income is £18 000.

First work out how much Sean earns above the £17 335 threshold and then calculate 9% of this amount.

Earnings above the threshold = £18 000 − £17 335 = £665

9% of £665 = 0.09 × £665 = £59.85

This is the amount Sean repays in a year, but the question asks how much he repays each month so you have to divide by 12.

£59.85 ÷ 12 = £4.98

He repays £4.98 per month.

Deductions from your salary are usually rounded down.

Activity 6

★ Check the figures for NI and the repayment of the student loan on the payslip.

Exercise 2B

1 Alan earns £4230 per month and has a tax-free allowance of £9000. Calculate

   a) his taxable income
   b) the amount of income tax that Alan pays in the year.

The standard personal allowance can be adjusted for various reasons so not everyone's tax-free allowance is £10 600.

2 For someone with a personal allowance of £10 600, the graph of their income tax and NI contributions for 2015–16 has this shape.

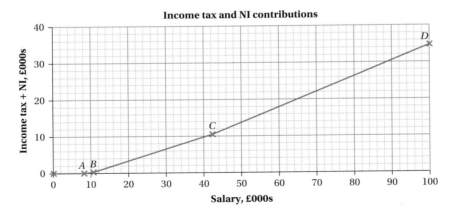

Income tax and NI contributions

a) Describe the significance of each of the points A, B and C in terms of income tax and NI thresholds.
b) For each of the points A, B, C and D, estimate what percentage of a person's salary is taken in income tax and NI.

3 During the year 2015–16, Ann earned £950 per week. She paid NI at the non-contracted out rate. Calculate the amount of NI she paid each week.

4 Misha earns £435 per week. She is contracted out for National Insurance contributions. Calculate the amount Misha pays weekly in NI contributions.

5 During the year 2015–16, Sasha was employed on a salary of £21 588. She had a personal tax allowance of £10 600 and paid NI at the non-contracted out rate. Throughout, give answers to the nearest penny.

    a) Calculate Sasha's monthly NI payments.
    b) Calculate the monthly income tax that she pays.
    c) Calculate the monthly repayments of Sasha's student loan.
    d) Find Sasha's monthly take home income after she has paid National Insurance, income tax, pension of £50 per month and student loan repayments.
    e) Copy and complete this payslip for Sasha.

Employee No.	Employee	Date	National Insurance No.		
01234	Sasha	31/09/2015	AB123456C		
Payments	**Units**	**Rate**	**Amount**	**Deductions**	**Amount**
Salary	1	☐	☐	PAYE Tax	☐
				PAYE NI	☐
				Personal pension	50.00
				Student loan	☐
				Total deductions	☐
Tax period 6	**Total gross pay** ☐			**Totals year to date**	
				Total gross pay TD	☐
				Tax paid TD	☐
				NI TD	☐
				Pension TD	☐
				Student loan TD	☐
Tax code 1060L	**Payment method: BACS**			**Net Pay**	☐

6 Jack was expecting to be paid £80 000 in the 2015–16 tax year. However, he was given a pay rise of 10% at the start of the tax year. He had a personal tax allowance of £10 600.

The questions below relate only to the extra charges levied as a result of this 10% pay rise.

Give all answers to the nearest penny.

    a) Calculate the additional non-contracted out National Insurance that Jack pays.
    b) Calculate the additional annual income tax that he pays.
    c) Jack has not yet paid off his student loan. How much extra annual pay did he receive from his 10% pay rise after the deductions due to National Insurance, income tax and student loan repayments?

(AQA, 2013)

The flows of money you receive and spend throughout your life will probably be very uneven. At times, you are likely to spend more than you earn, for example if you are at university, buying a house or retired. At other times you may earn more than you spend, for example if you are a young professional living at home or have a high enough income to allow you to save.

You will probably need to borrow money at some times and invest it at others to cope with these changes. It is important that you understand how to borrow sensibly and control your debt so that you don't get into difficulties paying back what you owe.

In the next few pages you will learn some important things about controlling debt. The theory is not too difficult; it's putting it into practice that can be hard!

Key point

If you want to borrow money you will have to pay it back with interest. This can be quite expensive, so:

- keep this cost (the interest rate) to a minimum
- make sure the cost is justified by the benefit you receive from having the money.

Different banks will have different interest rates and there may be other charges or conditions so if you need to borrow money it's worth doing some research first.

Controlling debt is as important in business as it is in personal finance. Many football clubs have extensive debt and their successes and failures in dealing with these debts provide useful lessons for everyone.

The debts incurred by Manchester United Football Club (MUFC) when the Glazer family acquired a controlling share in 2006 were a matter of great concern to fans.

One group of fans even broke away and founded another football club, F.C. United.

The bar chart shows the interest payments for these debts since 2006.

The estimates for 2015 and 2016 are based upon the outstanding debts before and after the refinancing which occurred in May 2015. There were two types of debt:

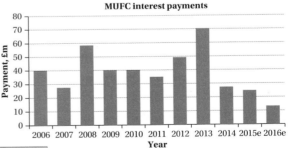

MUFC interest payments

	Bank Loan	US Bonds
2015	£206m 2.54% variable	£176m 8.375% fixed
2016	£147m 1.79% variable[†]	£278m 3.79% fixed*

Source: *the andersred blog*

[†]This rate depends upon both MUFC's revenue and LIBOR.

LIBOR is an inter-bank lending rate.

*Fixed until 2027.

There is no point in having debt at all unless you are putting the money to good use. In this case, it is sensible to borrow if it enables a player to be bought who significantly increases the club's chances of success.

For most people, the current ways of borrowing together with some typical interest rates are as shown. However, note that these rates can be extremely variable.

Payday loans	Too large to be considered. Some rates can be over 1000%
Credit cards	20–30%
Bank loan	4%
Mortgage	4%
Student loan	5.5%

Credit cards

Credit cards can be a sensible way to borrow money for the short term if you use them carefully. They may have very attractive gimmicks and introductory offers but the typical interest rates (excluding introductory offers) are too high to pay for an extended period.

If you have a lot of credit card debt to repay, following this flow chart is sensible.

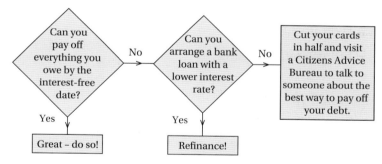

This will mean your remaining debt has a much lower interest rate and will be easier to control.

Student loans

Student loans have several features that make them significantly more attractive than you might think from the interest rate.

- Your repayments are determined by your income so they should always be affordable.
- The debt will be *totally* written off after 30 years.

Key point

To compare loans, compare their interest rates. The figure called APR is the best figure to use for this comparison and is explained in the next section.

Activity 7

★ Show that refinancing increased MUFC's debt.

★ Use the figures in the table to justify the two estimated bars of the bar chart.

★ Do you think MUFC's debt is under control?

Refinancing is the process of changing from one source of borrowing to another. This is usually done in order to improve the terms of the loans in one way or another.

It is expected that many students will never pay off their student loan debt. For these students the actual interest rate is likely to be negative!

2.5 Annual percentage rate (APR)

Loans can have all sorts of fees and special conditions which make them difficult to compare. The government therefore introduced a standard figure called the APR. You should always look for this number and ignore any other rates that you might be quoted. This example illustrates why.

> **Borrow £1000 today**
> Two simple repayments:
> £600 after 1 year
> £600 after 2 years

You have to repay £1200 so you might easily think of this as £100 interest per year so a 10% interest rate. However, you pay back £600 after 1 year so you only have access to the full £1000 for a single year. This means the actual interest rate, the APR, is higher than 10%.

In this case the APR is approximately 13%. Before you look at how you (or an accountant) might work this out, it is probably best to check that this percentage rate does match the borrowing in the advertisement in a way that agrees with common sense.

Example 5 Suppose you borrow £1000 at an annual interest rate of 13% and you pay back £600 after 1 year.

a) How much do you owe at the end of 1 year after making this payment?

b) To clear the loan, how much extra would you need to repay after a further year?

a) Calculate the total value of the loan after one year and then subtract the payment.

£1000 × 1.13 = £1130

You repay £600

So you still owe £530

> To calculate an increase of 13%, multiply the original value by 1.13 ⊞ 1060

b) After the second year you must repay the £530 plus the year's interest.

Total amount to repay = £530 × 1.13 = £598.90

As you can see, this is very close to the £600 of the advertisement.

This shows that the APR is indeed very close to 13% and not the lower 10% rate that advertisements like this used to quote before the government legislation.

There is a formula which is used to obtain the APR.

Assume that

- a loan of £C is taken out
- i is the APR expressed as a decimal (for example, if the APR is 18% then $i = 0.18$)
- the loan is repaid in m instalments of

$$£A_1 \text{ after } t_1 \text{ years}$$

$$£A_2 \text{ after } t_2 \text{ years}$$

...

$$£A_m \text{ after } t_m \text{ years}$$

Key point

The APR i is given by the equation

$$C = \frac{A_1}{(1 + i)^{t_1}} + \frac{A_2}{(1 + i)^{t_2}} + \cdots + \frac{A_m}{(1 + i)^{t_m}}$$

Example 6 A loan was taken out at APR 30%. £1000 was repaid after 1 year, a further £1000 was repaid after 2 years and a final payment of £1404 was made after 3 years. How much was the loan for?

Use the APR equation and substitute in the values. ⊕ 1186

$$C = \frac{1000}{1.3} + \frac{1000}{1.3^2} + \frac{1404}{1.3^3}$$

$$= 2000$$

The loan was for £2000.

You can also use the formula to work out the value of i (the APR).

For Example 5 putting all the numbers into the formula gives

$$1000 = \frac{600}{1 + i} + \frac{600}{(1 + i)^2}$$

This equation can be solved by various methods including quadratic equations, trial and improvement and an equation solver on a computer or calculator. With more complicated equations of this type, a good method is to use a spreadsheet so that you can quickly try different values of i.

Activity 8

In Example 5, the first repayment occurs after 1 year, so $t_1 = 1$.

Write down the values of C, m, A_1, A_2 and t_2.

The formula for APR is often written using the mathematical symbol sigma, Σ.

This stands for 'sum'. Each term of the formula has the form $\frac{A_k}{(1 + i)^{t_k}}$ for $k = 1$, $k = 2, \ldots, k = m$.

This gives the 'official' form of the formula for APR.

$$C = \sum_{k=1}^{m} \left(\frac{A_k}{(1 + i)^{t_k}} \right)$$

Activity 9

Choose one method and check that $i = 0.13$ approximately solves the equation to show that 13% is the actual APR for this example.

Example 7 Rachael borrows £4000.

a) One lender asks her to pay this back in two equal instalments of £2500, one at the end of the first year and a final instalment at the end of the second year.

Show calculations to confirm that the APR is 16.26%.

b) Another lender offers Rachael the same method of paying back a loan of £4000 in two equal instalments at the end of each of the first year and the second year but with an APR of 15%.

Calculate the value of each instalment.

(AQA, 2012)

a) Use the formula for APR and substitute in values for A_k, t_k and i.

$$C = \sum_{k=1}^{m} \left(\frac{A_k}{(1 + i)^{t_k}} \right) = \frac{A_1}{(1 + i)^{t_1}} + \frac{A_2}{(1 + i)^{t_2}}$$

$A_1 = 2500$, $A_2 = 2500$, $t_1 = 1$, $t_2 = 2$, $i = 0.1626$

So, $C = \dfrac{2500}{1.1626} + \dfrac{2500}{(1.1626)^2}$

> $i = 0.1626$ is 16.26% expressed as a decimal.

$\qquad = 3999.96$

This is extremely close to £4000 so this confirms that the APR is 16.26%.

b) Use the formula for APR again, but this time substitute in values for C, t_k and i.

$$C = \sum_{k=1}^{m} \left(\frac{A_k}{(1 + i)^{t_k}} \right) = \frac{A_1}{(1 + i)^{t_1}} + \frac{A_2}{(1 + i)^{t_2}}$$

$C = 4000$, $t_1 = 1$, $t_2 = 2$, $i = 0.15$

The two repayment amounts are equal so you can write A for each one.

$$4000 = \frac{A}{1.15} + \frac{A}{1.15^2}$$

Change the fractions to decimals and solve the equation to find A.

$4000 = 0.8696A + 0.7561A$

$4000 = 1.626A$

$\qquad A = 2460.02$

The value of each instalment is £2460.02

> $\dfrac{A}{1.15} = \dfrac{1}{1.15}A$
>
> $\qquad = 0.8696A$

Exercise 2C

1 a) If a loan at APR 15% is repaid in three equal annual instalments of £437.98, state the values in the APR formula of m, A_1, A_2, A_3, $1 + i$, t_1, t_2 and t_3.

b) Find the amount of the loan.

2

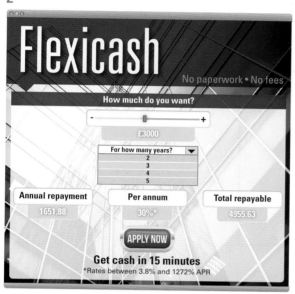

Check that the rate 'per annum' in this advertisement is actually the APR.

> The * indicates that you should read the small print. In this case the rate is clearly just an illustration and is not guaranteed.

3 When buying a camera costing £750, Asif is offered two different repayment methods.

Method A: paying back a single lump sum of £1000 at the end of 2 years.

Method B: paying back £450 after 1 year and a further £450 after another year.

a) Calculate the APR for repayment method A.

b) Repayment method B advertises an APR of 13.1%.

Show calculations to confirm that this is a good approximation.

(AQA, 2008)

4 Imogen considers borrowing £1500 from a lender offering loans at an APR of 25%. She would have to pay back the loan in two equal repayments, the first at the end of 1 year and the second at the end of 2 years.

a) Find the amount of each repayment.

b) A different lender offers the loan for a single payment of £2000 at the end of two years. What is the APR in this case?

5 Tom wants to borrow £1500. He wants to pay this back in two equal instalments, the first instalment at the end of 1 year and the second at the end of 2 years.

a) One lender offered the loan on the basis that Tom paid back £1000 for each of the two instalments. Show calculations that confirm that the APR in this case is approximately 21.5%.

b) A different lender offered an APR of 20%. In this case, how much would Tom pay back in each instalment?

(AQA, 2009)

2.6 Mortgages

Suppose you and a friend each have gross salaries of £27k per annum (this is roughly the national average) and that you want to buy a house together that costs £250 000 (this is roughly the cost of an 'average house' in many places).

> k is used to represent thousands so £27k means £27 000.

A typical mortgage calculator tool on the internet will show you what you might be able to borrow.

How much can I borrow?

No. of applicants	2
Applicant 1 salary	£27 000
Applicant 2 salary	£27 000
You can borrow	£216 000*

*These figures are for illustration only. All lending is subject to status, a full financial assessment and a survey of the property.

Activity 10

★ Can you spot how the amount 'you can borrow' has been calculated?

★ What percentage of the purchase price is the amount £216 000?

Q In practice, the percentage of the purchase price that you need to borrow can be an important consideration for you and the lender. Why is that?

Even if you are able to get a mortgage for such a large percentage of the house price at a reasonable interest rate there is another thing to consider. You will still need a large amount for the cash deposit. Even for the 'average' property in the example above £34 000 is needed on top of the £216 000 you can borrow. This is one of the major reasons young adults have for saving significant amounts of their incomes.

The repayment of a mortgage is usually based upon equal monthly payments to the lender. For example, suppose you have:

- a mortgage for £150 000
- repayments of £1000 per month
- gross annual interest of 5%, so that at the end of each year 5% of the amount owed at the beginning is added to the mortgage.

The initial amount owed, $£A_0$, is £150 000.

Activity 11

★ Why do lenders only allow people to borrow up to a fixed multiple of their salaries?

★ Why do they (in general) also expect the borrower to pay a large deposit?

After 1 year, you have repaid £12 000, but interest of $0.05 \times 150\,000 = £7500$ is added.

> 5% of £150 000 = 0.05 × £150 000.

The amount owing after 1 year, $£A_1$, is
£150 000 + £7500 − £12 000 = £145 500.

You have therefore paid £7500 in interest and only repaid £4500 of the mortgage.

The equation to carry out this calculation is $A_1 = A_0 \times 1.05 - 12\,000$.

Similarly, $\qquad A_2 = A_1 \times 1.05 - 12\,000$

$$A_3 = A_2 \times 1.05 - 12\,000$$

The general equation is

$$A_{n+1} = A_n \times 1.05 - 12\,000$$

> This is called a recurrence relation.

These are the results of the calculations for the first few years.

n	0	1	2	3
A_n	150 000.00	145 500.00	140 775.00	135 813.75

Activity 12

★ Explain where the 1.05 in the recurrence relation has come from.

★ Check the table of results given above and calculate the next few values in the table.

Exercise 2D

1 a) Set up a spreadsheet to carry out the calculations above for a large number of years.

   b) After how many years will this mortgage loan be repaid?

2 Suppose that monthly repayments of £1500 are made instead of £1000.

   When will the mortgage be paid off now?

3 In questions 1 and 2, roughly what are the totals of all the repayments in each case?

4 Harry takes a mortgage for £120 000. He repays £920 per month for the life of the mortgage. The mortgage has an interest rate of 0.6% per month which is fixed for the first 4 years.

The amount of the mortgage at the end of the nth month, $£A_n$, is given by

$A_n = 1.006A_{n-1} - 920$ where the amount at the start is $£A_0$ which is £120 000.

   a) Explain the significance of the number 1.006 in the recurrence relation.

   b) Use the recurrence relation $A_n = 1.006A_{n-1} - 920$ to give the amount of the mortgage debt outstanding at the end of each of the first six months.
   You should give your values correct to the nearest penny.

   c) Hence find the amount Harry will have paid off his mortgage in the first six months.

(AQA, 2013)

2.7 Savings and investments

Just as APR is the general figure that you should use to compare loans, there is a general figure that is designed to help you easily compare savings rates. This rate is called AER (annual equivalent interest rate). If an account pays interest only once every year then the AER is often just called the interest rate.

Example 8 Suppose you invest £2000 at a fixed interest of 3% per year.

a) Explain why the amount in the account after n years is given by £2000 $\times$ 1.03^n

b) How many years does it take for the amount to double?

a) The multiplying factor for a 3% increase is 1.03 and this must be applied n times to the amount of £2000.

b) 2000 $\times$ 1.03^n = 4000

Substitute different values for n in the equation. Use trial and improvement until you find one which gives a value close to 4000. ⊞ 1057

n = 20 gives 2000 $\times$ 1.03^{20} = 3612   (so 20 is too small)

n = 25 gives 2000 $\times$ 1.03^{25} = 4188   (so 25 is too big)

n = 23 gives 2000 $\times$ 1.03^{23} = 3947   (so 23 is too small, but it's very close)

n = 24 gives 2000 $\times$ 1.03^{24} = 4066   (so 24 is too big, but it's very close)

> After 23 years the amount has not quite doubled. In situations like this it is usual to give the answer that 'the amount has doubled after 24 years'.

By trial and improvement, 23 < n < 24. The amount will have doubled after 24 years.

Key point

After n years at an annual interest of r, an amount of £P will have grown to £$P(1 + r)^n$.

Note that the value of r should be expressed as a decimal. For example, for an interest rate of 3%, r = 0.03

$$£P(1 + r)^n = £P \times (1 + r)^n$$

Activity 13 ◯

★ Now suppose that interest is added to an account at, say, 0.2% per month.

★ What formula will give the amount in an account after 1 year if the initial deposit is £100?

★ Calculate this amount. Is it the same as if the account had received interest of 12 $\times$ 0.2% at the end of the year? If not, why not?

For an account paying interest monthly you might see any or all of three savings rates advertised. For example:

Monthly rate   0.2%

Nominal rate   2.4%

AER   2.43%

The nominal rate is just $12 \times 0.2\%$. It ignores the **compounding effect** that means interest obtained in earlier months earns interest itself in subsequent months.

> **Key point**
>
> **Financial calculation – AER**
>
> The general formula connecting the different rates is as follows.
>
> The annual equivalent interest rate (AER), r, is given by
> $$r = \left(1 + \frac{i}{n}\right)^n - 1$$
> where i is the nominal interest rate and n the number of compounding periods per year.
>
> Note that the values of i and r should be expressed as decimals.

The term 'simple interest' is sometimes used to refer to a quick method of calculating the interest on an account by ignoring the effect of compounding.

For the example on the left,
$n = 12$ since there are 12 months in a year
$i = 0.024$
$r = 0.0243$
The monthly rate as a decimal is $\frac{i}{n} = 0.002$

If you want to express the AER as a percentage, you will need to convert the decimal value of r to a percentage by multiplying by 100%. 1015

Exercise 2E

1 Pete deposits £3800 in an account for 12 months. It earns compound interest at the rate of 1.21%, paid every 6 months. Interest is rounded to the nearest penny.

a) Find the missing values A and B in the table.

b) Calculate the AER on this investment.

	Starting value (£)	Interest (£)	Final value (£)
First 6 months	3800.00	45.98	3845.98
Second 6 months	3845.98	A	B

2 Sheila deposits £800 in a Best Saver account for 9 months. It earns compound interest every 3 months. The nominal rate is 2% per annum.

a) What is the interest rate for each 3 month period?

b) Find the missing values A to E in the table above.

c) Calculate the AER for this account.

	Starting value (£)	Interest (£)	Final value (£)
First 3 months	800	4	804
Second 3 months	804	A	B
Third 3 months	C	D	E

3 Imran deposits £4000 in an account for 18 month. It earns 1.98% interest paid every 6 months.

a) How much money does Imran have in the account at the end of 18 months?

b) Find the nominal rate and AER for this account.

2.8 VAT and other percentages

Value added tax (VAT) is the third major tax revenue source for the UK government after income tax and National Insurance. The standard rate of VAT for 2015–16 is 20%.

The graph shows how the rate has varied.

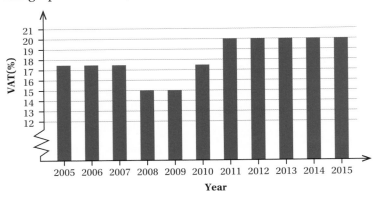

It is called a 'value added' tax because of the way the tax is collected.

Suppose that, without VAT, you buy a product for £300 from a business that had earlier bought the raw materials from a supplier for £100.

When you include VAT you have to pay £360 for the product and the business has to pay £120 for the materials.

The £60 extra that you have paid is collected according to the 'value added' at each stage.

The supplier pays tax of £20 (20% of the £100) and the business pays tax of £40 (20% of the £200 value it added to the product).

> **Key point**
> Using the VAT rate of 20%, gives two simple equations.
>
> $$\text{VAT} = 0.2 \times \text{Price excluding VAT}$$
>
> $$\text{Price including VAT} = 1.2 \times \text{Price excluding VAT}$$

The relationships between the different amounts are summarised in the diagram.

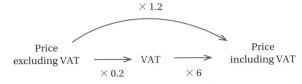

The same type of diagram used above for VAT is helpful for other questions involving percentage changes.

Activity 14

★ Explain the factor × 6 in the diagram.

★ You are given a price including VAT. How would you find

a) the amount of VAT

b) the price excluding VAT?

Key point

Many questions involving percentages are best solved using multiplying factors, for example,

A loss of 28%: New amount = 0.72 × Old amount

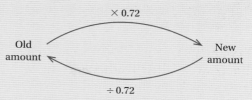

 1060

When you know the new amount and need to find the old amount, reverse the process. ⬢ 1073

Example 9 In 2015, the number of tickets sold for a festival was 209 520. This was a decrease of 12.7% on the number sold in 2014. How many tickets were sold in 2014?

A decrease of 12.7% means that the number sold in 2015 was 87.3% of the number sold in 2014.

Convert the percentages to decimals and show this on a diagram.

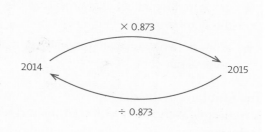

To find the number sold in 2014 divide the number of sales for 2015 by 0.873

$$\frac{209\,520}{0.873} = 240\,000.$$

Exercise 2F

1 A men's clothing store buys in ready-made suits from overseas and sells them with a mark-up of 50% for £270. What is the profit on each suit?

> Mark-up is another way of saying profit.

2 If you make a loss of 35% when you sell a car for £1625, how much did you pay for the car?

3 An antique dealer prices a table to give a profit of 150%. In a sale, the price of the table is reduced by 30%.

a) What is the dealer's percentage profit on the table?
b) The table was sold in the sale for £2100. What had the dealer paid for the table?

4 The price of a dining table was £1010.50, including VAT at $17\frac{1}{2}$%.

Later, VAT was increased to 20%. A notice was displayed in the showroom.

NOTICE

VAT has been increased by $2\frac{1}{2}$%.

The price of the table has been increased by

$$£1010.50 \times \frac{2\frac{1}{2}}{100} = £25.26$$

and will now cost £1035.76

Was the notice correct? Fully justify your answer.

2.9 Exchange rates

A country's currency has a value that fluctuates relative to other currencies. Banks buy and sell currencies from each other and these trades set a market price for each currency.

The diagram shows some currency exchange rates at one point during June 2015.

Activity 15

★ What mathematical relationship would you expect between the numbers in the diagram? Check if you are correct.

★ Work out an equivalent diagram for currency exchange in the reverse direction.

The numbers in the diagram are called 'mid-market rates'. There are charges associated with trading and so, at the time the rates were as shown above, the actual prices on one currency exchange market were:

	Bid	Ask
GBP/USD	1.5273	1.5284

A bank using these prices would receive $1.5273 for each £1, but would receive £1 for every $1.5284.

This 'bid-ask' spread for banks is relatively small. However, the exchange rates when a holidaymaker buys and sells a currency are considerably greater. For example, on 7th June 2015 the Post Office website was offering $1.469 for every £1 but was buying back dollars at $1.720 per £1.

> Some half a trillion dollars of trades take place each day just between pounds (GBP) and dollars (USD).

> The 'bid-ask' spread is one of the main ways the businesses providing you with cash make their profit.

Example 10

Exchange rate when buying dollars 1.469   Exchange rate when selling dollars 1.720

a) You plan to go to the States so you buy £300 worth of dollars. How many dollars do you get in exchange?

b) Your trip is cancelled and you exchange your dollars back. How much money have you lost in these transactions?

a) $300 \times 1.469 = 440.70$ (dollars)   For every 1 pound you exchange, you get 1.469 dollars.

b) First work out how many pounds you get back and then calculate how much money you have lost.

When changing back to pounds, you need 1.720 dollars to get 1 pound.

$$\frac{440.70}{1.720} = 256.22 \text{ (pounds)}$$

£300 − £256.22 = £43.78 so you have lost £43.78

Different methods of purchasing goods when abroad can make an enormous difference to your finances. In May 2015, a comparison website analysed the costs of five different methods of obtaining €1000.

Specialist Credit Card	£730[1]
Cash (Cheapest Bureau)	£736[2]
Cash (Post Office)	£754
Debit Card	£810
Cash (Airport)	£830

[1] But you must repay the card in full.
[2] But paying for this with a credit card incurs extra charges.

Best buys will change but there are some things you can learn from this list. It is a good idea to plan ahead and shop around for the best cash rates. Waiting until you get to the airport will almost always be much more expensive.

If you don't want to take a large amount of cash with you on holiday using a specialist credit card can be good. Using a debit card may be more costly and you need to carefully check what foreign exchange fees and transaction fees it charges.

If you are comparing prices yourself rather than relying on a comparison website, be careful to take everything into account and look out for extra charges.

Types of charges include:
- Commission
- Exchange rate spread
- Application fee, Topping-up fee, Annual fee, Inactivity fee (for prepaid cards)
- Transaction charges
- Foreign exchange fees
- ATM fees
- Delivery charges (for internet or telephone orders)

Exercise 2G

1 In the USA the price of a Kindle with free 3G is $190.

   The exchange rate is $1.528 to £1.

   Calculate the cost of the Kindle in pounds.

2 £500 is converted into euros using the exchange rate of £1 = €1.25

   The euros are then converted into dollars at the rate of €1 = $1.2

   How many dollars are obtained?

3 £400 is converted into euros using the exchange rate of £1 = €1.28

   An additional £10 commission is charged.

   a) How many euros are obtained and what is the total cost of these euros?

   b) What commission-free exchange rate would match this deal?

4 Given that the mid-market price for £1 is $1.5278 and 191.91 Japanese Yen, find the mid-market price for $1 in Yen.

5 A driver wanted to rent a car for one week in France. At the time, the cost was 275 euros and the exchange rate was €1 = 80 pence.

   a) The driver opted to pay in euros using her credit card, which charged a 2.75% fee. Calculate the cost of renting the car in pounds giving your answer to the nearest penny.

   b) The cost of renting this car if the driver had opted to use Dynamic Currency Conversion would have been £229.35. What exchange rate has been used to do this conversion?

   (AQA, 2009)

Dynamic Currency Conversion is a method used with cards where foreign currency prices are automatically converted into pounds. You have the advantage of knowing exactly what you will be charged but the exchange rates for this service tend to be poor and can add up to 4% extra onto the cost.

2.10 Inflation

Inflation is the one form of taxation that can be imposed without legislation.

Milton Friedman

Starting in the late 1990s, Zimbabwe had a long period of instability leading to its currency becoming worthless.

Zimbabwe now largely uses the US dollar as its currency.

Much less drastic examples of money losing its value over time can be found in all countries. For example, in the UK a house which could be purchased for say £2000 in 1950 might cost you £250 000 in today's money. For the UK, the official measure of how quickly the pound is losing its value is the Consumer Price Index (CPI). Each month, the average price of a 'basket' of 700 goods and services from over 100 000 retail outlets is calculated.

Activity 16

There is an inevitable debate about which items should be included in the representative 'basket', and other indices, such as RPI and RPIX are used for some purposes. Can you think of particular goods and services whose costs will be important to some people but not to others?

Q You may like to investigate the differences between some different indices.

The effects of a relatively gentle inflation with the costs of goods and services steadily increasing are that

- the economy is stimulated since consumers and businesses have an incentive to use money either by spending or investing
- debt is reduced in real terms both for individuals and, most importantly, for the state.

In 2015, the UK target rate for inflation as measured by CPI was 2% per annum.

In early 2015, house price inflation (not included in CPI) was 8.4%, illustrating how different indices can give different results for inflation.

In early 2015, the annual UK CPI inflation rate became zero for the first time on record. The main reasons for this unexpected rate were believed to have been falls in the price of oil and food. Whilst this might sound extremely good for most people (apart from North Sea oil producers) the fact that this was so far below the target rate was of concern to some economists.

The falls in both oil and food prices were, of course, largely not under the UK's control.

This graph shows just how much CPI has varied about the 2% level since 2005.

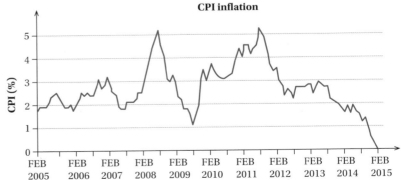

Data from ONS

Activity 17

★ The value of a house, currently worth £180 000, is projected to increase by 3% per year for the next 5 years. Find the house's value, £V, after 5 years.

★ The projected annual increase cannot be known exactly. Assuming that the annual increase could range from 2.5% to 3.5%, find an inequality satisfied by the value of the house after 5 years.

Inflation is important for your long-term financial planning. If, for example, you invest £10 000 and earn interest at 10% per annum, then after 7 years your investment will have nearly doubled to £10 000 $\times$ 1.1^7 = £19 487.17. However, if inflation was also at 10% during this period then the *value* of your investment has effectively been unchanged.

One way to reflect 'real' values is to do calculations relative to inflation. Newspaper articles will sometimes signal that they are doing this by phrases such as 'in real terms' or 'in 2000 prices'.

When Clare started work she had a student loan of £12 000. Each year she has to repay 9% of everything she earns in excess of £17 335. She started work on 1st January 2015 with an annual salary of £20 000.

To model her repayments, Clare assumed that the interest on the student loan would be 2% above inflation, that the threshold for repayments would keep pace with inflation and that her salary would increase at 4% above inflation. She set up a spreadsheet in '2015 money' to see when she would repay her loan.

	A	B	C	D
1	**End of year**	**Salary**	**Repayment**	**Loan remaining**
2	2015	20 000	239.85	12 000.15
3	2016	20 800	311.85	11 928.30
4				

Activity 18

★ Show the calculations that give the numbers in columns C and D on the spreadsheet.

★ Write formulae for cells A4, B4, C4 and D4.

★ Set up the spreadsheet and determine when Clare will repay her student loan.

In other words, with £19 487.17 you can only buy the same goods as you could have bought for £10 000 when you started the investment.

Clare's assumptions are reasonable.

- The government does try to increase various thresholds in line with inflation.

- Most people's salary increases do (at least initially) beat inflation as they learn and improve at their jobs.

Notice how by working in '2015 money' Clare does not even need to estimate what the inflation rate will be!

Consolidation exercise 2

1 Pete earns £3780 per month and has a tax-free allowance of £9600. Calculate his taxable income.

2 When water freezes its volume increases by 4%.

   What volume of water will make $1000\,\text{cm}^3$ of ice?

3 It is usually true that if you borrow a larger amount of money, then the APR you are charged will be smaller. In the past, the standard rates for bank loans were much higher than today.

   In February 2007, if you borrowed £4950 from a typical high-street bank and repaid the loan over 5 years you would pay £119.63 per month.

   If you borrowed £5000 from the same bank and repaid the loan over 5 years you would pay £98.28 per month.

   a) Calculate the total interest you would be charged over the 5 years for the £4950 loan.
   b) Calculate the total interest you would be charged over the 5 years for the £5000 loan.
   c) What should you do?

   (AQA, 2008)

4 Funds in the Platinum account earn interest at 2% per annum.

   a) If you invested £500 in the Platinum account and left it for 4 years without making deposits or withdrawals, how much would be in the account after 4 years? You can assume that the interest rate remains unchanged throughout the 4 year period.
   b) Find the nominal interest rate for the Platinum account if the interest is compounded monthly.

5 A shopkeeper normally sells coats at a price which includes a mark-up of 100% for profit. In a sale she wants to make a profit of at least 20% on each coat. By what percentage can she reduce the price of the coat in the sale?

6 The exchange rate spread can be very large for holiday currencies that are not as much in demand as euros and dollars. In this situation, getting the best exchange rate can often be better than looking for commission-free deals.

   Compare the following two possible sources of Turkish lira.

   A

Commission-free foreign currency		
	We sell	We buy
Turkish lira	3.7382	4.4941

   B

Great exchange rates			
	We sell	We buy	Commission
Turkish lira	3.8517	4.3282	£5

   a) What amounts of Turkish lira do you get with each company if you spend a total of £400? Which is the better deal?
   b) Use trial and improvement, perhaps with a spreadsheet, or solve the equation

   $$A \times 3.7382 = (A - 5) \times 3.8517$$

   to find the number of pounds at which the two deals are equally good.

7 a) Suppose you invest £S at a fixed rate of 0.17% per month. Then, after n years, the amount in the account will be £P where

   $$P = S \times 1.0017^{12n}.$$

   i   What is the amount in the account after 1 year if you invest £500?
   ii Hence find the AER for this investment.
   b) What is the formula for calculating the amount in an account after n years if you invest £S and the interest rate is 1.9% every 6 months?

8 The annual rate R, expressed as a decimal, at which an amount £P would increase to an amount £A after n years is given by the formula

$$R = \sqrt[n]{\frac{A}{P}} - 1.$$

An investment of £4000 has grown to £5000 after 8 years.

Find the rate of interest on this investment, expressed as a percentage.

> $\sqrt[8]{}$ stands for the 8th root. Make sure you know how to use your calculator to find roots like these.   ⊞ 1932

9 The table shows some average food prices, in pence, for the UK for 2000–2009.

	2000	2001	2002	2003	2004	2005	2006	2007	2008	2009
Milk (pint)	34	37	36	37	35	35	35	37	42	44
Potatoes (kg)	70	87	87	89	96	84	80	83	110	136
Sugar (kg)	55	57	62	69	74	74	74	77	84	96
Bread (loaf)	52	51	57	58	64	71	81	90	119	123

  a) Estimate how many loaves, pints of milk and kilograms of potatoes and sugar a 'typical' family might buy in a week.

  b) Hence calculate the weekly cost of your 'basket' of goods in each of the years 2000–2009.

  c) Find the annual inflation rates for the UK that are given by your index.

> Hint: Think about what the percentage increase is between years.

10 Sheila has a mortgage for £80 000. The annual interest rate is 4%.

  To pay off the mortgage quickly, she makes monthly payments of £1500.

  a) Show clearly that, at the end of the first year, Sheila owes £65 200.

  b) Let £A_n be the amount she owes after n years.

    Sheila uses the recurrence relation

    £$A_{n+1} = 1.04A_n - 1500$

    to find the amount owed at the end of each year.

    What mistake has she made? Correct her recurrence relation.

  c) A spreadsheet is set up as shown.

	A	B	C
1	n	A_n	
2	0	80 000	
3	1	65 200	
4	2	49 808	
5	3	33 800.32	
6			

    i What formula can be used for cell B6?

    ii Copy and complete the spreadsheet. When does Sheila pay off the mortgage?

11 The table shows the cost of a well-known burger in three countries in July 2015.

Canada	5.71 Canadian dollars
UK	£2.84
USA	$4.79

The exchange rates at the time were:

£0.65 = $1 = 1.23 Canadian dollars

In which country was the burger most expensive? Show your working fully.

12 To pay for a package holiday, Mary considers borrowing £600 from two different lenders.
   a) The first lender requires Mary to pay a single lump sum of £1000 at the end of 2 years. Calculate the APR charged on this loan.
   b) The second lender charges an APR of 30% and requires two equal repayments, the first repayment at the end of the first year and the second repayment at the end of the second year. Calculate the amount of each repayment.
   c) Although the two lenders charge similar APRs, the two total repayments are very different. Why is that?

13 The interest on a loan of £5000 is 1.9% per month. A formula which converts a monthly interest rate of r into an APR of R is

$$R = (1 + r)^{12} - 1$$

where both r and R are expressed as decimals.
   a) Use this formula to find the APR, expressing this as a percentage.
   b) Explain why the APR is greater than $12 \times 1.9 = 22.8\%$.

14 £1500 is invested in an account earning 0.85% interest every 6 months.
   a) Copy and complete the table.

	Start value	Interest	Final value
First 6 months	1500	12.75	1512.75
Second 6 months			
Third 6 months			

   b) What is the nominal interest rate on this account?
   c) Find the AER.

Review
After working through this chapter you should:

- be able to calculate the income tax, National Insurance and student loan repayments that will be deducted from your pay each month
- know how to compare loans using APR
- know how to compare savings rates using AER
- appreciate the effects of inflation on your savings
- be able to use spreadsheets and recurrence relations to find solutions to problems in financial contexts.

Investigation

In the news

For one or more of the stories shown below, research any necessary facts and then write a simple explanation of the issues and your conclusions about these topics.

Do politicians understand the difference between debt and deficit?

UK National Debt

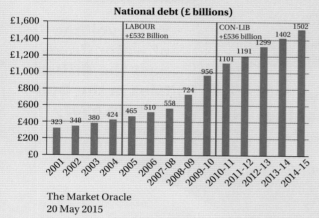

National debt (£ billions)

The Market Oracle
20 May 2015

'We are paying down Britain's debts'

Prime Minister David Cameron

UK National Deficit

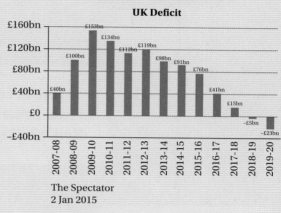

UK Deficit

The Spectator
2 Jan 2015

'The deficit has not come down'

Shadow Chancellor Ed Balls

These two quotes were made in the run up to the election of May 2015.

Yet another misleading quote on the figures in the graphs was that of Chancellor George Osborne saying 'The deficit has been halved.' He apparently meant that the deficit had been halved as a percentage of GDP.

Are houses really less affordable now?
June 5th 2015

Much is being made of reports that over the last 50 years UK house prices have risen on average by 2.7% in real terms.

This figure needs to be put into perspective. Wages have also risen much faster than inflation during this period.

Furthermore, houses now are of a much higher standard than in 1965. Central heating, double glazing and fitted carpets are now the norm whilst outside toilets are unheard of. Who today would be prepared to live in houses with as few modern conveniences as most houses of the 1960s?

Subprime mortgage misery
Sept 7th 2008

Two US government sponsored enterprises, Fannie Mae and Freddie Mac, have suffered huge losses and have had to be rescued by the federal government.

The boom in the US housing market is over. It was fuelled by high risk home buyers who had poor credit histories and no deposits. Many of these people will now lose their homes.

With the collapse of subprime lending, the demand for housing will fall further leading to negative equity, bankruptcies in the construction industry and widespread general recession.

3 Modelling and estimation

What is the land area of the British Isles?

What volume of water flows past the British Isles in the Gulf Stream every hour?

What is the radius of the Earth?

In this chapter you will learn how to answer questions like these. Some you will be able to answer accurately using good mathematical models and data. For others you will need to make simplifications and Fermi estimates of data that you do not know. This chapter gives you practice in modelling, and shows you some useful techniques which will help you to develop your modelling skills.

How many fish are there in the Earth's oceans?

How much water is there in the Earth's oceans?

If all the ice in the Antarctic ice cap melts, by how much will the oceans rise?

To model a situation you can use any relevant mathematics that you know – simple or advanced. The worked examples and questions in this chapter use these skills from GCSE mathematics, but you may find other ways of tackling them.

You should know how to

- estimate the answer to a calculation 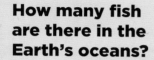 1005, 1043, 1969
- calculate the volume of a cuboid and cylinder 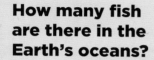 1137, 1138
- use the index laws 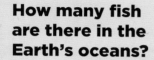 1951
- write large and small numbers in standard form 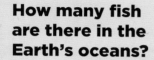 1049, 1051
- calculate percentages. 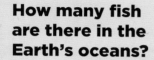 1030, 1031

You may have noticed from maps and photographs of the Earth that the British Isles are, roughly, as far north as Canada. The photograph of the polar bear shows a scene that is typical of Hudson Bay but not of the Irish Sea at a similar latitude. So, why do we have such a mild climate?

One major reason is the warm ocean current, called the Gulf Stream, flowing from Florida, across the Atlantic to Northern Europe. To find out why it has such a major effect on our climate, think about this question:

Example 1 How much warm water flows past the British Isles in the Gulf Stream every hour?

Activity 1

★ List the quantities that you will need to estimate to answer this question about the Gulf Stream.

★ What assumptions would you make when making these estimates?

One very important factor is the speed of the Gulf Stream. You could make these assumptions and estimates:

> Assume that the Gulf Stream has a constant speed all year.

> Assume that all the water in the Gulf Stream has the same speed, whether at the surface of the ocean or in the depths.

> Estimate the speed to be walking speed, 5 km/h.

Next you could model the cross-section of the Gulf Stream.

> Assume that the Gulf Stream has a rectangular cross-section.

You might be able to estimate the speed of the Gulf Stream based on your experience of currents when swimming in the sea. The accepted average speed for the Gulf Stream is about 6.4 km/h but the figure of 5 km/h is perfectly sensible and will give a reasonable answer to the question.

A cuboid is an easy-to-use simplification.

An average speed of 5 km/h means that in 1 hour the water flows 5 km.

Then the model for the water flowing past the British Isles in 1 hour is a cuboid like this:

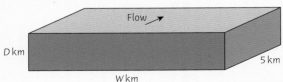

where D is the depth of the Gulf Stream and W is its width in kilometres.

Now you need to make assumptions or estimates for D and W.

> ▶ Assume that $D = 1$ and $W = 100$.

Values of half or double these would be just as sensible.

The volume of water flowing past the British Isles in 1 hour is therefore $1 \times 100 \times 5 = 500\,\text{km}^3$

Volume of cuboid $=$ length $\times$ width $\times$ height

Convert this into cubic metres, using
$1\,\text{km}^3 = 1\,000\,000\,000\,\text{m}^3$

$$500\,\text{km}^3 = 500 \times 1\,000\,000\,000\,\text{m}^3$$
$$= 500\,000\,000\,000\,\text{m}^3$$

You can find 'official' answers to questions like this on the internet, though you should be careful to search for reputable sources. If your answer is a long way out then you should review all the assumptions you made and see if, in hindsight, some of them were unrealistic. However, remember that estimates of some quantities are disputed by experts in the subject. You are only using rough and ready methods to obtain 'ball-park' figures.

In fact the rate of flow of the Gulf Stream varies at different times and in different locations.

Activity 2

Q Research on the internet:

- the average speed, width and depth of the Gulf Stream as it passes the British Isles. Do you need to change any of the assumptions made in the example? If so, do a new calculation.

- the amount of water flowing past the British Isles in the Gulf Stream in one hour. Is the estimate of $500\,000\,000\,000\,\text{m}^3$ per hour reasonable?

★ The California current flows south along the California coast. It is approximately $40\,\text{km}$ wide and $1\,\text{km}$ deep. Estimate the amount of water flowing per hour.

In the Gulf Stream problem, the key to the solution was the formula

Volume of water per hour $=$ Area of cross section $\times$ Speed

Using this simple formula, you can split an apparently difficult problem into two much more straightforward parts:

- estimating the area of cross section
- estimating the speed.

Later in this chapter you will see many other examples of splitting up problems which appear difficult into simpler parts, in order to solve them.

3.2 Standard form

On page 81 the estimate for the rate of flow of water in the Gulf Stream was 500 000 000 000 m³ per hour.

In standard form this is 5×10^{11} m³ per hour.

When estimating quantities, it is useful to be able to do calculations with both very large and very small numbers in standard form.

Standard form:
number between × power of 10
1 and 10

$$500\,000\,000\,000 = 5 \times 10^{11}$$

 1051

Key point
To multiply two numbers in standard form, multiply the numbers and powers of 10 separately. For example

$$4 \times 10^2 \times 2 \times 10^3 = 4 \times 2 \times 10^2 \times 10^3 = 8 \times 10^5$$

To multiply powers of 10, add the powers.
$$10^2 \times 10^3 = 10^5$$
1033

Example 2 In a vacuum, light travels 299 792 458 metres every second. Light takes approximately 497 seconds to reach Earth from the Sun.

Estimate the distance from the Earth to the Sun.

For Fermi estimation, you can round the numbers to 1 significant figure.

300 000 000 metres every second
500 seconds

Write the numbers in standard form.

3×10^8 and 5×10^2
The distance from the Earth to the Sun is
$3 \times 10^8 \times 5 \times 10^2$
$= 15 \times 10^{10}$
$= 1.5 \times 10^{11}$ m

15×10^{10} is not in standard form, because 15 is not between 1 and 10.

You may also need to divide one large number by another.

Key point
To divide two numbers in standard form, divide the numbers and powers of 10 separately. For example
$$\frac{6 \times 10^5}{2 \times 10^3} = 3 \times 10^2$$

To divide powers of 10, subtract the powers.
$$10^5 \div 10^3 = 10^2$$
1033

Example 3 How many movies could you store on a computer's hard disc?

State any assumptions that you make.

> ▶ Assume the hard disc can store 10^{13} bits of information.
>
> ▶ Assume that **all** this storage can be devoted to movies.
>
> ▶ Assume that each movie is 900 MB.
>
> ▶ $1\,MB = 8 \times 10^6$ bits.

Find or make estimates for the hard disc and movie size.

Convert 900 MB to bits and round to get an estimate.

$$900\,MB = 900 \times 8 \times 10^6$$
$$= 7200 \times 10^6$$
$$= 7.2 \times 10^9$$
$$\approx 7 \times 10^9 \text{ bits}$$

The number of movies is
$$\frac{10^{13}}{7 \times 10^9} = \frac{10^4}{7} = \frac{10\,000}{7} \approx 1500$$

$\approx$ means 'is approximately equal to'

You could store roughly 1500 movies.

You can write very small numbers in standard form. For example, the mass of a water molecule is approximately
0.000 000 000 000 000 000 000 000 03 kg.

$0.000\,000\,2 = 2 \times 10^{-7}$

⊞ 1049

In standard form this is 3×10^{-26} kg.

Example 4 How many water molecules are there in your body?

> ▶ Assume that your body is entirely composed of water molecules.
>
> ▶ Assume your weight is 70 kg.
>
> ▶ Assume 1 water molecule weighs 3×10^{-26} kg.

The human body is mostly water.

$$70 = 7 \times 10$$
$$10 \div 10^{-26} = 10^{1\,-\,-26} = 10^{27}$$

Number of water molecules is
$$\frac{7 \times 10}{3 \times 10^{-26}} = \frac{7}{3} \times 10^{27} \approx 2 \times 10^{27}$$

There are approximately 2×10^{27} water molecules in the body.

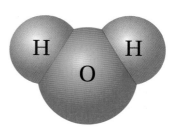

Exercise 3A

1   Write these numbers in standard form.

   a) 9000
   b) 0.000 002
   c) 85 000 000
   d) 0.000 015

2   Write these correctly in standard form:

   a) 24×10^3
   b) 360×10^5
   c) 0.8×10^3
   d) 0.03×10^5

3   Without using a calculator, work out

   a) $2 \times 10^7 \times 3 \times 10^4$
   b) $\dfrac{6 \times 10^7}{3 \times 10^4}$
   c) $(1.2 \times 10^7) \times (1.2 \times 10^4)$
   d) $(4 \times 10^7) \div (8 \times 10^5)$
   e) $6 \times 10^{-7} \times 3 \times 10^4$
   f) $(6 \times 10^7) \div (3 \times 10^{-4})$

4   a) By modelling different parts of the body as cuboids (or as cylinders), estimate the volume of your body.
   b) By using the fact that human beings can float in water (but only just!), use your weight to find a really good estimate of your volume.

   When an object floats, the weight of water displaced is equal to the weight of the object. 1 kg of water has volume $1000\,\text{cm}^3$.

   c) Compare your estimate from your model in a) with your better estimate in b). How good, or otherwise, was your model in a)?
   d) How could you improve your model in a)?

5   Planck's constant, h, is a number which is fundamental to the nature of matter and radiation in our universe.

   $h = 0.000\,000\,000\,000\,000\,000\,000\,000\,000\,000\,000\,662\,6$

   a) Express h in standard form.
   b) What approximation to h would you use in a Fermi estimation?
   c) Without using a calculator, use your answer to b) to write down an estimate for h^3.
        (h^3 is a number used in Niel Bohr's model of the hydrogen atom.)

6   a) Estimate how long it would take to walk from Land's End to John O'Groats.
   b) Carefully list all the simplifications and assumptions that you make.

7   Astronomical distances are expressed in *light years*. A light year is the distance travelled by light in 1 year.

   a) In a vacuum, light travels 299 792 458 metres every second. Express a light year in metres.
   b) The distance from the Sun to Alpha Centauri is 2.4 light years. Express this distance in metres.
   c) Voyager 1 is the furthest spacecraft from Earth and the first to enter interstellar space. It is travelling at approximately 6.1×10^4 km per hour. Travelling at this speed, how many years would it take to travel between the Sun and Alpha Centauri?

8   Blood is roughly 7% of the human body.

   a) Use your estimate for the volume of your body from Q4 to estimate the volume of blood in your body.

   In one minute, all of a person's blood is pumped through their heart.

   b) What volume of blood will your heart pump in a lifetime?

The remaining questions in this exercise are about water usage in one Tanzanian village, contrasted with water usage in the UK.

Grace Madeje lives with her husband and three children in a village in Tanzania.

There is no running water in their village. Each day, Grace has to walk to collect water and carry it back for herself and her family.

The nearest source of water is 11 km from her village.

9   Here are some questions you could ask based on this information:

- How far does she walk each time she goes to collect water?
- How much water could she carry at a time?

Write two more questions you could ask.

10 a) List some of the main uses for the water Grace collects.
    b) How much water does Grace need to collect each day for herself and for her family?

11 Estimate how many hours a day Grace spends collecting water.

What effect do you think this has on her daily life?

12 Roughly 1 in 9 people worldwide do not have access to safe water.

    a) Estimate the number of people who do not have access to safe water.

    b) Write a statement comparing your estimate to the number of people in the UK.

13 a) List some of the main uses for water in your home.
    b) How much water does your family use each day?
    c) How much water would a family of five people be likely to use each day in the UK?

In the UK, we are lucky to have easy access to some of the safest treated water in the world, just by turning on the tap. If you check your findings with figures from the internet, you find estimates of personal (domestic) usage vary a lot.

Activity 3

Q Domestic use of water in the UK is relatively small compared to other uses. List some of the major non-domestic uses of water in a developed country like the UK.

You modelled the Gulf Stream by comparing it with a more familiar speed – walking speed. This is a good example of a technique that is very useful for Fermi estimation.

Key point

A good way to solve some estimation problems is to estimate a quantity that

- can be scaled (up or down) to the required quantity and
- is either known or easier to estimate.

'Scaling' means multiplying or dividing both quantities by the same number.

Example 5 How many human heartbeats are there in the UK each year?

In this case you could start by estimating the number of heartbeats for one person in one minute. The time period 'one minute' seems suitable (rather than, say, one day or one millisecond) because heart rates are often given in beats per minute.

If you do not know an average figure for heartbeats per minute, it is easy to measure your own pulse over a minute. (Or instead of one minute, you could measure your pulse for 10, 15 or 30 seconds and scale up from that.)

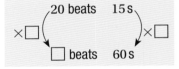

Pulse: 20 beats in 15 seconds
= 80 beats in 60 seconds

> Assume that one person has roughly 80 heartbeats per minute.

Number of heartbeats per year is $80 \times 60 \times 24 \times 365$

Round to 'nice' numbers that are easy to multiply. Use $\approx$ to show the calculation is approximately equal.

$$80 \times 60 \times 24 \times 365 \approx 100 \times 50 \times 20 \times 400$$
$$= 100 \times 1000 \times 400$$
$$= 10^2 \times 10^3 \times 4 \times 10^2$$
$$= 4 \times 10^7$$

80×60 in 1 hour

$80 \times 60 \times 24$ in 24 hours (1 day)

$80 \times 60 \times 24 \times 365$ in 1 year

> Assume that there are 6×10^7 people in the UK.

Paper 1

Total number of heartbeats is $6 \times 10^7 \times 4 \times 10^7$

$$= 24 \times 10^{14}$$
$$= 2.4 \times 10^{15}$$

There are approximately 2×10^{15} heartbeats in the UK each year.

This is a good example of scaling – a technique that is often useful for estimating quantities.

Example 6 Use this picture of the globe to help you estimate the radius of the Earth.

In this case you first need to decide which length 'you know or is easy to estimate'. For this problem, it would be useful to know that the straight-line length of the UK is very close to 1000 km. However, you can estimate this length using common sense and a little knowledge of travelling in the UK.

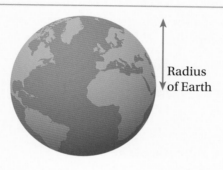
Radius of Earth

The radius is the distance from the centre of the Earth to the outer edge.

Activity 4

Think of a long distance journey you have made by car, mainly on main roads and motorways.

★ How long did it take?

★ Assume the car's average speed was 100 km/h. Use this to estimate how many km you travelled on your journey.

★ Find the start and end points of your journey on the map of the UK.

★ Compare the distance of your journey with the length of the UK. How many times longer than your journey is the length of the UK? Use your answers to estimate the length of the UK.

Length

Your answer for the length of the UK can now be used to answer the initial question about the radius of the Earth.

 Assume that the length of the UK is 1000 km.

From the globe, the radius of the Earth is roughly 6 times the length of the UK, so the radius of the Earth is approximately 6000 km.

You may like to find and use a larger map of the UK.

Scaling is a special case of a second technique: subdividing or breaking up a task into small solvable parts.

Key point
To solve some problems, start by subdividing the problem into smaller and more manageable parts.

Example 7 What is the land area of the British Isles?

To break this problem up into small solvable parts requires simple but orderly logical thinking. You could think of the process as a network of related activities.

The diagram shows one way of breaking this problem down.

Estimate the overall dimensions of the British Isles.

Approximate Great Britain by a triangle.

Approximate Ireland by a square.

Add the two areas.

> Estimate for length is 1000 km (see page 87).

▶ The length of Great Britain is about 1000 km.

Estimate the width, by comparing it to the length.

▶ Great Britain is (roughly) a triangle of height 1000 km and base 500 km.

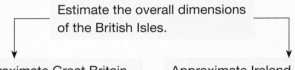

$\text{Area} = \dfrac{1}{2} \times \text{base} \times \text{height}$

$= \dfrac{1}{2} \times b \times h$

▦ 1129

Calculate the area of the triangle using $\text{Area} = \frac{1}{2} \times \text{base} \times \text{height}$.

Its area is approximately $\frac{1}{2} \times 500 \times 1000 = 2.5 \times 10^5 \text{ km}^2$.

▶ Ireland is (roughly) a square of side 250 km.

Its area is approximately $250 \times 250 \approx 6 \times 10^4 \text{ km}^2$.

Add the two estimates for area.

$2.5 \times 10^5 + 6 \times 10^4 \approx 3 \times 10^5$

The total area is approximately $3 \times 10^5 \text{ km}^2$.

> The calculation ignored all the other (over 6000) smaller islands that make up the British Isles and also ignored large areas of water, such as Lough Neagh in Northern Ireland. For Fermi estimation you need to make assumptions to simplify problems. Do not over-complicate them!

Paper 1

Example 8 How much water is there in the Earth's oceans?

The amount of water in the oceans means the volume of water in the oceans.

First plan a strategy to solve the problem. Don't worry about the calculations yet.

You can model the total amount of water as a prism (a 3D solid with the same cross-section all the way down).

Volume of prism = area of cross-section × height

Here is a possible strategy.

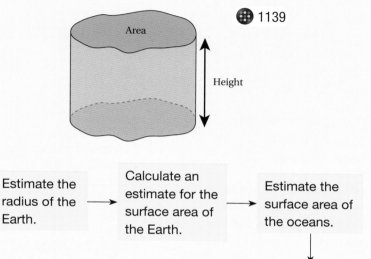

⬚ 1139

| Estimate the radius of the Earth. | → | Calculate an estimate for the surface area of the Earth. | → | Estimate the surface area of the oceans. |

| Estimate the depth of the oceans. | → | Calculate the volume of the oceans. |

The depth of the oceans might seem difficult to estimate. However, you only need an order of magnitude estimate.

Activity 5

Before reading on, decide how *you* would go about the five sub-parts to this problem.

★ Estimate the radius of the Earth.

★ Calculate an estimate for the surface area of the Earth.

★ Hence estimate the surface area of the oceans.

★ Estimate the depth of the oceans.

★ Hence calculate the volume of the oceans.

Compare your answers with the solution to Example 8.

> ▶ The radius of the Earth is roughly 6000 km.

surface area $= 4\pi r^2 \approx 4 \times 3 \times 6000^2 \approx 4 \times 10^8\,\text{km}^2$.

> ▶ 70% of the Earth's surface is covered with water.

Calculate 70% of 4×10^8.

$0.7 \times 4 \times 10^8 \approx 3 \times 10^8\,\text{km}^2$

Area of oceans $\approx 3 \times 10^8\,\text{km}^2$.

> ▶ The height of most dry land varies from sea level to
> approximately 8 km. Assume the mean depth of the sea
> is 4 km.

Volume = area of ocean × depth

$\qquad = 3 \times 10^8 \times 4 = 12 \times 10^8 = 1.2 \times 10^9 \approx 10^9\,\text{km}^3$.

There is approximately $10^9\,\text{km}^3$ of water in the Earth's oceans.

This estimate depended upon some knowledge of physical
quantities and the formula for the surface area of a sphere.
However, you could calculate reasonable estimates without
this knowledge. For example, you could model the Earth as a
cube and then estimate the Earth's surface area by reducing the
surface area of a cube to allow for the rounding of each corner.

Use the estimate for
radius of the Earth from
page 87.

The surface area of a
sphere of radius r is $4\pi r^2$

For a quick estimate you
can approximate π by 3.

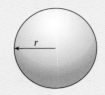

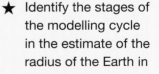 1122

There are very few
mountains over 8 km high.
The mean figure of 4 km
is probably far too big for
the average height of dry
land but is reasonable
for the depth of the sea
because the seabed has
deeper troughs and 'higher'
mountains than dry land.

Key point
It is more important to obtain an estimate by any means available
rather than be unable to proceed due to a lack of precise factual
knowledge.

Fermi estimation is an example of mathematical modelling. You
will recall the diagram used in Chapter 0.

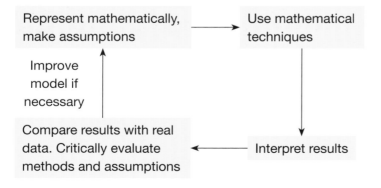

Activity 6
★ Identify the stages of
the modelling cycle
in the estimate of the
radius of the Earth in
Example 6.

★ Find real data to
compare with the
figures used in the
estimate.

★ Suggest some
ways that you could
improve the model.

Exercise 3B

1   Use the globe to estimate

   a)   the distance from the south to the north of Africa

   b)   the distance from the west to the east of Africa at its widest point

   c)   the length of the Gulf Stream, from Florida to the UK.

2   Count your pulse rate for 10 seconds. Use this to estimate how many times your heart will beat from birth until your 18th birthday.

3   Daisy used a picture of the globe to estimate the length of Greenland, from north to south. When she checked her answer she found it to be very inaccurate.

   a)   What might she not have noticed, when estimating the length from the globe?

   b)   Would this have made her estimate too small or too large?

   c)   Suggest how she could improve her method.

4   In the mid-19th century, American merchant ships could sail from England to New York in roughly 20 days. Estimate the average speed of these ships in km per hour.

5   In the mid-19th century, British ships tended to take significantly longer than American ships to sail from England to New York. The sailors did not know why this was.

   a)   What could have affected their journey speeds?

   b)   How had the American sea captains avoided this effect?

Before answering question **6** check your answer to question **5**.

6   Make a Fermi estimate of how much longer the British ships used to take to sail from England to New York.

Consolidation exercise 3 on page 99 has a variety of modelling problems. You could tackle some of these now to practise the techniques you have learnt so far. You might like to choose one where you will be able to relate each step to a stage in the modelling cycle and compare your model with reality.

3.5 Estimation technique 3 – stating assumptions

The first step in the modelling cycle is 'Represent mathematically, make assumptions.' In all the modelling you have done so far in this chapter, you have made assumptions and stated them clearly in your solutions. This is the third technique for solving Fermi estimation questions and this section will show you why it is so important in modelling.

Example 9 How many fish are there in the Earth's oceans?

According to the tagline in the film *Finding Nemo* there are 3.7 trillion fish in the Earth's oceans.

If you are faced with a problem like this, where there are data that you do not know and cannot find out, make the best assumptions you can and use them in your model. State your assumptions clearly. When you get to the fourth step in the modelling cycle, 'Compare results with real data. Critically evaluate methods and assumptions', you can then look back and perhaps improve your assumptions.

Use your estimates from Example 8. State your assumptions clearly.

> ▶ Assume that the volume of the oceans is $10^9 \, \text{km}^3$.
>
> ▶ Assume that the average depth of the oceans is 4 km.
>
> ▶ Assume that most fish live up to 20 m below the surface.

Convert the estimate for the volume of water to m^3.

Volume of water $= 10^9 \, \text{km}^3 = 10^{18} \, \text{m}^3$.

Estimate the volume of water containing fish.

The estimate for the volume assumed a depth of 4 km = 4000 m.

The zone for fish is only $\frac{20}{4000} = \frac{1}{200}$ of this.

Volume of water containing fish is $10^{18} \times \frac{1}{200} = 5 \times 10^{15} \, \text{m}^3$.

> ▶ Assume that each fish has its own territory. Model this as a cube with side length 10 m.

The figure given in the film *Finding Nemo* may be correct. However, it is usually wise to treat figures such as this with some suspicion unless you know how they have been calculated.

In Example 8 you saw that the total volume of the oceans is roughly $10^9 \, \text{km}^3$. However, it is still very difficult to estimate the number of fish because there are regions and depths where very few fish live and there are other regions which are densely populated with shoals of closely-packed fish.

For this problem, even experts disagree with each other's assumptions!

$1 \, \text{km}^3 = 1000 \times 1000 \times 1000 \, \text{m}^3 = 10^9 \, \text{m}^3$

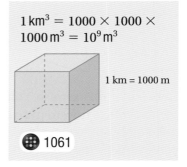

1 km = 1000 m

⊞ 1061

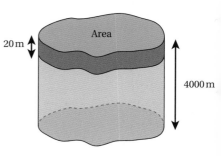

Calculate the volume of water for each fish.

$10 \times 10 \times 10 = 1000 = 10^3 \, m^3$

Then the number of fish is approximately

$5 \times 10^{15} \div 10^3 = 5 \times 10^{12}$.

There are roughly 5 trillion fish in the Earth's oceans.

The assumptions in Example 9 might appear to be very arbitrary. However, these assumptions are clearly stated and can be easily improved if more information is discovered or becomes available.

Key point

When you can only proceed by making an assumption for which you have little evidence, then make it anyway but state clearly what assumption you have made.

Using assumptions

Scientists have used this technique of making and stating assumptions to make important scientific breakthroughs.

One example is the equation $f = C\left(\frac{1}{b^2} - \frac{1}{a^2}\right)$ which was mentioned in Chapter 0 to this book. This equation gives the frequencies corresponding to lines in the spectrum of Hydrogen.

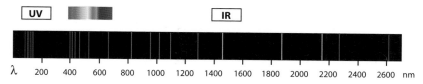

The frequencies corresponding to these lines had been well known for many years before a Swiss school teacher, Johann Balmer, recognised a numerical connection. By substituting integer values, 1, 2, 3, ... for a and b in the equation $f = C\left(\frac{1}{b^2} - \frac{1}{a^2}\right)$ and using the constant value 1.097×10^7 for C, now known as the Rydberg constant, Balmer's equation gave the frequencies corresponding to known lines and, remarkably, correctly predicted other lines.

Some 30 years later, Niels Bohr was able to explain the physical significance of this equation. He did so by making an unjustified assumption that the hydrogen atom could be modelled as a mini solar system.

The results from Bohr's model matched reality so well that it won his peers over. In subsequent years the model has been extensively refined but the success of the original model was a crucial breakthrough for quantum mechanics.

Activity 8

★ Show how you could find the volume of water containing fish more simply, using the surface area of the oceans and the depth of 20 metres.

Q Critically evaluate the results in Example 9, by

- using the internet to compare the figure of 5 trillion with accepted estimates for the number of fish in the oceans

- finding more detailed information about the distribution of fish in the oceans and developing an improved model.

Lines such as these have important uses in spectroscopy, for example in medical diagnosis and in astrophysics.

Spectroscopy is the study of the intensity of radiation at different wavelengths.

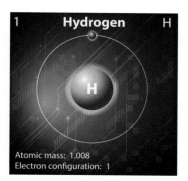

1   **Hydrogen**   H

H

Atomic mass: 1.008
Electron configuration: 1

3.6 Useful facts and formulae

There is no need to memorise these data, but you may find some useful when tackling modelling questions, especially those involving Fermi estimation.

The human body

Weight: 80 kg

Height: 1.7 m

Lifetime: 75 years

Heart rate: 80 beats per minute

Volume of blood: 5 litres

Sprinting speed (fastest): 10 m/s (or $10\,\mathrm{ms^{-1}}$)

Walking speed: 3 mph

Water

Density: 1 gram per cubic centimetre (or $\mathrm{g/cm^3}$)

The Earth

Radius: 6000 km

It spins on its axis once every day

It travels around the Sun once every year

Data storage

A bit is a binary digit, 0 or 1

1 byte = 8 bits

1 MB is 10^6 bytes

Equivalences

VOLUME

1 litre = $1000\,\mathrm{cm^3}$

DISTANCE

1 mile ≈ 1600 m

1 foot ≈ 30 cm

WEIGHT

1 pound ≈ 500 g

TIME

1 year ≈ 365 days

1 day = 24 hours

1 hour = 60 minutes

1 minute = 60 seconds

Useful formulae

For a cube of side length l:

Volume = Length of side^3 = l^3

Surface area = 6 × Length of side^2 = $6l^2$

For a sphere of radius r:

Volume = $\frac{4}{3}\pi r^3$

Surface area = $4\pi r^2$

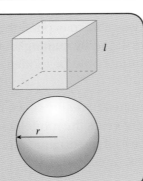

Activity 9

In modelling problems you may do calculations that are based on formulae, but that seem to be simply common sense.

Which formulae are used in the following examples?

Example 10 The Shangai Maglev Train has a top operating speed of 430 km/h. At this speed, how long would it take to travel the length of Great Britain? ⊕ 1121

The time is $\frac{1000}{430} \approx \frac{1000}{400}$ hours or roughly $2\frac{1}{2}$ hours.

Example 11 The Earth has mass 6×10^{27} grams and volume 1.1×10^{27} cm³. Find its density. ⊕ 1246

> The Earth is the most dense planet in the Solar System.

The density is $\frac{6 \times 10^{27}}{1.1 \times 10^{27}} \approx 5.5\,g/cm^3$.

The formulae used were

$$\text{Speed} = \frac{\text{Distance}}{\text{Time}} \quad \text{and} \quad \text{Density} = \frac{\text{Mass}}{\text{Volume}}$$

These two formulae are often written algebraically, for example, as

$$v = \frac{s}{t} \quad \text{and} \quad D = \frac{M}{V}$$

> When you are writing your own solutions you can explain your calculations using 'common sense' or word formulae or algebra.

In Chapter 0, you met the Drake Equation,

$$N = TShl$$

for the number of civilisations in our galaxy that could communicate with us using radio waves.

An equation, even a very simple one, is a very clear way of demonstrating the *structure* of your solution. For example, how the different variables are connected to give the final answer.

An equation or formula is also useful if you are going to set up a spreadsheet – perhaps to explore the effect of changing some of your assumptions about the values of different variables.

> In practice, many professions require people to understand simple formulae, to be able to substitute numbers into formulae and use spreadsheet formulae. In general, only mathematical and scientific professions require an ability to *manipulate* algebraic expressions.

3.7 Lessons from history

The fourth step in the modelling cycle is to critically evaluate and perhaps improve the model.

In an end-of-course examination you have very little opportunity to carry out a full cycle of the modelling process. Nevertheless, the unbiased evaluation of the model is a crucial stage of modelling.

As a case study in critical evaluation, here is a time-line showing how the properties of light were modelled over centuries.

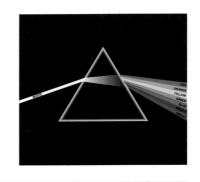

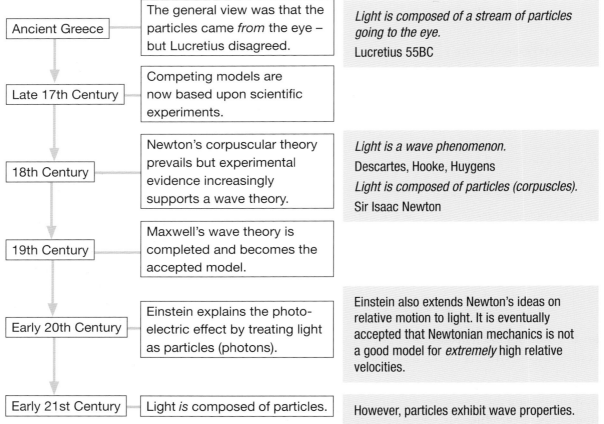

Ancient Greece	The general view was that the particles came *from* the eye – but Lucretius disagreed.	*Light is composed of a stream of particles going to the eye.* Lucretius 55BC
Late 17th Century	Competing models are now based upon scientific experiments.	
18th Century	Newton's corpuscular theory prevails but experimental evidence increasingly supports a wave theory.	*Light is a wave phenomenon.* Descartes, Hooke, Huygens *Light is composed of particles (corpuscles).* Sir Isaac Newton
19th Century	Maxwell's wave theory is completed and becomes the accepted model.	
Early 20th Century	Einstein explains the photo-electric effect by treating light as particles (photons).	Einstein also extends Newton's ideas on relative motion to light. It is eventually accepted that Newtonian mechanics is not a good model for *extremely* high relative velocities.
Early 21st Century	Light *is* composed of particles.	However, particles exhibit wave properties.

One perspective on this time-line might be that scientists are constantly getting things wrong. How, for example, could Maxwell's wave theory gain universal support in the 19th century, only to be overturned within 50 years? A remark by one of the physicists who worked on the new model in the early 20th century is revealing:

> *[Maxwell's wave theory] … remains for all time as one of the great triumphs of human intellectual endeavour.*
>
> Max Planck (1858–1947), winner of 1918 Nobel Prize for Physics

The scientists who came after Maxwell and saw further than him, progressed the model by 'standing upon his shoulders'. His wave model played a crucial role in preparing the way for further research. It is also worth noting that if all the experimental results had fitted Maxwell's wave theory, then there would have been *no* progress.

Critically evaluating your models

In your own work, a rough and ready model, perhaps using Fermi estimation, can help you to produce a better model. When you compare your initial model with reality, a critical evaluation might show you *how* you could improve it. Even if you conclude that the basis for your first model was completely wrong, this will have the positive effect of saving you time creating an elaborate but incorrect model.

The model used for Example 5 on page 86 was not 'completely wrong' but it did involve some big simplifications.

Activity 10

Critically evaluate the assumptions made in the solution to Example 5.

★ List all the simplifications and omissions that you can find.

Q Collect secondary data from the internet.

★ Write new assumptions and make any changes to the initial model.

★ Calculate an improved estimate of the number of human heartbeats in the UK each year.

★ Comment on the different estimates from the two models. For example, are they of the same order of magnitude?

Financial models

The performance of the economy has an immense and direct effect upon people's lives. However, governments have to base their policies on economic models for which there is rarely a consensus. In 2015, throughout the developed world there were opposing views on the appropriateness and the success or otherwise of austerity packages.

The following two statements were made on 11th February 2015, in the run-up to a general election. They both refer to expenditure for the whole of the UK. However, they paint very different pictures of the success or otherwise of the coalition government's attempt to reduce the annual deficit.

> *Perhaps most damagingly of all for the UK government's credibility, it has failed to meet its own deficit reduction targets.*

Nicola Sturgeon, First Minister of Scotland and Leader of the Scottish Nationalist Party

This is a good example of how producing one model can lead to further areas of investigation and eventually to a refined model which may differ significantly from the original one.

Secondary data is data collected by other people. Primary data is data you collect yourself.

Mathematical models and their critical evaluations are widespread in fields other than physics, especially in meteorology, medicine and economics.

A model that gives inaccurate predictions is soon evaluated unfavourably by the public! So an organisation whose reputation (and finance) depends upon how closely its model agrees with reality has to make a considerable effort to perfect the model.

Activity 11

In 2010, the CPI replaced the RPI as the UK government's main measure of inflation.

Use the internet to explore and evaluate the appropriateness or otherwise of using CPI when deciding the annual increases in income-related benefits and pensions.

This government has cut borrowing by £52bn from the level we inherited. We are vying with the United States for the strongest economic growth in the G7.

Spokeswoman for the Scottish Secretary

There is sometimes little consensus on the data itself. Chapter 2 introduced indices that measure economic inflation (the general increase in the level of prices). These indices can differ significantly. The figures for CPI and RPI in Dec 2014 were:

CPI   0.5%   RPI   1.6%

Meteorological models

One major use of computer time is to try to predict the weather. Although the weather in the UK tends to be less extreme than elsewhere in the world, weather damage can still be extensive. By the start of 2015, the accuracy of the Met Office in correctly forecasting rain three hours ahead was 73.3% compared to its target of 70%.

The Meteorological Office (Met Office) is the national weather service for the UK. They collect and analyse weather data from all over the world to provide weather and climate forecasts.

Activity 12

A different measure of the Met Office's success gives it an *equitable threat* score of roughly 50% for its three-hour rain forecasts.

Q Find out what this means.

★ Decide which measure the Met Office should use.

Activity 13

Q How is BMI calculated? What measurements does it use?

★ Use the internet to critically evaluate BMI as a model of obesity. What 'better' measures have been suggested?

A critical evaluation of BMI should consider how accurate the BMI is in identifying people who are obese.

Review

After working through this chapter, you should:

- know the steps in the modelling cycle

- be able to use Fermi estimation to set up a model when you do not know all the data
- know how to use three important techniques:
    - scaling a quantity that you know or which is easy to estimate
    - subdividing a problem into smaller, solvable parts
    - making and stating your assumptions and simplifications
- be able to evaluate the accuracy of your model.

Consolidation exercise 3

The questions in this exercise are grouped by theme.

- Some can be solved in a rough and ready way using Fermi estimation.
- For others you might need to research or collect data.

> You can tackle these questions with little technical knowledge as long as you are confident enough to make appropriate simplifying assumptions.

Do questions of both types, on themes that interest you. Make sure you carry out all the stages of the modelling cycle on at least one problem.

- First, use a simple model with quick Fermi estimates of quantities that you need.
- Then evaluate your model by comparing it with the reality of the situation that you are modelling.
- Finally, improve your model by using more accurate data or by developing a more sophisticated model.

Human beings

1 How many hairs are on a person's head?

2 How long will you spend on the phone in your lifetime?

3 How far can a person throw a stone?

4 How many cells are there in the human body?

Meteorology

5 What volume of rain water falls on the UK each year?

Food and drink

6 What weight of food is consumed in the UK each year?

7 What fraction of a swimming pool would be filled by the amount of liquid that you will drink in your lifetime?

8 How many eggs are consumed each day in the UK?

Science

9 The density of sea water is 1.03 grams per cubic centimetre. What is the total mass of salt in the oceans?

10 What is the speed of Oxford as the city rotates about the Earth's axis?

11 What is the speed of Oxford in its rotation about the Sun?

Economics

12 What weight of domestic rubbish is collected each year by councils throughout the UK?

13 How much will you earn in your lifetime?

14 What is the annual cost of petrol used in cars in the UK?

Technology

15 On a CD-ROM, the bits of data are stored using a series of tiny pits. What is the distance between two adjacent pits?

> Perhaps assume that the pits are arranged in a square array.
>
> This rough and ready model will give a sensible answer for the distance between adjacent pits.

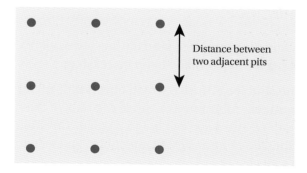

Distance between two adjacent pits

Miscellany

16 Throughout the world, how many people are airborne at any one time?

17 How many bricks are there in the UK?

18 What weight of grass clippings is collected each year from a lawn?

Investigation

The Antarctic ice cap

Recent trends in changes to the amount of Antarctic ice have led to a number of confusing reports.

Sea ice around Antarctica increasing

On 22 September 2014 sea ice surrounding Antartica reached its maximum extent.

National Snow and Ice Data Center (NSIDC)

Antarctic ice shelves melting

US scientists have found that the floating ice shelves around Antartica have lost 310 km³ of ice every year. They studied satellite data from an 18-year period.

Sea ice around Antarctica is expanding

While the amount of land ice has been decreasing, the amount of sea ice has actually been increasing.

The land ice is the ice which covers 98% of the land of the continent of Antarctica. Since the bulk of the Antarctic ice is land ice, the net effect is a loss of ice. This melting of ice is expected to have significant effects, one of which is a rise in sea level.

A detailed survey of Antarctica, called Bedmap2, was published by the British Antarctic Survey in 2013. This used over 25 million measurements to compile an accurate map of Antarctica and the ice depth across the continent. The survey found that the depth of ice varied from 0 to 4000 metres.

The melting of sea ice does not greatly affect sea level. Think about a drink with ice cubes in it. As the ice melts, the level of drink in the glass does not change. When an iceberg melts there is a very small increase in sea level because fresh water (from the iceberg) is less dense than sea water.

The two major ice sheets are Greenland and Antarctica. Between them, they contain about 75% of the world's fresh water.

For the past 60 years, British Antarctic Survey (BAS) has carried out most of the UK's scientific research in Antarctica. They have five research stations, two Royal Research Ships and five aircraft in and around Antarctica.

Here is a simple relief map of Antarctica.

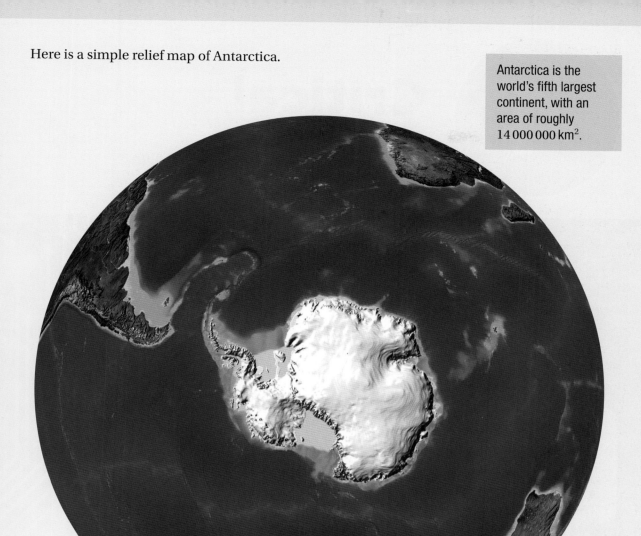

Estimate the total amount of ice in Antarctica, showing full details of all your assumptions and calculations.

Suppose all this ice were to melt. Use your estimate of the amount of ice to calculate what effect this would have on sea level.

Write a report or prepare a presentation on your findings.

Include:

- a critical evaluation of your assumptions and simplifications
- comments on the significance of your conclusions.

4 Critical analysis

Is emigration damaging the UK?

Has the NHS wasted millions on failed attempts to tackle obesity?

Evidence ⟶ Reas

In this chapter you will study how to analyse and judge the merits of a person's conclusions as well as looking at the way these conclusions have been justified. This chapter should also help you improve in the way you communicate your own ideas.

Should green belt land be released for housing?

Can animals predict earthquakes?

Who should you believe about climate change?

ing → Conclusion

In this chapter you will apply general reasoning skills. You will use relatively little mathematics but you should understand and be familiar with the content of Chapters 1, 2 and 3.

In your future profession, as well as in your everyday life today, you will need to think in a logical and reasoned way in order to decide to what extent you should believe what you read or are told.

Critical analysis involves exploring arguments logically. You will need to be sceptical and yet open to persuasion by a well-constructed argument.

A well-constructed argument should involve

- robust evidence
- correct reasoning
- an appropriate conclusion.

Each section of this chapter will contain a number of short passages for you to read and reflect upon and answer some specific questions. After tackling each one, you should check your thoughts with those given at the back of the book.

Concern over the number of women in positions of authority within our organisation misses the point. No individual talented person working here has their professional progress inhibited because of their gender.

CEO of a public authority

In May 2015, the Health Secretary, Jeremy Hunt, said that one in five 11-year-olds was obese and that tackling this would be a priority. Some arguments for and against the ways successive governments have tackled obesity were later given in *The Times*.

Question A

What evidence has been used to back up the CEOs statements? Do you think this is a well-constructed argument?

NHS accused of wasting millions

Professor Modi took over this month as president of the Royal College of Paediatrics and Child Health.

'There have been millions directed to trying to intervene once obesity is established, and yet they've all been failures, so we need to tackle causes,' Professor Modi told *The Times*.

In a stinging assessment of healthy eating messages, Neena Modi said that government anti-obesity campaigns directed at adults and young people 'have all been failures'. Efforts should be devoted to pregnant women and babies instead, she said.

Programmes to train toddlers to like healthy food and make breastfeeding 'trendy' should be tested to find ways of stopping people becoming fat in the first place. I know from my own research that [obesity] actually starts in utero and in infancy,' Professor Modi said.

'An obese mother with gestational diabetes is going to have a baby which is more adipose, has more fatty tissue. That baby has been set on a trajectory which is leading towards obesity in childhood and adulthood. What we don't know yet is how we can intervene.'

Extract from *The Times*, May 26th, 2015

As part of the National Health Service Five Year Forward View, published on 23 October 2014, Simon Stevens, the head of the NHS England, launched a £5 million lifestyle intervention programme to encourage people to lose weight. Other promotions of healthy living include the £11 million-a-year Change4Life campaign and a £155 million programme for school sports.

Question B

a) Very briefly, state
   - what evidence Professor Modi is using
   - what conclusion she draws.

b) Can you think of some possible flaws in her argument as it is presented here?

The reasoning in many arguments depends on mathematical, especially statistical, ideas. The following extract is typical in this sense but hopefully it is not typical in terms of its flawed logic!

> *An X-ray provides no evidence of the disease in approximately one out of five patients known to have the disease.* } Evidence

C

> *If we take no action on the basis of a negative X-ray then there is a 20% chance that we are delaying treatment of a very serious disease.* } Conclusion

The author of the article seems to assume that the conclusion follows immediately from the evidence. To check if this reasoning is correct, consider the results for a sample of 1000 patients.

	Disease	No disease
X-ray positive	8	9
X-ray negative	2	981

Question C

a) Is the statement in the first sentence of Article C correct? Justify your answer.

b) Is the statement in the second sentence of article C correct? Justify your answer.

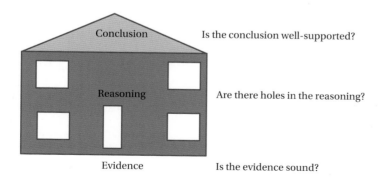

Conclusion — Is the conclusion well-supported?

Reasoning — Are there holes in the reasoning?

Evidence — Is the evidence sound?

When considering how good a test is at diagnosing a medical condition, two terms are often used.

A **false positive** result is one where the test has incorrectly indicated the presence of the disease. In this trial there are 9 false positive results.

A **false negative** result is one where the test has incorrectly not indicated the presence of the disease. In this trial there are 2 false negative results.

Key point

An argument will typically consist of a conclusion that is supported by reasoning on the basis of some evidence. This is illustrated in the diagram.

If you cannot follow an argument, it may be because it is not expressed clearly enough. Consider, for example, the following extract.

Asthmatic athletes must be allowed to take Salbutamol, even though it is blatantly obvious that they are enhancing their performance. No sane person could think it sensible to allow one athlete to take a performance enhancing drug but then ban other performance enhancing drugs. However, the death of so many cyclists shows that Lance Armstrong's ban is fully justified.

D

Question D

a) List examples from this extract of the use of
   - emotive language
   - vague language.

b) What knowledge does the author, perhaps unjustifiably, assume the reader has?

c) In what way is the argument given in this extract contradictory?

d) • What assumption does the author appear to be making about Salbutamol which is never actually stated?
   • Similarly, what is being implied about the deaths of cyclists?

Now suppose that a friend has returned from a trip to the Democratic Republic of Congo and shows you a photograph of an animal you have never seen before, an okapi.

Activity 1

Which of the following deductions, if any, can you make from the evidence of this photograph?

a) All okapis have brown bodies.

b) Some okapis have brown bodies.

c) At least one okapi has a brown body.

d) At least one okapi has at least one brown side.

Whatever you decided in answer to Activity 1, it would be *reasonable* to assume that 'some okapis have brown bodies'. This conforms to a general knowledge about animals, for example that they do not have different coloured sides.

Q The potential overlap of a drug being needed for medical purposes for some athletes and performance enhancing for others is problematic. You could look up official guidelines and try to do better than the author of Extract D in forming a logical opinion on this tricky issue.

Key point

When critically analysing an article you must decide if the author is making reasonable assumptions about the subject matter and the reader's knowledge in order to keep the article to an appropriate length. The article may lack clarity if too many assumptions are being made. This may be done deliberately by the author.

Not making any assumptions could make an article very long and frustrating to read.

Clarity may be damaged by:

- emotive language
- vague phrases
- unjustified assumptions
- self-contradiction.

Key point

Whether a vague phrase such as 'in the next few years' is appropriate will, of course, depend on the context. Phrases you use in everyday conversation might be totally inappropriate in an academic journal.

In the next few activities you can practise applying the ideas you have met so far in this chapter. Article G is quite typical of many letters printed in local newspapers.

It is damaging to the UK for people to retire abroad, taking their accumulated wealth and skills with them. Employers must encourage people to stay on at work beyond pensionable age. In any case, people are living much longer and will need to continue working to avoid poverty in old age.

Question E

a) What is this extract's conclusion?

b) What main reason is given? Does it support the conclusion?

c) Is there anything self-contradictory in this extract?

It is wrong not to eat everything on your plate whilst there are people starving in Africa.

Question F

a) What is the conclusion of this statement and what is given as the reason for this belief?

b) Does the conclusion follow from the reason?

The inappropriate building of houses on green belt land must be stopped before all our countryside disappears. Over 5% of the beautiful land surrounding our town has already been built upon and so we must act quickly to protect what little remains of the once extensive areas of unspoilt fields and woodland. No reasonable person could therefore support the building of a large number of over-priced 'executive' homes on the site of the old dairy. This development will make the developers rich whilst doing nothing to help the young of our town find the houses they so badly need.

G

In 1955, the Minister of Housing urged local authorities to protect the countryside around their towns by labelling some areas 'green belts'. These are rings of countryside where building is to be resisted 'for the foreseeable future'. Because of the green belt policy, land is continuing to be protected for agriculture, forestry and leisure, and urban sprawl has been controlled.

Question G

Comment fully on this letter. Make sure you consider all the key points of this section.

4.3 Selectivity of data

Zener ESP cards each show one of five different symbols.

The idea is that a subject tries to guess the symbol on the card chosen by another person. There is, of course, a 20% likelihood that a guess will be correct by chance. Despite a mass of negative evidence for ESP, stories such as the following often crop up.

ESP proved

In extensive tests of 10 000 high school students in the USA, one student (who has asked to remain anonymous) correctly predicted ten Zener cards in a row. Dr Reuss reported this remarkable performance in a television interview, saying that the odds were only 1 in 10 million that this could have occurred by chance.

ZENER
CARDS
suitable for
TELEPATHY

ESP stands for extra-sensory perception.

Question H

Show that the odds of a person being correct every time when predicting ten Zener symbols is indeed 1 in 10 million.

Although this result sounds remarkable, it is not remarkable at all. In the experiment, each student actually had roughly 500 separate opportunities to score 10 in a row. So there was a total of 500 × 10 000 = 5 million opportunities for someone to get 10 in a row.

If this large-scale experiment happened twice there would be 10 million opportunities to score 10 in a row. As the odds of doing this are 1 in 10 million you would actually *expect* this to happen roughly once every second time the experiment was run!

The following extract about Zener cards is amusing in that following this 'advice' would make *not* showing psychic skills virtually impossible!

Hint: There are 5 possibilities for the first card, 5 × 5 possibilities for the first two cards and so on.

Key point

One individual piece of data can seem very convincing but, ideally, you should consider a range of information. In particular, reporting one success is misleading if millions of failures are ignored.

Develop your psychic skills with Zener cards

- The score due to chance is 20%. Scores above 20% show psychic skills. Scores below 20% show reverse psychic skills and these are also important.
- Find the person you get the best results with.
- Find your best time for getting good results.
- You should do the test at least 100 times under each set of conditions.

Very much less amusing is the fact that important ideas about statistics can be poorly understood by lawyers and medical experts. As a result, there have been a number of high-profile miscarriages of justice.

You could look up details of cases such as those of Donna Anthony, Angela Cannings (featured in the BBC drama *Cherished*) and Sally Clarke.

The precise details of actual cases can be quite complicated but a simple made-up example illustrates the ludicrous nature of some of the arguments used in these cases.

Cheating lottery winner jailed

Lottery winner Fluky Fred has been jailed for cheating in the National Lottery. It is not known how he carried out this fraud but the statistical evidence was convincing. At his trial, it was pointed out that the probability of his winning by chance was only 1 in 14 million.

Question I

Explain to a non-mathematician what is wrong with the evidence used to convict Fred.

Selectivity of data has naturally occurred whenever you read a newspaper article about someone's research. In the following extract this significantly alters the conclusion.

Male professors pick male applicants

Only 18% of biology professors are female and the proportion is even lower in other sciences. Now we know why. Professors of biology, chemistry and physics were asked to rate applicants for a post. Each professor got the same application, but some received it from 'John' and others from 'Jennifer'.

Guess what! Not only were the professors more likely to choose John, but any Jennifer lucky enough to be selected was offered a lower starting salary!

Science faculty's subtle gender biases favour male students

Despite efforts to recruit and retain more women, a stark gender disparity persists within academic science. In a randomised double-blind study ($n = 127$), science faculty from research-intensive universities rated the application materials of a student – who was randomly assigned either a male or female name – for a laboratory manager position. Faculty participants rated the male applicant as significantly more competent and hireable than the (identical) female applicant. These participants also selected a higher starting salary and offered more career mentoring to the male applicant. The gender of the faculty participants did not affect responses, such that female and male faculty were equally likely to exhibit bias against the female student.

Corinne Moss-Racussin et al, 'Science faculty's subtle biases favour male students', *PNAS*, Vol 109 no 41, 21 August 2012

The symbol n is typically used to refer to the size of a sample.

A faculty is a group of academic departments and (in the United States) the term is also used to refer to the *members* of the departments.

Question J & K

a) In what significant way has the newspaper article misrepresented the primary data?

b) What precisely did the writer of the newspaper article do to cause the misrepresentation?

c) Which of the two articles did you find easier to read? List a few of the reasons for this.

The research paper is the **primary source**.

The newspaper article is the **secondary source**.

Look up the purpose of 'double-blind' studies.

You saw in Chapter 1 that sampling is an important aspect of the selection of data. Whether you have to test the durability of a piece of furniture, the safety of a drug or the opinions of the electorate, the time and cost involved are good reasons for testing as small a sample as possible.

In the case of durability, the testing often results in the destruction of the object and so widespread testing would not be sensible.

Key point

When using just a small sample you should take steps to obtain a representative sample so you can draw sensible conclusions. For a good survey you should always include information about the size of the sample and how it was selected.

If a particular group was *not* represented in the sample, then any conclusions cannot be applied to members of that group.

Activity 2

A health authority commissioned a survey of the over-55s in its area in order to plan for future needs. The sample was obtained by selecting people in several local care homes in order that 25% of participants were from each of the age groups 55–65, 65–75, 75–85 and 85+.

a) In what ways could this sample be said to represent the over-55s in this health authority?

b) In what ways is the sample likely to be not representative?

In the biblical book of Daniel, there is a very early description of a 10-day trial of a diet of vegetables and water. The sample size of the people on this diet was only 4, but there was a larger control group on the standard diet of meat and wine.

The size of the sample you use depends on what you intend to use the results for. Opinion polls may be based on a few thousand people whereas initial clinical trials sometimes have very small sample sizes. However, clinical trials often give different results for people of different ethnicities, ages and medical conditions so they have to be repeated for different target groups

Some initial ideas concerning the mathematical theory of sampling are given in Chapter 6. Even if you are not going to study that chapter, it is useful to understand the way that the results from trials and surveys are often reported. For example, following an opinion poll, support for the Traditional Party was reported as:

Key point

Representative samples are used to enable testing to take place with minimal cost and time.

Traditional Party $38.1 \pm 3\%$ ($n = 1811$, 95% confidence level)

A 95% confidence level means that 95 out of 100 polls of this type would give the Traditional Party's support at between 35.1 and 41.1%.

Of course figures such as these refer to a poll *on that day*. The 2015 General Election showed that there can be significant changes in voting intentions just before the election itself.

In your everyday life you have to be careful not to slip into the habit of judging things on the basis of very small samples or on the basis of a biased sample.

Gladys always drank a carton of orange juice every day and she lived to be 103.

Xavier Gonzales had not been looking forward to moving when his company relocated him to the Isle of Wight. Even though he had previously been living on a similar sized Spanish island he was surprised by the courtesy and politeness of his new British neighbours.

Question M

What sampling error do you imagine Xavier might have made which led him to be so pleasantly surprised by the people on the Isle of Wight?

Trialling is very important for judging things such as the effectiveness of drugs and working practices. However, it is also important that you take various precautions when conducting these trials.

At one time, it was thought possible that female hormones might be useful in preventing altitude sickness. One particular trial was conducted by a team of male doctors in the Himalayas. One third of the team took a standard altitude-sickness medication called Acetazolamide, another third took a cocktail of female hormones and the remaining third were given pills containing no active ingredients. No-one on the trip knew who was receiving which treatment. For ethical reasons, members of the trip did know what the three possibilities were. Concern amongst team members about possible side effects of their treatments led to the trial being abandoned.

Question N

a) Research the concepts of placebos and control groups. What were the control groups in this trial? What was the purpose of the pills with no active ingredients?

b) Why was it felt ethical to inform the doctors of the types of treatment?

Key point

Patients can experience beneficial effects simply by believing they are receiving treatment. Control groups taking a placebo can be used to check that the effect of a trialled drug is greater than this placebo effect.

Question L

What is the significance of this example of long life?

At a height of 3600 metres, there are roughly 40% fewer oxygen molecules in each breath. Endurance athletes often take advantage of this and may spend a few weeks training at between 2000 and 3600 metres to improve the efficiency of their oxygen use during an event.

However, if a person ascends to these types of height too quickly, they can suffer from altitude sickness. This involves symptoms such as severe headaches, nausea, lack of sleep and exhaustion. The symptoms usually disappear quickly if a person descends and they can then make a slower ascent later.

4.5 Misleading with data

Diagrams are a very effective way to display data and are therefore widely used in marketing and the media. This diagram dramatically shows the effect of journey length on fuel economy. However, it would look much less dramatic if the vertical scale were to go from 0 to 1.1.

There is no reason to think that this diagram was meant to deceive and even small changes in fuel economy are extremely important. However, the superficial visual effect of this diagram would influence many people reading the article accompanying it.

Reporters sometimes deliberately choose the scales and the starting points of axes to mislead and enable headlines to be more dramatic than is justified. The examples in Activity 3 are all based on misuses of diagrams that have actually occurred.

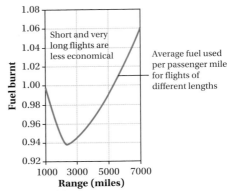

Activity 3

For each one of the following diagrams comment on any unusual features and how it may have been meant to mislead the reader.

(a)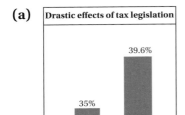

(b)
Viewers undecided

1s FIFA corrupt?

42% Very likely
34% Likely
31% Not likely

(c)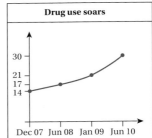

(d)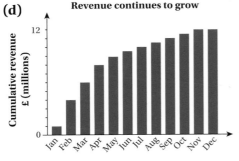

(e)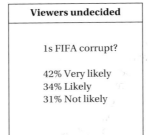

(f)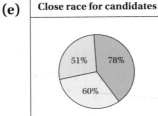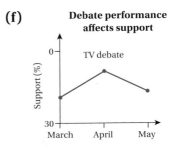

Key point

When interpreting a graph, always check carefully that what it appears to be showing is actually the case. The standard conventions are to use uniform scales, starting from 0 on both axes. There are sometimes good reasons for not following these conventions but this should be clearly flagged to the reader.

Some more complicated examples where you need to criticise data and the way it is used are given below.

Casualties soar in 20 mph zones

Official statistics show that this year there has been a 20% increase in deaths in 20 mph zones.

Question O

The official statistics were as shown.

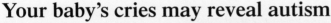

Fatalities	20 mph zones	30 mph zones
2010	6	563
2011	7	636

a) Comment on the newspaper article and its headline.

b) Why might the number of deaths in 20 mph zones have increased even if each 20 mph zone has stayed just as safe?

Q The Royal Society for the Prevention of Accidents has a policy statement *20 mph Zones and Limits* which puts a strong, statistics based, case for 20 mph zones.

Your baby's cries may reveal autism

Autism spectrum disease (ASD) is a poorly understood condition. There is no cure for ASD but behavioural intervention can be useful and should be carried out as soon as possible. This intervention can involve extensive practical, emotional and financial commitment for a family.

Researchers at the University of Pittsburgh may have had a break-through in the early diagnosis of ASD. These findings could enable families to diagnose ASD in children as young as 6 months old. A small number of 6-month-old babies in the study were later diagnosed with ASD. The pain-related cries of these babies were among those which had the highest sound frequency measured in hertz.

'We definitely don't want parents to be anxiously listening to their babies cry,' said psychologist Dr Stephen Sheinkopf. 'The differences in cries were detected by sophisticated technology and not people. It's unclear if the human ear is sensitive enough to detect this.'

P

Question P

a) What criticism could you make of the newspaper's headline?

b) What vague words in the newspaper article might make you wonder about the value of the research project?

c) In the small study, one baby who was *not* later diagnosed with ASD had a pain related cry with a sound frequency which was significantly higher than that of any other baby in the study. What would have been the drawback of behavioural intervention for this baby?

d) The only relevant finding of the research paper was that in a sample of 12 babies who made pain-related cries, the only baby who was later diagnosed with ASD had the second highest frequency cry. Comment on the value of the research paper findings and the newspaper article.

4.6 Coincidence?

Making mental connections between things is a natural human activity. Just as babies learn that some actions get them instant attention, adults eventually realised that smoking causes lung cancer. When two things appear to be linked they are said to be **correlated**.

In May 2015, Sally, Alex and Kate met for the first time on holiday in Turkey. After a very mild earthquake, Sally remarked that she had visited each of the affected regions of Peru, New Zealand, Japan and Nepal shortly before each country experienced a major earthquake in 2007, 2011, 2012 and 2015 respectively. Sally was surprised to discover that Alex and Kate had also visited all four of the affected regions. Like Sally, Alex had gone to the regions before the major earthquakes, whereas Kate had always been there shortly after the earthquakes.

Activity 4

Can you think of some reasons why the various correlated events of this true story were not as remarkable as they might at first seem?

Hint: Think about what careers some of these people may have had which would take them to these places.

Key point

If two events, A and B, are correlated then

- this could be coincidental
- A could cause B
- B could cause A
- a third factor could cause both A and B.

Activity 5

In each of the following scenarios decide whether you think there is a causal link between A and B or whether a third factor might have caused both A and B.

a) After Clare moved into their road the residents noticed an increase in the number of pigeons [A] and an increase in the number of rats [B].

b) Students who score highly in the first question on an examination paper [A] gain a good grade [B].

c) Countries with good university research facilities for Mathematics [A] have high Gross Domestic Products (GDP) [B].

d) As ice-cream sales increase [A] so does the crime rate [B].

e) Rising atmospheric CO_2 levels [A] are correlated with obesity levels [B].

The possible causation behind observed correlations is of great importance to scientific investigations. When scientists trial a new drug they look for correlations between taking the drug and recovery rates and between taking the drug and possible side effects.

Newspaper articles often feature strange correlations. Often, these articles imply a causal connection which is not justified by the evidence.

Optimists are luckier than pessimists

A scientific analysis of the personalities of lottery winners has convincingly demonstrated that optimists outnumber pessimists in the ratio of 2:1.

Question Q

a) Can you think of a possible causal link for this result? How would you check your idea?

b) What words in the extract are designed to convince you that optimists are actually luckier than pessimists?

Animals do predict earthquakes!

Many people, from the time of ancient Greece onwards, have reported odd animal behaviour before an earthquake. Seismologists have treated these claims with scepticism. However, these reports have now been justified with data from VLF (very low frequency) receiving stations in Peru. They have shown that there were large irregularities in VLF signals before the magnitude 7 Peruvian earthquake of August 2011. This gives us an explanation for the animal behaviour. Just as you can create static electricity by rubbing objects together, so subterranean movements of rocks can cause static electricity.

R

VLF signals are radio frequencies in the range 3 kHz to 30 kHz (between 3000 cycles per second and 30 000 cycles per second).

These frequencies are impractical for speech transmission but are used for some navigational, meteorological and military communications.

Question R

a) Anecdotal reports, for example of odd animal behaviour, may indicate things to investigate but do not provide evidence in their own right. Why is that?

b) List the steps in the chain of reasoning leading to 'This gives us an explanation for the animal behaviour'.

For each step, comment fully on whether the article has made a good case.

For each step, describe experiments that could be carried out or data that could be gathered to check if the explanation of animal behaviour suggested in the article is sensible.

In Chapter 3, you considered and critically evaluated a variety of models from physics, meteorology, medicine and economics. In this section you will analyse models connected with the hotly-debated topic of climate change.

The diagram below represents a simple model but thinking about this model will help you understand many of the issues connected with climate change.

A major source of evidence concerning factors involved in the warming and cooling of the Earth are the records of atmospheric CO_2 and temperature in the Earth's past. On the graph below, each trough is an ice age and each peak is an inter-glacial period such as the present.

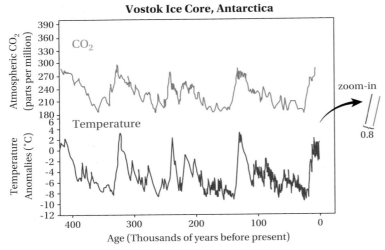

J R Petit et al., 'Climate and atmospheric history of the past 420,000 years from the Vostok ice core, Antartica', *Nature*, Vol 399 no 6735, 3 June 1999

The two graphs show how closely the changes in atmospheric CO_2 and temperature match each other. With a much finer horizontal scale, it is interesting that for much of the time there is a lag between the two effects with CO_2 levels following approximately 800 years after temperature changes.

In the film '*An Inconvenient Truth*', Al Gore stressed the importance of the correlation between the graphs and concluded:

> '*There is one relationship which is far more powerful than all the others. When there is more CO_2 in the atmosphere the temperature gets warmer.*'

(S)

Activity 6 💬🔍

a) Think of some factors that could affect the amount of energy received from the Sun.

b) Think of some factors that could affect the amount of energy radiated from the Earth.

Temperatures and concentrations of CO_2 for the last 650 million years have been obtained using ice cores from Vostok, Antarctica.

If the two graphs are superimposed then, for most of the time, the red graph lags very slightly behind the blue graph.

Question S

a) What is the major flaw in Al Gore's argument for explaining historic temperature changes in terms of atmospheric CO_2.

b) Can increases in CO_2 cause climate change?

If you are interested in the historic shifting between warm periods and ice ages you should research Milankovitch cycles in the Earth's orbit. These are also correlated with variations in temperature and suggest a model along the lines of:

Natural variations in the Earth's orbit

Changes in the radiant energy received from the Sun

Changes in the temperature of the Earth

Changes in atmospheric CO_2
(warm oceans release CO_2 to atmosphere)

There are perhaps two major issues for you to bear in mind when considering these arguments.

1. People on both sides of the debate can be highly selective about the data they use. The temperature of regions such as Antarctica, the USA and deep oceans are interrelated in a very complicated way and in the short term, small changes in one area are probably due to other factors rather than any overall climate change.

2. This graph shows that the temperature of an affected region can fall for long periods whilst the general trend is still upwards.

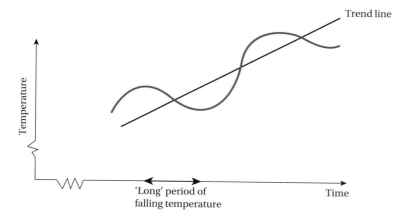

Because of human activity, the amount of atmospheric CO_2 has risen well above the levels seen in the Vostok ice cores and is now roughly 400 parts per million. What is much less clear is precisely how strong an effect this will have on the temperature.

Activity 7 Q

The internet is a rich source of arguments for you to analyse about how much the Earth is warming. For example, you could consider

★ Professor Richard Lindzen of MIT responding to Tim Yeo, Chairman of the UK Energy and Climate Committee on 28th January 2014.

★ Dr Don Easterbrook claiming that 'Global warming stopped in 1998', at the Senate hearing of 26th March 2013.

Key point

Temperature changes (or a lack of them) in a particular region and/ or over periods of a few years have no relevance in this debate. Many people (even scientists) are unwilling to suspend judgement. Once you have made your mind up it is often too easy to grasp at supporting evidence and ignore contradictory evidence.

The potential environmental repercussions of not acting early enough are enormous.

This does not excuse lazy thinking, but does explain why many people feel that they do not have the luxury of waiting any longer before expressing an opinion.

Consolidation exercise 4

1

National debt doubles
£
2009

£

2014

Criticise this visual representation of the national debt.

2 *It is only fair that students should pay tuition fees at university. Many studies have shown that graduates earn more than non-graduates on average and so going to university increases earning power.*

   a) State the evidence and the conclusion of this extract.

   b) If you assume that graduates do earn more than non-graduates on average, does it follow that 'going to university increases earning power'? Carefully justify your answer.

3 In Aysha's year at college there are 250 students. 70 of these students study Mathematics at A level. To determine the general attitude of students to their choice of course, Aysha gives a questionnaire to 16 students in her Maths A level set and to 9 other students. The results are summarised in the table.

	Dissatisfied	Satisfied	Very satisfied	Good	Excellent
Studying maths	1	4	3	6	2
Not studying maths	3	3	2	0	1
Total	4	7	5	6	3

   a) Comment on the categories Aysha has used for this study.

   b) Aysha reports that 84% of the students in her year are happy with their choice of course. Explain how she arrived at this figure.

   c) Give two reasons for doubting the figure of 84%.

   d) Assume that the students in Aysha's set are representative of the students studying Mathematics A level and that the 9 students are representative of the other students. Calculate a better estimate of the percentage of students who are satisfied with their course.

4

Councils waste £10 billion per year

Councils have a £50 billion annual procurement budget. However, a report commissioned by the government has found that councils can save up to 20% on their spending without affecting any of their services simply by shopping around. The management consultancy firm they employed looked at spending in two areas.

	Annual spending (£k)	Proposed saving (£k)
Mobile phones	600	133
Solicitors	6400	642

Procurement is obtaining or buying goods or services. Councils procure things such as care, education and transport services. These may be more or less expensive depending on who provides the services.

£k means thousands of pounds. k is a common abbreviation for thousands.

a) Calculate the percentage savings that can be made in each of the two areas of spending.
b) Describe the evidence and reasoning that might have been used to reach the conclusion of the newspaper headline.
c) Explain why the newspaper headline is *not* supported by the evidence.
d) It is suggested that the management consultancy firm should have worked out the saving on the *total* spending in the two areas considered. Carry out this calculation and find how much this would suggest could be saved from the £50 billion budget.
e) Do you think that your answer to d) is reasonable for the savings that could be made? Justify your answer.

Critical analysis of your own work

The analysis you have learnt to apply when studying the work of other people should be applied to your own work so that any summarising or report writing you do does not contain the many flaws that you have seen in this chapter.

Review

After working through this chapter you should:

- be able to recognise the evidence, reasoning and conclusions of an article
- know that the clarity of an argument can be damaged by emotive language, vague phrases, unjustified assumptions and self-contradiction
- be aware that each side of a debate may be *very* selective about the evidence they use
- know the importance of studying visual data carefully in case the presentation is misleading
- understand the difference between correlation and causation
- be able to apply your knowledge to the critical analysis of articles and mathematical models.

Investigation

Global warming

View the Lindzen–Yeo debate mentioned in Activity 7 on page 117. Summarise Professor Lindzen's arguments and conclusions in particular those relating to the question:

'Is the decade 2000–2010 the hottest on record? Is it? Yes or No?'

Can you write a brief report putting Professor Lindzen's answers in a form that members of the UK Energy and Climate Committee might have understood?

5 The normal distribution

What proportion of students have an IQ of over 130?

What proportion of people in the UK can run 5 k in under 21 minutes?

In this chapter you will learn how to use the normal distribution to find the likelihood of events in a range of applications.

What is the chance that a tyre will be within legal guidelines after 25 000 kilometres?

You should know how to:

- express an amount as a percentage of a whole ⊕ 1302
- convert between fractions and decimals ⊕ 1015
- find means and standard deviations (Chapter 1)
- plot histograms (Chapter 1).

5.1 Features of a normal distribution

Whether you are designing a kitchen, a car or a work station for an office or factory, you will have to pay great attention to 'typical' human dimensions such as height, weight and lengths of arms.

You will not be surprised that histograms of many of these sorts of measurements have a bell-shaped outline with most measurements in the centre of the distribution.

What is perhaps more surprising is *precisely* how similar some of these histograms are. Many have the following features.

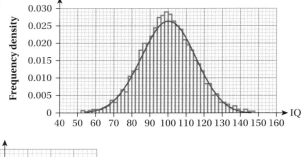

- The distribution is symmetrical about the mean and bell-shaped.
- About $\frac{2}{3}$ or 67% of the data are within 1 standard deviation (s.d.) of the mean.
- About 95% of the data are within 2 s.d. of the mean.
- Virtually all the data are within 3 s.d. of the mean.

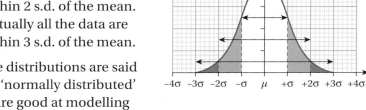

These distributions are said to be 'normally distributed' and are good at modelling not just the dimensions of humans and other animals but also many other variables such as IQ scores, manufacturing errors, weights of mechanically-filled packets, life spans of animals and various products.

> Standard deviation (σ) is a measure of spread. Look back to Chapter 1 if you need a reminder.

> The normal distribution is the single most important idea in statistics. It occurs naturally in a broad variety of situations.

Activity 1 💬

The amount of household waste from each household in a town can be modelled by a normal distribution with mean 80 litres per week and standard deviation 5 litres. The council is considering giving every household a 90-litre bin for household waste.

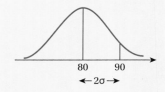

Out of 100 000 households one week, how many would you expect to have

a) at least 10 litres of spare capacity in their bin

b) too much waste for their bin?

You should think of the normal distribution as a 'mathematical model' for the actual distribution. This is illustrated in Example 1 where a particular normal distribution models the data well in some ways, but less well in others.

> The histogram is based on the speeds of 400 million observed cars, many travelling at well over the speed limit.

Example 1

a) What percentage of cars were travelling at over 80 mph?

b) The speeds can be modelled by a normal distribution with mean 70 mph and standard deviation 10 mph.

What proportion of cars does the model predict to have been travelling at over 80 mph?

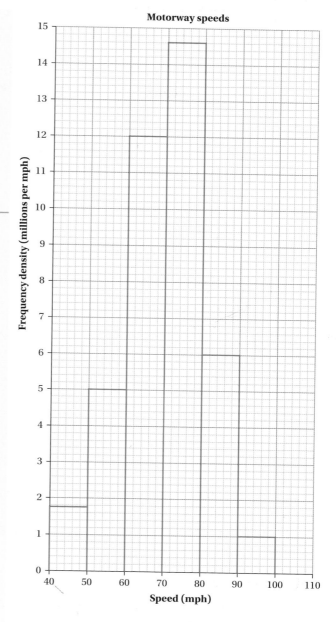

a) Calculate the number of cars travelling over 80 mph.

$6 \times 10 + 1 \times 10 = 70$ million

Work out what percentage of the whole number of cars this is. ⊕ 1302

$\frac{70}{400} \times 100 = 17.5\%$

> In a histogram the areas of the bars represent the number of cars.

b) Over 80 mph is everything more than 1 s.d. above the mean. For a normal distribution about $\frac{2}{3}$ of the data are within 1 s.d. of the mean so the other $\frac{1}{3}$ of the data is equally split between values above 80 mph and values below 60 mph.

Over 80 mph is approximately $\frac{1}{6}$ or 16.7% of the distribution. This is reasonably close to the figure in part **a**, obtained from the histogram of data.

Suppose a distribution of lengths has mean 10 cm and standard deviation 2 cm. You should be able to switch quickly between lengths and standard deviations above the mean, as in this table.

Key point

For a normal distribution of given mean and standard deviation, think in terms of numbers of standard deviations above (or below) the mean.

Length (cm)	14	11	8	15	7
No. of s.d. above mean	2	0.5	−1	2.5	−1.5

It is often convenient to use negative numbers of standard deviations when a value is below the mean.

Exercise 5A

1 A distribution of weights has mean 60 kg and standard deviation 8 kg. Complete this table of values.

Weight (kg)	64	44			56
No. of s.d. above mean			−1	2.5	

2 State whether the distributions shown below appear to form a normal distribution. If the distribution is not normal give the reason.

a)

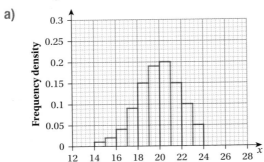

b)

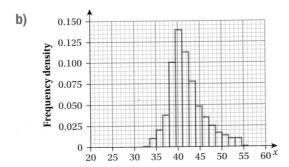

b)

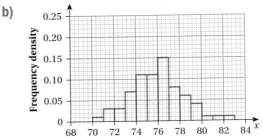

c)

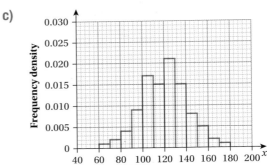

d)

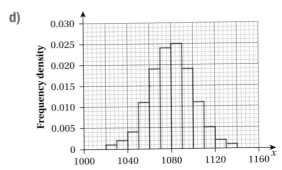

3 Estimate the mean and standard deviation of each of the normal distributions.

a)

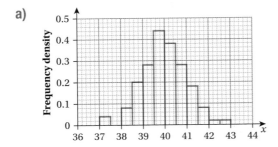

4 Adult female heights are normally distributed with mean 165 cm and standard deviation 6.5 cm. Adult male heights are normally distributed with mean 173 cm and standard deviation 7.5 cm.

a) Comment on the differences between the distributions of female heights and male heights.

b) Explain why the proportion of women taller than 178 cm is the same as the proportion of men taller than 188 cm. What is this proportion?

c) In a random sample of 1000 women, how many would you expect to be taller than 178 cm?

5 The daily returns on the FTSE 100 (a portfolio of shares) have a mean of 0.03% and a standard deviation of 0.24%.

> These figures are based upon historic performance over the last 30 years and are not guaranteed for the future.

a) Comment on the significance of the value of the mean.

b) Comment on the value of the standard deviation compared to that of the mean.

c) Assume that the daily returns are normally distributed.

   i On how many days in a year would you expect the daily return to be less than −0.21%?

   ii On how many days in a year would you expect the daily return to be greater than 0.51%?

> Normal distributions play a major role in modelling customer demand. This is vital for many businesses so that the right amount of stock is kept in store.

6 a) Why do most businesses keep a stock of products for which they do not yet have orders?

b) Why would a business not want to keep a larger than necessary amount of stock?

c) A car dealership models its sales per year by a normal distribution with mean 500 cars and standard deviation 25 cars. According to this model, how likely is it that one year's sales will be less than 450 cars?

7 The weights of packets of biscuits are normally distributed with mean 200 g and standard deviation 2.5 g. A random sample of 200 packets is selected and each packet is weighed.

a) How many packets would you expect to have a weight outside the range from 195 g to 205 g?

b) How many packets would you expect to have a weight in the range from 197.5 g to 202.5 g?

c) How many packets would you expect to have a weight in the range from 197.5 g to 200 g?

8 The number of calls per minute to a telephone answering service at peak time can be modelled by a normal distribution with mean 20 and standard deviation 5.

a) How likely is it that the number of calls per minute will be more than 35?

b) How likely is it that the number of calls per minute will be less than 15?

c) How likely is it that the number of calls per minute will be between 15 and 30?

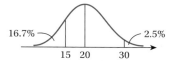

The standard normal distribution is a mathematical description of the bell-shaped distributions that appear in nature. It is denoted by the symbol N(0, 1).

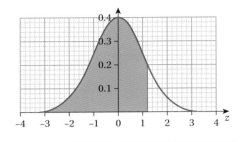

N(0, 1) has mean 0 and s.d. 1.

The area under the curve is 1.

Since the area under the entire curve is 1, the shaded area shown above will equal the proportion of the distribution which has $z < 1.2$. This is denoted by $\Phi(1.2)$.

> N stands for normal.
> 0 is the mean and 1 is the square of the standard deviation.

Activity 2

a) State the approximate values of $\Phi(-2)$, $\Phi(-1)$, $\Phi(0)$, $\Phi(1)$ and $\Phi(2)$. Think about the symmetry of the graph.

b) Estimate the value of $\Phi(1.2)$ from the graph shown above.

> z is conventionally used for the variable of N(0, 1).
> Φ is the capital Greek letter phi.

As you have seen in Section 5.1 and Activity 2, you can find the approximate proportion of a normal distribution that lies above and below certain multiples of the standard deviation. For other values, tables of results for $\Phi(z)$ have been created and one is given in the back of this book. Even more conveniently, the values of Φ are stored in many calculators.

> You should find out whether you can use *your* calculator to find these values.

Example 2 Find the proportion of a standard normal distribution which lies between $z = -0.2$ and $z = 1.5$.

You need to find $\Phi(1.5) - \Phi(-0.2)$.

You can use values from your calculator or from the tables.

$\Phi(1.5) = 0.93319$

If you are using the tables, you will need to find $\Phi(-0.2) = 1 - \Phi(0.2)$.

$\Phi(-0.2) = 1 - 0.57926 = 0.42074$

$\Phi(1.5) - \Phi(-0.2) = 0.93319 - 0.42074 = 0.51245$

so 51.245% of the distribution lies between $z = -0.2$ and $z = 1.5$

> Tables only show values for positive z.
>
> To find the value for a negative number you have to use the fact that $\Phi(-z) = 1 - \Phi(z)$.
>
>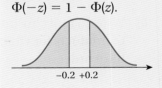

One way of looking at the result of Example 2 is that a randomly chosen member of the distribution has a 51.245% chance of lying between -0.2 and 1.5. The mathematical term for chance is **probability**. You can say that the probability that $-0.2 < z < 1.5$ is 0.51245. You can write this as $P(-0.2 < z < 1.5) = 0.51245$.

> Remember that probability is a number between 0 and 1.

Activity 3

What are the following probabilities?

a) $P(z = 1)$

b) $P(-0.2 \leqslant z \leqslant 1.5)$

c) The probability of an impossible event.

d) The probability of a certain event.

Example 3 For the distribution N(0, 1)

a) Find

  i $P(z < 1.3)$ **ii** $P(-1 < z < 1.3)$ **iii** $P(z > -1.2)$

b) Find the number a such that $P(z < a) = 0.70194$

c) Find a symmetric interval about 0 such that the probability of a randomly chosen member of the distribution being in this interval is 0.9

a) i $P(z < 1.3) = \Phi(1.3)$. Use your calculator or the tables to find $\Phi(1.3)$.

    $\Phi(1.3) = 0.90320$

> If you have a suitable calculator, the answers to part **a)** can be written down with just minimal explanation.

ii $P(-1 < z < 1.3) = \Phi(1.3) - \Phi(-1)$

    Remember that $\Phi(-1) = 1 - \Phi(1)$

    $\Phi(1.3) - \Phi(-1) = 0.90320 - (1 - 0.84134)$

                   $= 0.74454$

iii $1 - \Phi(-1.2) = \Phi(1.2) = 0.88493$

> From the symmetry of the normal distribution, $P(z > -1.2) = P(z < 1.2)$.
> (You could draw a diagram to help you see this.)

b) This time the problem is reversed and you already know the probability. You can use your calculator or use the tables the other way around.

  $\Phi(a) = 0.70194$ so $a = 0.53$

c) The interval is symmetric so it goes from $-b$ to b.
You have to find the value of b.

  $P(-b < z < b) = 0.9$

The other 0.1 is equally split between values less than $-b$ and more than b.

So $P(z < b) = \Phi(b) = 0.95$

As in part **b)**, you already know the probability so solve the reverse problem.

$\Phi(b) = 0.95$ so $b \approx 1.645$

The interval is $(-1.65, 1.65)$

Exercise 5B

1 a) Find

    i $P(-1 < z < 1)$ **ii** $P(-2 < z < 2)$

    iii $P(-3 < z < 3)$

  b) Compare your answers with 95%, $\frac{2}{3}$ and 1.

2 Find

  a) $P(z > 1.7)$ **b)** $P(z < -1.3)$

  c) $P(1 < z < 2.5)$

In general, quantities that you are studying are unlikely to have a mean of 0 and a standard deviation of 1. However, all normal distributions have a simple connection with the standard normal distribution.

$\mu = 0, \sigma = \frac{1}{2}$

This has a smaller s.d. and so is less spread out. By doubling the height, the area is kept as 1.

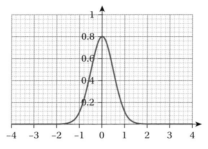

$\mu = 0, \sigma = 1$

The standard normal distribution.

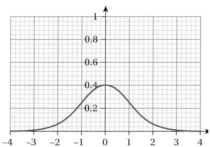

$\mu = 0, \sigma = 2$

This has a larger s.d. and so is more spread out. By halving the height, the area is kept as 1.

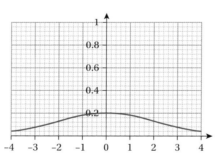

This means that *every* normal distribution is obtained from a standard normal by

- shifting in the x-direction until the mean is at μ
- stretching/shrinking from the y-axis to make the s.d. correct
- shrinking/stretching from the x-axis to keep the area equal to 1.

Key point

For *any* normal distribution, all that matters for calculation of probabilities is how many standard deviations a measurement is from the mean.

Example 4 Suppose that the heights of 13-year-old boys are normally distributed with a mean of 159 cm and standard deviation of 11 cm.

a) What is the probability that a 13-year-old boy will have a height of less than 165 cm?

b) How many of a random sample of 1000 13-year-old boys would you expect to be less than 165 cm tall?

---

a) The first step is to convert all numbers into 'number of standard deviations away from the mean'. Then carry out the calculation for the standard normal distribution.

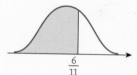

$$165 = 159 + 6$$
$$= \mu + \frac{6}{11}\sigma$$

165 is $\frac{6}{11}$ standard deviations away from the mean.

So $P(h \leqslant 165) = P\left(z \leqslant \frac{6}{11}\right)$
$$= \Phi(0.545)$$
$$\approx 0.71$$

$\sigma = 11$ and so $6 = \frac{6}{11}\sigma$.

$\frac{6}{11}$ or 0.545 is known as the 'z-value'. ⊕ 1063

b) The probability of a boy being less than 165 cm tall is approximately 0.71

To find the expected number in a sample, multiply the sample size by the probability.

So you would expect approximately $1000 \times 0.71 = 710$ of the boys to be less than 165 cm tall.

Key point

The notation for a normal distribution with mean 159 and s.d. 11 is $N(159, 11^2)$ or $N(159, 121)$.

The second number is the *square* of the s.d. and is called the **variance**.

Key point

In the distribution $N(159, 11^2)$, the value 165 produced a 'z-value' of 0.545

In general, if a value x is taken from $N(\mu, \sigma^2)$ then the z-value is given by the formula $z = \frac{x - \mu}{\sigma}$.

This process of finding the z-value is called **standardising** the variable.

Activity 4 💬

A condition of membership of Mensa is that you should have an IQ of over 130. The Mensa website states that Mensa is a society for the top 2% of the population. Use the fact that many IQ tests are designed to have the distribution N(100, 15^2) to check if having an IQ over 130 is the same as being in the top 2% of the population's IQs.

Example 5 The weights of the contents of jars of honey are normally distributed with mean 454 g and s.d. 4 g.

a) Find the z-values of   **i**   450 g   **ii**   455.5 g.

b) Find the probability that the contents of a jar of honey have a weight between 450 g and 455.5 g.

---

a) Use the formula $z = \dfrac{x - \mu}{\sigma}$.   ⬚ 1186

   i $\dfrac{450 - 454}{4} = -1$

   ii $\dfrac{455.5 - 454}{4} = 0.375$

b) $P(450 < \text{weight} < 455.5) = P(z < 0.375) - P(z < -1)$

$$= \Phi(0.375) - \Phi(-1)$$
$$= 0.64617 - (1 - \Phi(1))$$
$$= 0.64617 - 0.15866$$
$$\approx 0.49$$

Example 6 You know that the length of time a particular pizza firm takes to deliver a pizza on a Friday evening has distribution N(30, 64). At what time should you phone for a pizza if you want a 95% chance of it arriving by 7 pm?

Use your calculator or use the tables backwards to work out the z-value for 95%.

$\Phi(1.64) = 0.95$

There's a 95% chance of it taking less than 1.64 standard deviations above the mean.

$30 + 1.64 \times 8 \approx 43$ minutes

You should phone at 6:17 pm.

> Remember that the standard deviation is $\sqrt{64}$.

Example 7 The resting heart rate of a randomly selected person has distribution N(70, 12^2).

a) Find the probability that a person has a resting heart rate of over 100.

b) Find the heart rate below which 90% of the population lie.

---

a) Work out how many standard deviations 100 is above the mean of 70.

$100 = 70 + 30 = 70 + 2.5\sigma$

$P(z > 2.5) = 1 - P(z < 2.5)$

> $P(z > 2.5) + P(z < 2.5) = 1$ as the area under the normal curve is 1.

$\qquad\qquad = 1 - \Phi(2.5) = 0.00621$ or 0.6%

b) Use your calculator or tables to find the z-value which gives 90%.

$\Phi(1.28) \approx 0.9$

You need to find 1.28 standard deviations above the mean.

The required heart rate is $70 + 1.28 \times 12 = 85.4$

Exercise 5C

1 State the value of the standard deviation for each of these distributions.

   a) $N(5, 7^2)$ **b)** $N(5, 36)$ **c)** $N(17, 40)$

2 Calculate z-values for each of these.

   a) A value of 9 in a normal distribution with mean 6 and standard deviation 5.

   b) A value of 7 in a normal distribution with mean 8 and standard deviation 16.

   c) A value of -1 in $N(2, 3^2)$.

   d) A value of 6 in $N(4, 9)$.

   e) A value of 4 in $N(3, 7)$.

   f) A value of 3 in $N(5, 5)$.

3 For a standard normal distribution, find these probabilities.

   a) $P(z < 1.14)$

   b) $P(z > 2.03)$

   c) $P(z < -1)$

   d) $P(z > -3)$

   e) $P(1.5 < z < 2.5)$

   f) $P(-1 < z < 2)$

   g) $P(-2.41 < z < -1.56)$

4 **a)** For a standard normal distribution explain, using a diagram, why
$P(z < -a) = P(z > a)$,
for any positive number a.

   b) Hence explain why
$P(-a < z < a) = 2P(z < a) - 1$

5 For a standard normal distribution, find the value of the number a for each of these.

   a) $P(z < a) = 0.556$

   b) $P(z > a) = 0.01$

   c) $P(z < a) = 0.1$

   d) $P(-a < z < a) = 0.95$

   e) $P(-a < z < a) = 0.34$

6 Assume that pulse rates, R, are normally distributed with mean 68 and variance 10. Find numbers $68 - c$ and $68 + c$ for each of these probabilities.

   a) $P(68 - c < R < 68 + c) = 0.34$

   b) $P(68 - c < R < 68 + c) = 0.754$

> Hint: The answer to **6 a)** is the z-value of $68 + c$.
>
> Hint: First find the value of a such that $P(-a < z < a) = 0.34$

7 Even with measurements of such quantities as heights and weights, the distributions for some populations will not be normally distributed. Suggest possible reasons for distributions of heights to have the shapes shown below.

   a)

   b)

Consolidation exercise 5

1 The weights of babies when born follow a normal distribution with a mean of 3.5 kg and a standard deviation of 0.5 kg. New-born babies are classified as small, medium or large, according to their weight.

Babies with a weight of more than 4.3 kg are classified as large.

The probability of a new-born baby being classified as small is 0.137

   a) Calculate the probability that a new-born baby is classified as large.
   b) Calculate the maximum possible weight of a baby that is classified as small.

2 Body temperatures may be assumed to be normally distributed with a mean of 98.3 °F and a standard deviation of 0.8 °F.

Calculate the probability that a person, chosen at random, has a body temperature greater than 100.5 °F

3 The heights of boys aged 11 years may be assumed to be normally distributed with a mean of 149.3 cm and a standard deviation of 12.7 cm. Calculate the probability that a boy, chosen at random, has a height

   a) greater than 152 cm
   b) less than 154 cm
   c) less than 145 cm
   d) between 148 and 151 cm

4 Each day, Margot completes the crossword in her local morning newspaper. Her completion times, X minutes, can be modelled by a normal random variable with a mean of 65 and a standard deviation of 20.

Determine

   a) $P(X < 90)$ b) $P(X > 60)$

(AQA, 2010)

5 Electra is employed by E and G Ltd to install electricity meters in new houses on an estate. Her time, X minutes, to install a meter may be assumed to be normally distributed with a mean of 48 and a standard deviation of 20.

Determine

   a) $P(X < 60)$
   b) $P(30 < X < 60)$
   c) the time, k minutes, such that $P(X < k) = 0.9$

(AQA, 2007)

6 Assume that the wrist circumference of the male population of the UK is normally distributed with a mean of 184.5 mm and a standard deviation 13.6 mm.

   a) A manufacturer designs a watch for the UK market. It is designed to cater for wrist circumferences up to 205 mm. What percentage of the UK male population is catered for by these watches?
   b) A watch designed for the Japanese market will fit wrists with circumferences from 145 mm to 190 mm. Find the percentage of the UK male population that this watch will fit.

(AQA, 2005)

7 The rest pulse rate of a randomly-selected person can be assumed to be normally distributed with mean 68 and standard deviation 13. Find the probability that

   a) i a person has a pulse rate over 75
      ii a person has a pulse rate between 58 and 72
   b) Find the pulse rate below which 90% of the population lie.

8 A set of ear plugs claims to have noise reduction of 25 dB, if used properly. At low frequency (125 Hz), the plugs are found to have a mean attenuation of 31.6 dB with standard deviation 4.3 dB. Assume the attenuations are normally distributed. What is the probability that a set of plugs chosen at random will have attenuation below the advertised 25 dB?

9 An engineering firm receives an order to supply 10 000 nails. The firm's machines produce nails with lengths having mean 55 mm and standard deviation 1.5 mm. Quality control rejects a nail if it is less than 52 mm long or more than 57 mm long. To be sure of making enough good nails to fulfil the order the firm decides to produce 11 000 nails. Assuming a normal distribution for the lengths of nails, how many nails which will not be rejected is the firm likely to produce?

10 a) Assume that Japanese adult males have foot lengths which are normally distributed with a mean of 24.9 cm and standard deviation 1.05 cm.
       Calculate the probability that a Japanese adult male has a foot length greater than 27 cm.
    b) Assume that Japanese adult females have foot lengths which are normally distributed with a mean of 22.8 cm and standard deviation 0.89 cm.
       Calculate the percentage of the Japanese adult female population that has foot lengths between 22 cm and 25 cm.

       (AQA, 2011)

11 The heights of sunflowers may be assumed to be normally distributed with a mean of 185 cm and a standard deviation of 10 cm.
    Determine the probability that the height of a randomly chosen sunflower
    a) is less than 200 cm
    b) is more than 175 cm
    c) is between 175 cm and 200 cm.

       (AQA, 2006)

12 When a particular make of tennis ball is dropped from a vertical distance of 250 cm on to concrete, the height, X centimetres, to which it first bounces may be assumed to be normally distributed with a mean of 140 and a standard deviation of 2.5
    a) Determine
        i $P(X < 145)$
        ii $P(138 < X < 142)$
    b) Determine, to one decimal place, the maximum height exceeded by 85% of first bounces.

       (AQA, 2008)

Review

After working through this chapter you should:

- know the main features of a normal distribution
- be familiar with the notation $N(\mu, \sigma^2)$
- be able to find z-values for any normal distribution
- be able to use either tables or a calculator to find the probability of an event that is normally distributed.

Investigation

The central limit theorem

Abraham de Moivre
(1667–1754)

Pierre-Simon Laplace
(1749–1827)

Carl Friedrich Gauss
(1777–1855)

These three mathematicians all made significant contributions to the understanding of the normal distribution.

From the time of Galileo (1564–1642), it had been known that there were errors in astronomical measurements caused by both human error and imperfections in the instruments. In 1809, Gauss developed the theory of the normal distribution and was able to apply it to these errors.

Some aspects of the central limit theorem had been observed very much earlier by de Moivre in a context which is easy to investigate.

1 Take some coins and throw them repeatedly. Count the number of heads each time and then draw a histogram of your results. Like the one shown here, your histogram is likely to show some, but not all, of the features of a normal distribution.

Q *Why* the normal distribution should appear in this context (and many others) was explained in 1812, when Laplace discovered the central limit theorem. This theorem has been described as explaining the beautiful regularity in nature which underlies even the wildest confusion. You can research the central limit theorem on the internet.

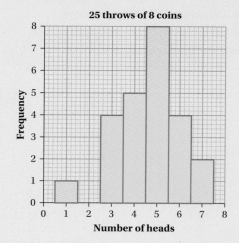

25 throws of 8 coins

The normal distribution

Taking a large number of coins, such as 100, and throwing them a large number of times is impractical. However, you can find plenty of simulators for this online.

2 Use a standard simulator (or write a program of your own) to investigate large numbers of throws and coins.

3 Check that for large numbers the histograms have a normal distribution shape. Can you discover how the mean and standard deviation of the normal distribution depend upon the number of coins?

A quantity such as your height depends on many random events, ranging from your genetic make-up to chance environmental factors. The central limit theorem explains why the random outcomes of all these events, across the whole population, give a remarkably smooth normal distribution of heights.

6 Confidence intervals

How does modern quality control depend upon the mathematical theory of the normal distribution?

Why is a baker's dozen greater than 12?

What do e marks mean on packages?

In this chapter you will learn about the distribution of the mean of a sample from a normal distribution. In particular, you will discover what the mean of a sample tells you about the mean of the population.

How does a trading standards officer check claimed weights for goods?

How are the number of polar bears in the Arctic estimated?

You should know how to:

- calculate the mean of a set of data 1200
- find z-values for any normal distribution (Chapter 5)
- use tables or a calculator to find the probability of an event that is normally distributed (Chapter 5).

6.1 Quality control

The number 13 is known as a 'baker's dozen'. In the past bakers would add an extra roll or loaf to a batch to avoid being fined for selling bread weighing less than the statutory limit.

Quality control is just as vital today. With products such as nut and bolts, the manufacturer's main issue is to reduce variability (the standard deviation). For products sold in closed packets or jars it is important to satisfy the legal obligation that comes from stating the weight of the contents on the container.

Depending on the product, the manufacturer might be concerned about the variability of the size, shape, weight or quality.

Real Mayonnaise
500 ml ℮
(475 g)

Pure Clear Honey
454 g ℮

Spiced Prunes
300 g ℮

Yeast Extract
250 g ℮

Italian Artichokes
260 g ℮

Wholegrain mustard
175 g ℮

If the content of a packet or jar is below what the label states, the manufacturer risks being prosecuted for short measure.

Activity 1

a) What would the manufacturer of the Real Mayonnaise need to do if they decided to use the 'baker's dozen' solution to avoid being prosecuted for short measure? Why is this type of solution not popular with manufacturers?

b) One possibility for the manufacturers of the Pure Clear Honey is to use an extremely accurate filling machine. Why is this solution used for some products and yet not for others?

Extremely accurate machines are used for the active ingredients of medicinal drugs.

Example 1 Suppose that the yeast extract bottling line is set up so that the weights of the jar contents are normally distributed with mean 260 grams and standard deviation (s.d.) 8 grams.

a) What is the probability that a randomly selected jar will be underweight?

b) Comment on your answer to part **a)**.

a) The contents have the distribution $N(260, 8^2)$.

Standardise the variable to find the 'z-value' for 250.

$z = \frac{250 - 260}{8} = -1.25$

Use your calculator or tables to find $\Phi(-1.25)$.

$\Phi(-1.25) = 1 - \Phi(1.25) = 0.10565 \approx 10.6\%$

In Section 5.3 you learned that if a value x is taken from $N(\mu, \sigma^2)$ then the z-value is given by the formula $z = \frac{x - \mu}{\sigma}$

Remember:
$\Phi(-z) = 1 - \Phi(z)$

b) Even though the manufacturers are putting an average of 10 g extra yeast extract into each jar, there is still a greater than 10% chance that a randomly chosen jar will be underweight.

Example 1 shows that looking at a single sample of one item from a population tells you very little. As an even more extreme example, consider sampling from the population of digits $\{0, 1, 2, 3, 4, 5, 6, 7, 8, 9\}$.

This population has mean 4.5 but a sample of one digit is equally likely to be any digit and so this digit is likely to be a long way away from the mean. However, consider what happens if you take a sample of size 12.

	A	B	C	D	E	F	G	H	I	J	K	L	M
1	Random digit	9	6	5	1	1	9	5	2	9	0	6	8
2	Mean	5.08											

In this spreadsheet, the random digits are generated by putting =RANDBETWEEN(0,9) in cell B1 and then copying this command across the row. The mean of the sample of 12 random digits is found by entering =AVERAGE(B1:M1) in cell B2.

Activity 2 💬

a) Set up a spreadsheet like the one above. What do you notice about the mean of your sample?

b) How could you make the mean of a sample even more likely to be very close to 4.5, the mean of the population? Amend your spreadsheet to check your idea.

> The manufacturers of the yeast extract in Example 1 were, on average, giving consumers 4% more than was claimed on the jars and yet you could easily find that a single jar is underweight. In general, it is not sensible to draw conclusions from a sample of size 1.

> In Section 1.1, you learnt that a random sample (also known as a simple random sample) is one where each member of a population has an equal probability of being chosen.

> **Key point**
> Quality control often depends upon taking a random sample from the population of *all* items that have been produced. The mean of a random sample is called a **point estimate** for the mean of the whole population.

Exercise 6A

1 An inspector weighs the contents of five randomly selected jars of wholegrain mustard and obtains values of 181 g, 176 g, 172 g, 178 g and 171 g. What point estimate for the average weight of contents does this give for the company's jars of wholegrain mustard?

2 A second inspector measures the weights of the contents of another 25 randomly selected jars and obtains a point estimate of 174.5 g. Which of the two point estimates is likely to be the most accurate? Justify your answer.

3 Use the results of questions **1** and **2** to obtain a point estimate based on both samples.

4 Suppose that a wholegrain mustard bottling line is set up so that the weights of the jar contents are normally distributed with mean 180 grams and standard deviation (s.d.) 3 grams. What is the probability that a randomly selected jar will be underweight?

6.2 The sample mean

Suppose that the weights W of chocolates on a production line are normally distributed with mean 10 g and standard deviation 2 g.

If samples of 25 chocolates are taken, then the mean weights $\overline{W}$ of these samples will themselves have a distribution. From what you saw in Activity 2, you will expect that sample means $\overline{W}$ will themselves have mean 10 g, but that they will be more tightly grouped than W. Mathematicians have proved that the distribution of $\overline{W}$ will itself be normal with mean 10 g.

They also proved that the standard deviation of the sample can be found by dividing the standard deviation of the population by the square root of the sample size. So, in this example the standard deviation of the sample $= \frac{2}{\sqrt{25}} = 0.4$ g.

Key point

If random samples of size n are taken from a normally distributed population, $N(\mu, \sigma^2)$ then the sample means have distribution $N\left(\mu, \frac{\sigma^2}{n}\right)$.

The standard deviation, $\frac{\sigma}{\sqrt{n}}$, is called the **standard error** of the mean.

Example 2 The weights of chocolates are normally distributed with mean 10 g and s.d. 2 g.

a) What is the probability that the weight of a randomly chosen chocolate lies between 9.8 g and 10.2 g?

b) What is the probability that the average weight of a chocolate in a box of 25 lies between 9.8 g and 10.2 g?

a) Work out the z-values for 9.8 and 10.2

$z = \frac{9.8 - 10}{2} = -0.1 \qquad z = \frac{10.2 - 10}{2} = 0.1$

So, $P(9.8 < W < 10.2) = P(-0.1 < z < 0.1)$

Use your calculator or tables to calculate the probability.

$P(-0.1 < z < 0.1) = \Phi(0.1) - \Phi(-0.1)$

$= 0.07966$

You may have been able to spot the z-values by observing that

$9.8 = \mu - 0.1\sigma$

$10.2 = \mu + 0.1\sigma$

b) The standard error is $\dfrac{\sigma}{\sqrt{n}} = \dfrac{2}{\sqrt{25}} = 0.4$ g.

$\overline{W}$ has distribution $N(10, 0.4^2)$

Work out the z-values for 9.8 and 10.2

$9.8 = \mu - 0.5\sigma$ and $10.2 = \mu + 0.5\sigma$

$P(9.8 < \overline{W} < 10.2) = P(-0.5 < z < 0.5)$

$= \Phi(0.5) - \Phi(-0.5)$

$= 0.38292$

Example 3 The weights of male students in a college are distributed normally with mean 70 kg and s.d. 5 kg.

a) What is the probability that the combined weight of 4 male students is less than 260 kg?

b) What assumptions have you had to make in answering part a)?

---

a) First work out the distribution of the sample mean.

A random sample of 4 students has mean weight distributed like

$$N\left(\mu, \frac{\sigma^2}{n}\right) = N\left(70, \frac{25}{4}\right)$$
$$= N(70, 2.5^2)$$

The mean of the weights must be less than $\frac{260 \text{ kg}}{4} = 65$ kg so you need to find $P(\overline{X} < 65)$. ⊕ 1200

$$P(\overline{X} < 65) = \Phi\left(\frac{-5}{2.5}\right) = \Phi(-2) = 0.02275$$

b) Each of the 4 students must have been randomly selected so that each weight is independent of the others.

Activity 3

The heights of a variety of sunflower are normally distributed with mean 2 m and s.d. 40 cm. 100 random samples of 50 flowers are collected for measurement. How many samples would you expect to have a sample mean of greater than 210 cm?

If you know the total for a given number of items you can easily find the mean and vice versa.

The z-value for 65 is $\frac{65 - 70}{2.5}$

Exercise 6B

1 A random sample of size 15 is taken from a normal distribution with mean 60 and s.d. 4. What is the probability that the mean of the sample is

a) equal to 62

b) greater than 62?

> Think about the area under a point.

2 If a random sample of size 5 is taken from a distribution N(100, 80), find the probability that the sample mean

a) is greater than 107

b) lies between 101 and 109.

3 The number of miles a motorist drives each day has an $N(90, 90^2)$ distribution.

a) Show that the probability that the motorist drives more than 840 miles in a week is less than 20%.

b) State any assumptions you had to make to answer part a).

4 The body length of a species of insect is normally distributed with mean 5.1 mm and s.d. 0.2 mm. What is the probability that

a) a randomly chosen insect has a body length less than 5.08 mm

b) the mean body length of a sample of 10 insects is less than 5.08 mm

c) the mean body length of a sample of 100 insects is less than 5.08 mm?

5 The heights of a certain species of plant follow a normal distribution with mean 20 cm and s.d. 3 cm. A botanist selects a sample of n plants and says that the probability that the mean height, $\overline{H}$, will lie between 19 cm and 21 cm is 0.68268

a) Explain why $P(\overline{H} < 21) = 0.84134$

b) Hence show that $21 = 20 + \frac{3}{\sqrt{n}}$

c) How large a sample did the botanist take?

6.3 Confidence intervals

A survey in the year 2000 of polar bears in the McClintock Channel area of the Arctic was widely reported in the press as producing an estimate of around 280 polar bears in that area.

In the actual survey, the results were stated as

'The number of polar bears was 284 ± 118 (95% confidence interval).'

Giving a range of possibilities rather than a single point estimate is common in opinion polls and also in scientific work.

Activity 4 Q

Why do you think the range for the polar bears from 166 to 402 is so wide? How are polar bear estimates obtained?

The ranges of possibilities are called **confidence intervals** because they are associated with a probability. For samples from normal distributions, you can find these probabilities precisely using the standard error of the mean.

For example, you know that, for any normal distribution, 95% of the distribution lies within 1.96 × standard deviation of the mean. Therefore, for samples from a normal distribution, 95% of the distribution lies within 1.96 × standard error of the mean.

The most commonly used confidence intervals for a standard normal distribution are as follows:

99% of the distribution lies in the range $[-2.58, 2.58]$

95% of the distribution lies in the range $[-1.96, 1.96]$

90% of the distribution lies in the range $[-1.64, 1.64]$

Example 4 The weights of the contents of boxes of a particular breakfast cereal can be assumed to be normally distributed with mean μ g and standard deviation 1 g.

A random sample of 200 boxes has a mean weight of 340 g. Calculate a 95% symmetric confidence interval for the mean weight of boxes of this cereal.

First calculate the standard error.

The standard error is $\dfrac{\sigma}{\sqrt{n}} = \dfrac{1}{\sqrt{200}}$.

The 2000 survey resulted in a hunting ban and the number of polar bears is now *estimated* to be increasing in this area.

The notation $[a, b]$ is used to mean the range of numbers from a to b **including** a and b.

You can use the methods you learnt in Chapter 5 to work out any confidence interval you require. However, the ones given here are used so often that it would be helpful for you to know them.

The interval is called **symmetric** because the mean is at the centre of the interval. You can assume that you will always be finding a symmetric confidence interval even if this is not stated.

Using the information about 95% confidence intervals, you know that there is a probability of 0.95 that μ lies within $1.96 \times \dfrac{\sigma}{\sqrt{n}}$ of the mean of the sample.

A 95% symmetric confidence interval for μ is therefore
$$\left[340 - 1.96 \times \frac{1}{\sqrt{200}}, 340 + 1.96 \times \frac{1}{\sqrt{200}}\right] = [339.86, 340.14]$$

As you might expect, there is a trade-off between the interval width and the confidence that you can have in your estimate. The greater the level of confidence, the wider the interval.

95% confidence intervals are very often used, so you will find it particularly useful to remember the formula $1.96\,\dfrac{\sigma}{\sqrt{n}}$

Different applications of mathematics may use different percentages for the confidence intervals. You might, for example, expect that near total confidence should be demanded for the safety of a medicinal drug, but be quite happy with 95% confidence in an opinion poll.

Key point

If a random sample of size n from a population $N(\mu, \sigma^2)$ has mean m, then a 95% confidence interval for μ is
$$\left[m - 1.96\,\frac{\sigma}{\sqrt{n}}, m + 1.96\,\frac{\sigma}{\sqrt{n}}\right]$$

It might seem strange that you will always be assuming that you know σ but not μ. In fact, σ will often not be known in advance and also has to be estimated, but this is not something you need to know for this course.

Activity 5

In a large consignment of tomatoes, you know that the number of damaged tomatoes per crate has a normal distribution $N(\mu, 4.8^2)$.

There are 201 damaged tomatoes in a random sample A of 9 crates.

There are 362 damaged tomatoes in a different random sample B of 16 crates.

Obtain 90% confidence intervals for μ from

a) sample A   **b)** sample B   **c)** samples A and B combined.

The number of tomatoes is a discrete variable. However, the continuous normal distribution is very useful in modelling these sorts of discrete variables.

⊕ 1248

This activity illustrates the connection between sample size and interval width. This relationship is caused by the important result that the standard error is $\dfrac{\sigma}{\sqrt{n}}$

Example 5 A biscuit manufacturer claims that the mean weight of biscuits in a packet is 450 g and the standard deviation is 5 g. A trading standards officer selects 10 packets and finds the weight of biscuits in each packet. The results, to the nearest gram, are

442, 450, 447, 446, 453, 449, 444, 454, 443, 457

Assume that the weight of biscuits in each packet is normally distributed with standard deviation 5 g. Find a 99% confidence interval for the mean weight of biscuits in a packet. Comment on the answer.

The point estimate for the mean is 448.5

The standard error is $\dfrac{\sigma}{\sqrt{n}} = \dfrac{5}{\sqrt{10}} \approx 1.58$

A 99% confidence interval is $448.5 \pm 2.58 \times 1.58$ which is $[444.4, 452.6]$.

The claimed mean is within this range so, on the basis of just this one sample, there is no reason to suspect the claim of the manufacturer.

Consolidation exercise 6

1   The weight, X grams, of the contents of a tin of beans can be modelled by a normal random
    variable with a mean of 421 g and a standard deviation of 2.5 g.
    a)   Find
             i $P(X = 421)$ ii $P(X < 425)$ iii $P(418 < X < 424)$.
    b)   Determine the value of x such that $P(X < x) = 0.98$

(AQA, 2013)

2   A machine produces circular discs which have an area of
    Y cm^2. The distribution of Y is normal with mean μ cm^2 and
    variance 25 cm^2.

    A random sample of 100 such discs is selected. The mean
    area of the discs in this sample is calculated to be 40.5 cm^2.

    Calculate a 95% confidence interval for μ.

> Remember that standard
> deviation is the square root
> of variance.

(AQA, 2011)

3   The weight of adult guillemots may be modelled by a normal distribution with mean μ g
    and s.d. 69 g. The total weight of a random sample of 9 adult guillemots is 8514 g.

    Construct a 98% confidence interval for μ based on these data.

> This question requires you to work out the z-values for the confidence interval, since the 98% confidence
> interval is not one listed earlier.
>
>
>
> 98%
>
> The area in the right hand tail is only 1% and so you need to use the tables to obtain
> $\Phi(2.33) = 0.99$.

4   The weight, X kg, of sand in a bag can be modelled by a normal random variable with
    unknown mean μ kg and known standard deviation 0.4 kg.

    The sand in each of a random sample of 25 bags from a large batch is weighed.

    The *total* weight of sand in these 25 bags is found to be 497.5 kg.
    a)   Construct a 98% confidence interval for the mean weight of sand in bags in the batch.
    b)   Hence comment on the claim that bags in the batch contain an average of 20 kg of sand.

(AQA, 2013)

5   The pears from a local supplier have weights that can be modelled by a normal distribution
    with standard deviation 8.3 g.

    A supermarket requires the mean weight of the pears to be at least 175 g. William, the fresh-
    produce manager at the supermarket, suspects that the latest batch of pears delivered does
    not meet this requirement. He weighs a random sample of 6 pears, obtaining the following
    weights in grams.

    160.6   155.4   181.3   176.2   162.3   172.8

    a)   Find a 95% confidence interval for the mean weight of pears in this batch.
    b)   Comment on William's suspicion.

(AQA, 2013)

6 Rice that can be cooked in microwave ovens is sold in packets which the manufacturer claims contain a mean weight of more than 250 g of rice.

The weight of rice in a packet may be modelled by a normal distribution with standard deviation 1.94 g.

A consumer organisation's researcher weighed the contents of each of a random sample of 50 packets. She found the average weight to be 251.1 g.

a) Construct a 96% confidence interval for the mean weight of rice in a packet, giving the limits to one decimal place.

> The 'limits' of a confidence interval [a, b] are the two numbers a and b.

b) Hence comment on the manufacturer's claim.

(AQA, 2011)

7 The volume of orange juice in a carton is normally distributed with a standard deviation of 10 ml.

The mean volume of juice should be 750 ml but the manufacturer is concerned that the cartons are being over-filled in the factory, He takes a random sample of 12 cartons and records the volumes in ml, as follows:

> 763   769   746   765   756   755   756   750   758   758   765   755

Calculate a 95% symmetric confidence interval for the mean volume and comment on whether the manufacturer should be concerned.

8 a) The variable L represents the length of an adult male grass snake.

L may be modelled by a normal distribution with mean μ and variance σ^2. A sample of size n is taken from the populations of adult male grass snakes. Write down the distribution of the sample mean $\bar{L}$.

b) The length of an adult male grass snake in England may be assumed to follow normal distribution with mean μ and variance $22 \, \text{cm}^2$.

A random sample of 10 adult male grass snakes has been collected in England. Their lengths, in centimetres, were measured and are recorded below.

> 102   87   109   93   98   112   86   105   97   89

Use the data above to calculate a 95% symmetric confidence interval for μ.

(AQA, 2013)

Review

After working through this chapter you should:

- know that the mean of a sample is a point estimate for the mean of the population
- know that the accuracy of a point estimate is likely to be improved by increasing the sample size
- be able to calculate the standard error, $\dfrac{\sigma}{\sqrt{n}}$
- be able to find confidence intervals for populations of known variance.

Investigation

e marks

To protect consumers from short measure, there are complicated mathematical regulations that manufacturers must follow in order to be allowed to put e marks on their produce.

Regulations Q

This investigation is based upon The Weights and Measures (Packaged Goods) Regulations 2006 guidance for businesses.

Worked example

To illustrate the process, this example applies these regulations in the case of a manufacturer who intends to put the mark 200 g e on his jar labels. The filling carried out by the production line being used has known standard deviation of 5 g. Samples of size 5 are taken every half an hour during each $2\frac{1}{2}$ hour production run.

Step 1: Calculate the tolerable negative error (TNE)

Nominal quantity(Q) in g or ml	Tolerable negative error (TNE)
5 – 50	9%
50 – 100	4.5 units
100 – 200	4.5%
200 – 300	9 units
300 – 500	3%
500 – 1000	15 units
1000 – 10 000	1.5%
10 000 – 15 000	150 units
Above 15 000	1%

Source: The Weights and Measures (Packaged Goods) Regulations 2006

The manufacturer wants to use the 200 g e mark so this first step is easy.

Reading across the table, Q = 200 tells you TNE = 9. In this case, the units are grams.

The e mark (sometimes called the estimated sign) shows the product meets the requirements of a particular European Directive. A product granted this mark has free access to all European Economic Area markets without needing to satisfy any further weights and measures regulations of individual countries.

This is a table which manufacturers are given and have to use to calculate the TNE for their own e mark.

Step 2: Calculate the target quantity (Q_T)

The manufacturer has to set up his production line to fill the jars with a target quantity which is at least as large as Q. The critical factor for calculating this amount depends on the standard deviation (s) of the filling process.

The target quantity (Q_T) is based upon the largest of the following three quantities.

$$Q \qquad Q - TNE + 2s \qquad Q - 2TNE + 3.72s$$

In this example, $Q = 200$, $s = 5$ and $TNE = 9$

The three quantities (in grams) are therefore

$$200 \qquad 200 - 9 + 10 = 201 \qquad 200 - 18 + 3.72 \times 5 = 200.6$$

The largest of these is 201.

However, adjustments are sometimes needed depending on the size of the sample taken.

In this example, 25 jars are sampled in each production run. This is judged to be a small sample size and so the regulations state that the target quantity has to be increased by $0.2s = 1$. Therefore

$$Q_T = 202$$

Step 3: Calculate the confidence interval for the sample mean

For most consumer protection purposes the lower limit of the confidence interval is the more important one. However, the upper limit matters for some health and safety issues (for example for filling aerosols) and for some customs and excise duties.

The confidence interval for e marks uses the standard error, $\dfrac{s}{\sqrt{n}}$, in the formula

$$Q_T \pm 3\frac{s}{\sqrt{n}}.$$

Now $\Phi(3) \approx 0.999$ and so the chance of a sample mean being more than 3 standard errors below the mean of a correctly targeted Q_T is 1 in 1000.

In the example, $n = 5$ and $Q_T \pm 3\dfrac{s}{\sqrt{n}} = [195.3, 208.7]$.

The government regulations give a table with results depending on the process variation.

Process variation, s(g)	Target quantity (g)	Minimum sample mean (g)
4		
5	202.0	195.3
6		

> This is the real crux of the e mark regulation. It uses the ideas of this chapter including the standard error and confidence intervals. You should note the use of 3 in the formula. This ensures that there is very little probability that a sample from a correctly set up procedure will be outside tolerance (since virtually all jars will be within 3 standard errors of the target quantity). You should contrast this with the answer to Example 1.

> The regulations use the phrase process variation instead of standard deviation. Be careful not to confuse process variation with variance.

> Check the calculations in this article and copy and complete the table for the other values of s.
>
> For a product of your choice, research how the manufacturer satisfies the government guidelines on sampling.

7 Correlation and regression

In what way are house prices related to earnings?

How is the wingspan of an airliner related to its length?

What is the connection between the amount spent abroad by people from the UK and the number of times they go to other countries?

In this chapter you will learn how to find and measure the strength of relationships between variables. These methods will help you to make predictions and assess how reliable these predictions are likely to be.

Does strong correlation mean that changes in one of the variables will cause the other to change?

Is a school's performance closely related to pupil absences?

How closely are football clubs' point scores linked to the number of goals they score?

You should know how to:

- recognise correlation 1213
- calculate the mean of a dataset 1200
- use and interpret scatter graphs 1213
- find the gradient of a straight line graph
- plot straight line graphs from their equations. 1396

When two variables are related you can show this on a **scatter graph** and draw a **line of best fit**. When the line of best fit is straight, the relationship is said to be **linear**. ⊕ 1213

Key point

To draw a line of best fit for a linear relationship

- Use sensible scales to draw x and y-axes that include the maximum and minimum values. Then use the data to plot points.
- Calculate the mean value of x and the mean value of y. Plot the mean point $(\bar{x}, \bar{y})$.
- Draw a straight line 'by eye' to pass through the mean point and as near as possible to the other points. Aim to follow the trend of the data with an equal number of points above and below the line.

The x and y- axes do not need to start at the origin.

Remember, you can find the mean using the in-built function on your calculator or

$$\bar{x} = \frac{\Sigma x}{n}$$

$$= \frac{\text{sum of values}}{\text{number of values}}$$

⊕ 1200

Activity 1 💬

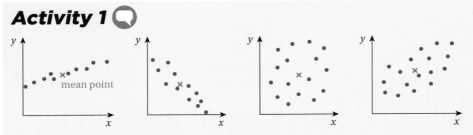

mean point

- ★ Describe the correlation in each of the four scatter graphs.
- ★ Suggest possible variables for x and y.
- ★ Use tracing paper to copy the axes and draw a line of best fit for each scatter graph.

A line of best fit allows you to use the value of one variable to predict the value of the other.

When these values lie within the range of the known data, this is called **interpolation**.

When the values lie outside the range of known data, this is called **extrapolation**.

Key point

Predictions made using extrapolation are much less reliable than those made using interpolation because the trend of the data may have changed.

A point that has extreme values or that lies far from the line of best fit is called an **outlier**. This may be due to an error or because the data is unusual in some way. It is often sensible to exclude outliers.

When an outlier occurs, try to identify the reason for it. If you suspect that an error has occurred or the values are so extreme that they will dominate results (for example, when calculating the mean) you can omit the outlier, but you should say so and give the reason. You may sometimes be given a definition of an outlier in terms of the quartiles and interquartile range (or the mean and standard deviation in a normal distribution).

Example 1 A delivery van takes goods from a factory to a different shop each day in a two-week period.

The table shows the distances to the shops and the times taken to reach them.

Day	Week 1		Week 2	
	Distance (miles)	Time taken (minutes)	Distance (miles)	Time taken (minutes)
Mon	135	160	128	157
Tues	106	135	204	246
Wed	226	273	117	130
Thurs	184	213	218	254
Fri	138	296	143	168

a) The delivery van was held up by a road accident on one of the days.
    Use a scatter graph to identify the day on which this happened.

b) Draw a line of best fit. State whether you include the data from the day of the accident and give a reason.

c) A new shop is 150 miles from the factory.
    Predict the time needed for the delivery van to reach this shop.

d) Find and interpret the gradient of the line of best fit.

---

a)

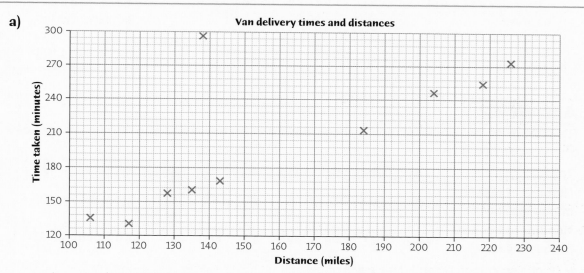

At the point (138, 296) the time taken is much higher than the trend suggests.
The accident was on the first Friday.

b) It is sensible to exclude the data from the day of the accident when finding the mean point.

Mean distance = 162 miles (3 sf), mean time = 193 min (3 sf).

The point (138, 296) has been excluded because the mean time would be affected by the unusually long time taken because of the accident.

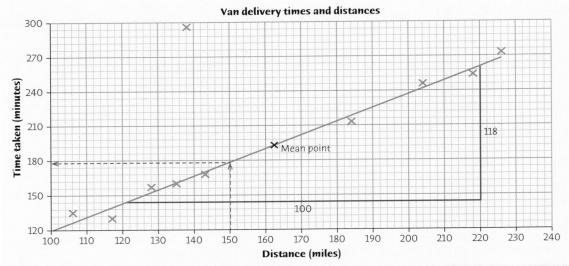

Van delivery times and distances

c) To predict how long it takes to travel 150 miles draw a vertical line from 150 miles to reach your line of best fit and then draw a horizontal line to find the time taken.

For a distance of 150 miles, approximately 180 minutes will be needed.

d) To find the gradient choose two points on the graph and draw a triangle to find the change in y-values and the change in x-values.

Gradient $= \frac{118}{100} = 1.18$ minutes per mile.

This suggests the van will need about 1.2 minutes for each extra mile.

> Gradient
> $= \dfrac{\text{change in } y\text{-value}}{\text{change in } x\text{-value}}$
>
> The gradient is the rate of change of y with respect to x.
>
> You can also think of the gradient as the change in y when x increases by 1 unit.

Activity 2

The table gives the number of mobile phone stations in ten towns and also the number of babies born in those towns last year.

Town	A	B	C	D	E	F	G	H	I	J
Number of mobile phone stations	46	36	61	38	25	63	54	63	73	62
Number of babies born last year	1227	1021	2297	1453	991	4033	2807	3673	3120	1951

★ Plot a scatter graph and draw a line of best fit.

★ List five other variables in the towns that are likely to be related to population size.

You should find that the two variables have positive correlation, but this does not mean that mobile phone stations cause more babies to be born.

Sometimes correlation is caused by a third variable, called a **confounder**. In this case the confounder is population size. Large towns have more mobile base stations and more newborn babies than small towns.

Key point

Correlation does not necessarily imply causation.

This means that there may be strong correlation between two variables but this does not mean that changes in one must cause changes in the other.

Exercise 7A

1 For each pair of variables select the most likely scatter graph from those below and describe what the correlation means in terms of the variables.

a) Maximum daily temperature and sales of ice creams.
b) Height and IQ.
c) Time a marathon runner spends training and the time taken to run a marathon.
d) Extension of a vertical spring and the mass attached to the lower end (assuming the upper end is fixed).
e) Engine size and time taken by a car to accelerate to 60 miles per hour.
f) Height of a horse chestnut tree and the circumference of its trunk.

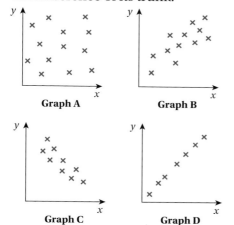

Graph A   Graph B

Graph C   Graph D

2 For each of the following pairs of variables state whether you expect the correlation to be positive or negative and justify your answer.

a) Height and shoe size.
b) Consumption of butter and consumption of margarine by households in England.
c) Weight of a loaded lorry and the time it takes the lorry to accelerate to 50 miles per hour.
d) Height from which a ball is dropped and the height of its first bounce.
e) Average speed and the time taken to travel between two towns.

3 Each year a savings bank carries out a survey to find out how much pocket money children get each week. The average amount for each age from 5 to 16 years in 2015 is shown in the table.

Age (years)	Amount (£)
5	3.28
6	4.00
7	3.71
8	4.02
9	4.88
10	4.74
11	6.71
12	7.36
13	8.13
14	9.72
15	9.13
16	10.27

a) Draw a scatter graph and a line of best fit.
b) Describe the correlation.
c) How much extra pocket money do children get for each year they get older?

4 Ahmed and Kayleigh have both used the same data to draw a scatter graph and a line of best fit.

Ahmed's graph and line of best fit

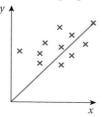

Kayleigh's graph and line of best fit

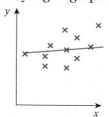

Describe how these students can improve their graphs and lines.

5 Each month the manager of a baker's shop records the price the shop has charged for loaves of bread and bread rolls and the numbers sold. The table gives the results.

Month	Price of loaf (pence)	Number of loaves	Price of roll (pence)	Number of rolls
Jan	93	895	15	492
Feb	94	910	16	480
Mar	105	840	17	512
Apr	110	836	19	542
May	106	890	18	528
Jun	98	900	16	530
Jul	89	926	15	520
Aug	88	934	14	472
Sep	87	945	13	468
Oct	89	918	16	502
Nov	95	922	17	506
Dec	98	910	18	540

a) i What relationship would you expect between the price of a loaf and the number sold?

ii Draw a scatter graph to check your answer to part i. Include a line of best fit.

iii How many loaves would you expect the baker to sell for £1.04?

b) i Draw a scatter graph of the number of rolls against the number of loaves sold. Include a line of best fit.

ii Describe the correlation and suggest a possible reason for it.

iii Predict the number of rolls that will be sold in a month when 870 loaves are sold.

c) i Draw a scatter graph of the price of rolls against the price of loaves. Include a line of best fit.

ii Suggest a third variable that may be the cause of the correlation.

6 The area and population of countries in Central America are given in the table.

Country	Area (square kilometres)	Population (thousands)
Belize	22 966	340.84
Costa Rica	51 100	4 755.23
El Salvador	21 041	6 125.51
Guatemala	108 889	14 647.08
Honduras	112 090	8 598.56
Mexico	1 964 375	120 286.66
Nicaragua	130 370	5 848.64
Panama	75 420	3 608.43

Source: CIA World Factbook 2014

a) A value that is more than 1.5 × interquartile range above the upper quartile is an outlier. Show that Mexico's area and population are both outliers.

b) Plot a scatter graph, omitting Mexico.

c) Draw a line of best fit and describe the correlation.

7 Investigate the correlation between these variables.

Animal	Body mass (kilograms)	Brain mass (grams)	Maximum life span (years)
Cat	3.3	26	28
Cow	470	420	30
Deer	15	98	17
Donkey	190	420	40
Goat	28	120	20
Horse	520	650	46
Pig	190	180	27
Rabbit	2.5	12	18
Sheep	56	180	20

Comment on your findings.

8 Sonja is investigating relationships between employment, earnings and house prices. She finds this table of information for regions of the UK in 2014 and draws a scatter graph.

	Employment rate (%)	Average annual earnings (£thousands)	Average house price (£thousands)
North East	69.5	24.876	152
North West	70.0	25.229	168
Yorkshire & Humber	72.0	24.999	172
East Midlands	73.8	25.027	181
West Midlands	70.2	24.920	193
East	75.9	26.830	269
London	72.3	35.069	485
South East	76.4	28.629	320
South West	76.1	25.571	238
Wales	70.1	24.384	166
Scotland	73.4	27.045	187
Northern Ireland	68.1	24.020	132

Data from ONS

a) Sonja draws this graph.
Sonja wants to draw a line of best fit.
What would you advise Sonja to do about the outlier? Explain why.

b) i Draw a scatter graph of average house price against average annual full-time earnings.

ii Describe the correlation.

iii Suggest other variables that could affect house prices.

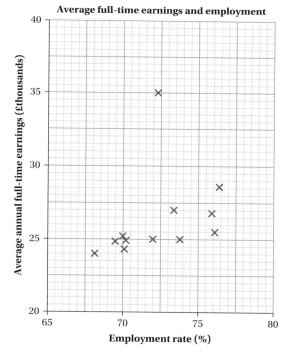

9 Car manufacturers' websites give a lot of information about the cars they sell. The data often includes car length, width and height, engine size, top speed, fuel consumption, fuel tank capacity and boot capacity. Other websites or car magazines give information about second-hand car prices and insurance and other costs.

Choose variables that you think may be related and investigate the correlation.

Write a summary of your findings.

Drawing a good line of best fit 'by eye' is sometimes difficult and different people often draw different lines. However, there is a method for calculating the equation for a line of best fit. The method is complicated, but it is so useful that calculators and spreadsheets have an in-built function for carrying it out. A line of best fit calculated in this way is called a **regression line**.

Key point

To find the regression line of y on x:

- enter the data into a calculator or spreadsheet
- use the in-built function to find the equation.

> To find the regression line on a spreadsheet, draw the scatter graph, add a **linear trendline** and display its equation.

For example, you can find regression lines using this information from weather stations in England in July 2014.

Station	Total rainfall (mm)	Total sunshine (hours)	Mean maximum daily temperature (°C)
Bradford	101.4	136.1	22.2
Camborne	70.0	183.2	20.3
Eastbourne	59.6	236.3	22.3
Heathrow	84.6	178.8	25.8
Lowestoft	94.8	207.8	21.7
Oxford	90.3	185.7	24.9
Ross	95.2	171.9	23.9
Sheffield	123.8	143.9	23.2
Whitby	100.1	192.5	20.8

Source: Met Office

Entering the total rainfall values for x and the total sunshine values for y gives the regression line $y = 290.03 - 1.188x$

Key point

In the equation $y = a + bx$

- the constant term, a, is the value of y when $x = 0$
- b is the gradient of the regression line. This gives the increase in y when x increases by 1 unit.

> Some calculators give the regression equation as $y = ax + b$

The equation $y = 290.03 - 1.188x$ suggests that

- there would be about 290 hours of sunshine in a month when there is no rain (but this is an extrapolation so it may be unreliable)
- an extra millimetre of rainfall indicates a reduction of about 1.2 hours of sunshine.

Activity 3

★ Find out how to use *your* calculator to find the equation of the regression line of total sunshine on total rainfall. Also find a regression line for total sunshine on mean maximum daily temperature.

★ Enter the weather data into a spreadsheet. Draw a graph of total sunshine against total rainfall and a graph of mean maximum daily temperature against total sunshine. Add linear trendlines with their equations and use these to check the equations found earlier.

💬 Discuss your findings.

> When asked for a 'regression line' in this chapter, use linear regression. In some situations a curve is more appropriate than a straight line, but the study of regression curves is not included in this course.

Sometimes you may be given a graph and asked to draw the regression line on it.

Example 2 The table and scatter graph show the number of visits abroad by people from the UK and the total amount they spent in other countries in each month of 2014.

Month	Visits (*n* thousands)	Expenditure (£*P* million)
Jan	3873	2410
Feb	3523	2196
Mar	3687	2374
Apr	4990	2718
May	5689	3074
June	6062	3416
July	6047	3563
Aug	8099	5050
Sept	6634	4201
Oct	5350	3300
Nov	3760	2050
Dec	3220	1710

Source: ONS (International Passenger Survey)

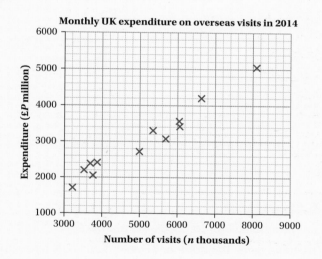

a) Find the equation of the regression line of P on n.
b) Plot the line on the scatter graph.
c) Use the regression equation to predict the total amount that will be spent abroad in a month when there are 7 million visits.
d) Explain why the equation should not be used to predict the expenditure when n is 2000 or 10 000.

---

a) First use your calculator to find the equation of the regression line.

$y = -185.6 + 0.6284x$

The regression line is $P = -186 + 0.628n$ (to 3 sf)

> Replace y by P and x by n. Round sensibly.

b) Use your calculator to find the mean point.

The mean point is (5078, 3005).

Calculate at least one other point on the line.
Use a value of x from the range shown on the graph.

When $n = 8000$, $P = -185.6 + 0.6284 \times 8000 = 4842$

Plot the points on the graph and join with a straight line.

> This predicts that in a month when there are 8 million visits, the expenditure will be about £4840 million.

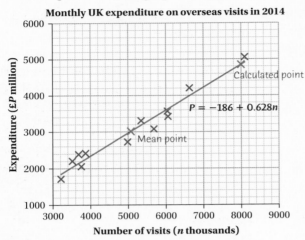

Monthly UK expenditure on overseas visits in 2014

Calculated point

$P = -186 + 0.628n$

Mean point

Expenditure (£P million)

Number of visits (n thousands)

c) 7 million = 7000 thousands.

$n = 7000$ gives $P = -185.6 + 0.6284 \times 7000 = 4213.2$

The regression equation predicts expenditure of £4210 million (3 sf).

> Use more than 3 sf in the working but round the answer.

d) Compare $n = 2000$ and $10\,000$ with the given data.

2000 and 10 000 are outside the given data.

Extrapolating may not give reliable predictions.

> Sometimes extrapolating gives silly results.
> For example, $n = 0$ gives $P = -185.6$.

Activity 4

Use your calculator to check the answers given in the example above.

The table gives the number of visits to the UK by people from abroad and the total amount they spent here in each month of 2014.

Month	Jan	Feb	Mar	Apr	May	Jun	Jul	Aug	Sep	Oct	Nov	Dec
Visits (n **thousands**)	2444	2187	2444	2970	3276	3260	3524	3628	3025	3010	2620	2410
Amount spent (£P **millions**)	1410	1194	1329	1532	1794	2112	2655	2554	2163	1760	1540	1690

Source: ONS (International Passenger Survey)

Find the equation of the regression line of P on n.
Draw a scatter graph showing the data and the regression line.

💬 Compare the results with those in the example.

Exercise 7B

1 In a memory test a volunteer memorises 100 words. Afterwards the volunteer is tested each week to see how many words she remembers. The table gives the results.

Time elapsed (x weeks)	Number of words remembered (y)
1	97
2	94
3	87
4	84
5	76
6	73
7	67
8	61

a) Draw a scatter graph.
b) i Find the equation of the regression line of y on x.
   ii Plot the line on your scatter graph.

2 The table gives the marks achieved by students in an oral test and a written test. The total mark for each test was 40.

Student	Oral mark, x	Written mark, y
Ann	18	21
Baz	32	30
Carl	36	32
Daisy	23	27
Ed	34	abs
Fran	28	26
George	37	27
Helen	24	31
Ian	31	33
Jack	24	22
Kay	16	23
Liam	27	23
Meera	17	14

a) Find the equation of the regression line. Ignore Ed as his results are incomplete.
b) Draw a scatter graph of the data and plot the line.

c) Ed got 34 marks on the oral test, but was absent for the written test.
   i Use your graph to predict the mark he might have achieved on the written test.
   ii Use the equation of the regression line to check your answer.

> Note that you do not need to draw a graph to answer question 3.

3 The table gives the cheapest and most expensive season tickets for Premiership football clubs in the 2013–14 season.

Club	Cheapest season ticket (£x)	Most expensive season ticket (£y)
Arsenal	1014	2013
Aston Villa	335	615
Burnley	499	685
Chelsea	595	1250
Crystal Palace	550	720
Everton	544	719
Hull	501	572
Leicester	365	730
Liverpool	710	869
Man City	299	860
Man Utd	532	950
Newcastle	383	710
QPR	499	949
Southampton	608	853
Stoke	459	609
Sunderland	400	525
Swansea	449	499
Tottenham	795	1895
West Brom	349	459
West Ham	640	910

a) Find the equation of the regression line of y on x.
b) Find the value of y when $x = 0$ and explain why the answer does not give any useful information.
c) Interpret the gradient of the regression line.

4 Amy measures the direct distance from Sheffield to other cities on a map, then finds the road distance from the internet. She puts her results in a table and draws a scatter graph.

Sheffield	Distance	
	Direct (x km)	By road (y km)
Birmingham	106	146
Bristol	232	289
Cambridge	170	198
Cardiff	245	320
Edinburgh	307	403
Leeds	51	56
Manchester	53	67
Newcastle	182	209
Norwich	206	234
Oxford	186	229

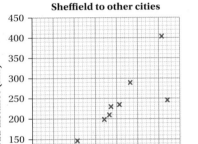

Direct and road distances from Sheffield to other cities

a) Amy thinks one of her points is incorrect. Identify the point and describe the error Amy has made.

b) Find the equation of the regression line of y on x.

c) Draw a correct scatter graph, including the regression line.

d) Interpret the gradient of the regression line.

e) The direct distance from Sheffield to Aberdeen is 426 km.
Explain why Amy should not use the regression equation to predict the distance by road.

5 The table shows average earnings for public and private sector jobs in regions of the UK in 2013.

Region	Median (£/week)	
	Public x	Private y
North East	426.40	349.20
North West	446.40	364.10
Yorkshire & Humber	418.10	366.40
East Midlands	438.90	371.40
West Midlands	429.80	379.20
East	436.50	396.30
South East	446.40	425.20
South West	425.30	364.70
Wales	429.10	343.40
Scotland	466.00	382.40

Source: ONS

a) The equation for the regression line of y on x is $y = 38.5 + 0.769x$ (to 3 sf).
Draw a scatter graph including this regression line.

b) Which region is represented by the point that is furthest from the regression line?

c) i What would the equation of the regression line be if the average weekly earnings in the public and private sectors were equal?

ii In what ways would the line differ from that shown on your graph?

6 The cost of car insurance usually depends on the driver's age.

The table shows the average cost of insurance for drivers living in a city.

Age (n years)	Insurance (£C)
18	1315
22	795
27	583
35	417
45	306
57	238
70	214

a) i Find the equation of the regression line of C on n.

ii Interpret the gradient of the regression line.

b) i Use the equation of the regression line to predict the cost of insurance for a 40-year-old driver.

ii Draw a scatter graph to show the data and line.

iii Use parts **b) i** and **b) ii** to comment on the use of linear regression in this context.

7 The table shows the monthly output of a factory and total cost of production.

Month	Output x (thousand)	Total costs y (£thousand)
Jan	15.3	69.5
Feb	18.4	71.3
Mar	23.7	85.9
Apr	35.6	119.2
May	32.9	106.7
Jun	38.7	123.8
Jul	29.6	96.7
Aug	31.3	107.8
Sep	25.1	92.4
Oct	28.0	95.9
Nov	21.7	81.7
Dec	19.6	76.9

a) Draw a scatter graph including the regression line of y on x.

b) Interpret
i the gradient of the regression line
ii the constant term in the regression equation.

c) The factory manager expects to produce 27.5 thousand units next month.
Use the regression equation to estimate the total costs next month.

d) The factory manager wants to increase production to 50 thousand units per month.
Give a reason why the regression equation may not give a reasonable estimate for the total costs when the factory reaches this level of production.

e) The selling price of each unit of output is £3.60.
Estimate, to the nearest thousand, the output at which the total income equals the total costs.

8 **a)** To find the relationship between stride length and height:

- measure the heights and average stride lengths of a number of people of varying sizes
- use these results to find the equation of the regression line of height on stride length.

b) In 1978 an expedition found footprints at Laetoli that were over $3\frac{1}{2}$ million years old. The largest set of footprints had a stride length of approximately 43 cm. The shortest set had an average stride length of approximately 29 cm.
i Use your regression equation from part **a)** to estimate the height of the individuals who made these footprints.
ii Give reasons why your answers to part **i** should be treated with caution.

c) Use the internet to find out more about the Laetoli and other ancient footprints.

There are a variety of ways of measuring the strength of the correlation between two variables. One of the most frequently used is a coefficient invented by the British mathematician Karl Pearson.

Pearson's product moment correlation coefficient (often abbreviated to pmcc) can be worked out using the formula

$$r = \frac{s_{xy}}{s_x s_y}$$

where s_x is the standard deviation of the x-values,
  s_y is the standard deviation of the y-values
and $s_{xy} = \sum \frac{1}{n}(x - \bar{x})(y - \bar{y})$ (called the covariance of x and y).

Scientific calculators and spreadsheets have an in-built function to calculate r.

You do not need to know this formula – just note that it is linked to the mean and standard deviation of both x and y.

Key point
To find Pearson's product moment correlation coefficient
- enter the data into a calculator or spreadsheet
- use the in-built function to find r.

On a spreadsheet use the formula PEARSON.

Activity 5
Find out how to use *your* calculator to find the product moment correlation coefficient.
Use the data from Example 2 and Activity 4.
Use a spreadsheet to check your results.

Activity 6
Q Use the internet to find out more about Karl Pearson and his contributions to mathematics and science.

Key point
The pmcc always lies in the range $-1 \leqslant r \leqslant +1$. The sign is the same as that of the gradient of the line of best fit and the size depends on the strength of the correlation.

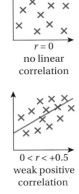

$r = +1$
perfect positive
correlation

$r = 0$
no linear
correlation

$r = -1$
perfect negative
correlation

Activity 7
For each of the following values of r, describe the correlation and draw a sketch to show what you would expect the scatter diagram to look like:
$r = 0.7$ $r = -0.9$ $r = 0$
$r = -0.2$ $r = 0.18$

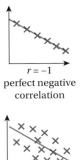

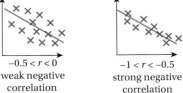

$+0.5 < r < +1$
strong positive
correlation

$0 < r < +0.5$
weak positive
correlation

$-0.5 < r < 0$
weak negative
correlation

$-1 < r < -0.5$
strong negative
correlation

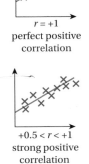

Example 3 The table gives the length and wingspan, in metres, of ten airliners.

Airliner	Length l (m)	Wingspan w (m)
Airbus A300–600	54.08	44.84
Airbus 320	37.57	34.09
Antonov An-38	15.67	22.06
Boeing 737–900	42.11	34.31
British Aerospace Avro RJ85	28.60	26.21
Canadair CRJ-700	32.41	23.01
Dornier 328	21.22	20.98
Embraer EMB120 Brasilia	20.00	19.78
Ilyushin Il-62	53.12	43.20
Tupolev Tu-154	47.90	37.55

Data from www.airliners.net

a) Find the equation of the regression line of w on l
  Write down the product moment correlation coefficient (pmcc).
b) Draw a scatter diagram to show the data and your regression line.
c) i Describe the correlation between the length and wingspan of the airliners.
  ii For light aircraft the value of the pmcc for length and wingspan is 0.625
  Compare this with your answer to part **a)** and comment on the correlation.

---

a) Use your calculator to find the equation of the regression line and r.
  $w = 7.7942 + 0.6467l$
  $w = 7.79 + 0.647l$ (3 sf)
  $r = 0.960$ (3 sf)

b) Use your calculator to find the mean point.
  $\bar{x} = 35.268$ and $\bar{y} = 30.603$

  Substitute a value for l into the regression equation.
  When $l = 50$, $w = 7.7942 + 0.6467 \times 50 = 40.129$

  Draw the scatter graph and join (35, 31) and (50, 40) to give
  the regression line.

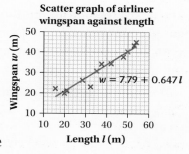

Scatter graph of airliner wingspan against length

c) i The wingspan and length of the airliners have strong positive correlation.
  This means that they are strongly related with wingspan increasing as length increases.

  ii 0.625 is less than 0.960 This means that the wingspan and length are less strongly
  correlated for light aircraft than for airliners.

Activity 8

★ Work through Example 3 using your calculator.

Q Use the internet to find other aircraft data. Use the data to investigate the correlation
between pairs of variables.

Exercise 7C

1 The table gives average house prices and rents in regions of England in 2011.

Region	House price (£P thousands)	Rent (£R/week)
North East	153	65.78
North West	175	68.65
Yorkshire & Humber	171	66.20
East Midlands	179	72.08
West Midlands	189	72.47
East	256	81.87
London	401	97.46
South East	301	89.94
South West	232	76.04

Source: ONS

a) Find
  i the equation of the regression line of R on P
  ii the value of the product moment correlation coefficient (pmcc).
b) Draw a scatter diagram to show the data and your regression line.

2 The table gives the amounts of coal and gas used in the UK in the months of one year. Both quantities are measured in million tonnes of oil equivalent.

Month	Coal (x)	Natural Gas (y)
Jan	4.1	10.9
Feb	3.4	9.5
Mar	3.6	9.6
Apr	2.4	7.9
May	2.8	7.1
Jun	2.7	5.6
Jul	2.9	5.5
Aug	2.2	5.2
Sep	2.8	5.8
Oct	3.5	8.2
Nov	4.0	9.0
Dec	4.6	10.6

Source: www.gov.uk (Department of Energy and Climate Change)

a) Find
  i the equation of the regression line of y on x
  ii the value of the product moment correlation coefficient (pmcc).
b) Draw a scatter diagram to show the data and your regression line.

3 Amelia finds this information about a model of car she likes on the internet.

Age (n years)	Mileage m	Price £P
0	250	8995
0.5	745	8990
1.0	2048	8495
1.5	14032	7275
2.0	35022	6500
2.5	14896	6995
3.0	25864	6490
3.5	14829	6495
4.0	9000	6540

Amelia wants to know whether the car's price is more closely related to its age or its mileage.

a) i Find the equation of the regression line of P on n.
  ii Write down the value of the product moment correlation coefficient (pmcc).
  iii Draw a scatter diagram to show the data and your regression line.
b) Repeat part a) using m instead of n.
c) Describe what your results tell you about the relationships between the price, age and mileage. Include comments about
  • correlation
  • what the gradients of your lines suggest
  • the prices implied for a new car.
d) Do you think linear regression is a good model in this context? Explain your answer.

Note that you are not required to draw a graph when answering questions **4** and **5**.

4 The table gives the bust and hip measurements in inches of size 12 and size 14 dresses at 8 shops.

Shop	Dress size 12		Dress size 14	
	Bust	**Hips**	**Bust**	**Hips**
A	35.4	39	37.4	42.9
B	35.9	39	37.8	41
C	35.4	37.6	37	40
D	36.2	38	38.2	40
E	36.2	38	38.2	40
F	37	40.2	39	42.3
G	37	40.2	38.6	41.7
H	37.8	40.2	40.2	42.3

Which has the stronger correlation: the bust and hip measurements for size 12 or the bust and hip measurements for size 14?

5 The UK extracts, exports and also imports oil and gas. The table gives the quantities in millions of tonnes for the years from 2000 to 2012.

Year	UK Extraction	Exports	Imports
2000	233.0	113.5	58.3
2001	221.2	116.8	62.7
2002	218.2	118.8	65.1
2003	207.7	102.9	68.9
2004	190.5	97.0	89.4
2005	171.8	87.4	91.5
2006	155.5	82.6	96.9
2007	147.8	79.3	101.5
2008	140.4	78.9	97.5
2009	127.2	78.8	102.4
2010	119.4	84.8	113.3
2011	96.7	81.5	117.6
2012	83.0	78.2	117.9

Source: www.gov.uk (HM Revenue and Customs)

Use regression and correlation to compare the relationships between UK extraction and exports and imports of oil and gas.

6 Amber and Harry have used the same data to draw these graphs:

Amber's scatter graph

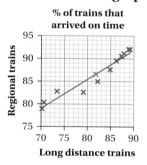

% of trains that arrived on time

a) What information is given on Amber's graph that is not given on Harry's graph?
b) What information is given on Harry's graph that is not given on Amber's graph?
c) Describe how correlation is shown on each graph.
d) Use one of the graphs to estimate the product moment correlation coefficient.

Harry's time series graph

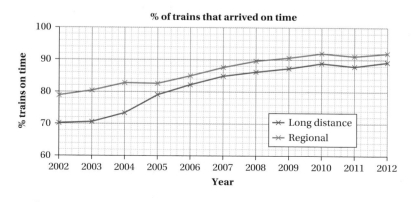

7   Oliver is investigating the relationship between the use of cars, buses and rail. The table gives the distances, in billions of kilometres, travelled by people in Great Britain in the years listed.

Year	Car	Bus	Rail
1952	58	92	38
1962	171	74	37
1972	327	60	34
1982	406	48	31
1992	583	43	38
2002	673	41	48
2012	642	42	70

Source: www.gov.uk (Department of Transport)

a) Find the product moment correlation for distances travelled by
    i   bus and car
    ii   bus and rail
    iii rail and car.
b) Draw a scatter diagram for the pair of variables with the strongest correlation. Include the regression line.
c) Write a paragraph that Oliver could use in his report. This should describe what the answers to parts a) and b) tell you about the relationships between the different ways of travelling.

8   The table gives the length, wingspan and mass of 12 seabirds.

Seabird	Length l cm	Wingspan w cm	Mass M g
Arctic Tern	34	70	108
Black-headed Gull	36	105	300
Common Gull	41	120	390
Fulmar	48	107	805
Gannet	94	173	3000
Great Shearwater	47	109	833
Guillemot	42	69	990
Herring Gull	56	87	1093
Kittiwake	39	103	400
Lesser Black-backed Gull	58	143	810
Puffin	28	55	400
Razorbill	38	65	660

a) i   Find the equation of the regression line of w on l.
    ii   Draw a scatter diagram to show the data and your regression line.
b) i   Write down the value of the product moment correlation coefficient (pmcc).
    ii   Describe the correlation between the length and wingspan of the seabirds.
    iii For garden birds the value of the pmcc for length and wingspan is 0.932
         Compare this with your answer to part b) i and explain what this tells you about the correlation of wingspan and length.
c) Gina says that there may be strong correlation between the mass and the length of the seabirds. Rashid says mass depends on volume and volume depends on length cubed, so mass and length cubed will be more strongly correlated.
    Is Rashid correct? Show how you decide.

9 A study of body measurements from 17-year-old students gave the following results. All measurements were recorded in millimetres.

Regression of y on x	Male		Female	
	Regression equation	pmcc	Regression equation	pmcc
Hand width on hand length	$y = 36.79 + 0.273x$	0.541	$y = 38.73 + 0.212x$	0.427
Foot width on foot length	$y = 43.86 + 0.224x$	0.574	$y = 31.46 + 0.251x$	0.491
Hand length on height	$y = 22.17 + 0.095x$	0.676	$y = 52.66 + 0.072x$	0.516
Foot length on height	$y = -24.53 + 0.165x$	0.743	$y = 63.57 + 0.106x$	0.561
Foot length on hand length	$y = 32.96 + 1.225x$	0.779	$y = 84.51 + 0.887x$	0.661

a) State whether you think each of these statements is true or false and justify your answer.

  i The correlation between each pair of variables is positive.

  ii In each case the correlation is stronger for males than females.

  iii For both males and females, foot length is more strongly correlated with hand length than with height.

  iv For females an extra 1 cm in foot length predicts an increase in foot width of about $\frac{1}{4}$ cm.

  v For males an extra 10 cm in height predicts an increase in hand length of about 1 cm.

  vi The regression line of foot length on height for males has a negative gradient.

b) What other body measurements do you think are correlated? Carry out your own investigation to find out whether you are correct.

10 The table gives the UK energy consumption E (million tonnes of oil equivalent) and the greenhouse gas emissions G (million tonnes of CO_2 equivalent) between 2000 and 2012.

Year	Energy Consumption, E	Greenhouse gas emissions, G
2000	241.1	750.4
2001	246.2	760.7
2002	240.3	744.0
2003	244.8	756.1
2004	247.1	757.5
2005	247.5	755.9
2006	242.9	746.3
2007	237.8	738.3
2008	232.7	718.7
2009	218.9	662.3
2010	225.3	680.2
2011	212.2	642.0
2012	214.8	656.3

Source: ONS

a) Use the data for energy consumption and greenhouse gas emissions to find

  i the product moment correlation coefficient.

  ii the equation of the regression line of G on E.

b) Draw a scatter graph to show the data and line.

c) The UK government has set a target for greenhouse gas emissions in 2027. Explain why using the regression line may not give a reliable prediction for the energy consumption that will achieve the target.

11 Find out about other measures of correlation (for example Spearman's rank correlation coefficient). Write a summary of your findings.

Consolidation exercise 7

1 A teacher thinks that the nearer students sit to her desk, the better they learn. The table gives the distance from the teacher's desk to each student and the mark they achieved in a test out of 80.

a) Plot a scatter graph.
b) Draw the line of best fit by eye.
c) Describe the correlation and comment on the teacher's theory.

Learner	Distance (m)	Mark
Chris	5.2	28
Sarah	5.6	24
Tom	6.3	16
Laura	5.6	29
James	4.3	48
Lucy	4.8	32
Sophie	3.3	66
John	3.8	40
Ryan	2.7	61
Holly	4.3	43
Zoe	7.2	13
Luke	5.2	35
Chloe	3.3	73

2 Sam uses data about dogs to draw scatter graphs and to calculate regression equations and correlation coefficients.

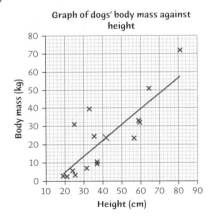

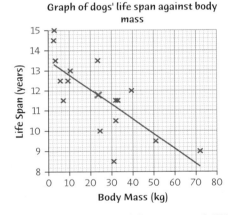

$y = -12.4 + 0.864x \quad r = 0.820$ $y = 13.4 - 0.0720x \quad r = -0.775$

Suggest comments that Sam can make to interpret his results.

3 The table gives the average daily time spent sleeping by babies of different ages.

a) i Plot a scatter graph of t against n.
   ii Some of the sleep time has been omitted when calculating the total for one baby. Identify which baby this is.

b) i Find the equation of the regression line of t on n, ignoring the incorrect data.
   ii Use the equation to predict how long a 10-month-old baby will spend asleep during a day.

Baby	Age (n months)	Average time asleep (t hours)
Amelia	6	14.7
Ben	12	12.9
Charlie	18	12.3
Daisy	9	11.0
Emily	24	10.5
Finley	1	15.8
Grace	15	12.6
Harry	3	14.6
Isla	4	14.7
Jack	21	11.3
Kai	2	15.4

iii Explain why you should not use the regression equation to predict how long a 30-month-old child will spend asleep during a day.

iv Use the regression equation to find t when $n = 0$. Comment on the result.

v Interpret the gradient of the regression line.

4 The table gives information about average annual household income, lack of qualifications and unemployment as well as the percentage of 18 year old students who go to university.

Region	Average household income (£thousands)	% of adult population with no qualifications	Unemployment rate %	18-year-old entry to university rate %
North East	29.479	17.8	9.8	27.2
North West	34.532	16.8	7.6	31.6
Yorkshire & Humber	32.256	17.7	8.2	29.6
East Midlands	35.639	16.3	6.1	28.0
West Midlands	33.999	18.1	7.5	29.8
East of England	39.560	14.1	5.3	30.7
London	48.824	12.4	7.5	34.8
South East	43.949	11.7	4.8	31.0
South West	38.515	12.5	4.9	26.7

Data from ONS, www.ucas.com (UCAS 2014 Application Cycle: End of Cycle Report)

a) Katie says 'High unemployment causes household incomes to be low.'

i Use a scatter graph, regression equation and correlation coefficient to investigate the relationship between average annual household income and the unemployment rate.

ii Comment on Katie's statement.

b) Will says that pupils are likely to go to university when household income is high. Stacy says that entry to university and unemployment are more strongly related. Carry out calculations to see if Stacy is right and comment on your findings.

5 A car manufacturer tests the braking distance of a new car travelling at different speeds.

a) Write down the regression equation of y on x and the product moment correlation coefficient.

b) Draw a scatter graph. Include the regression line.

c) Use your regression equation to predict the braking distance from a speed of

i 85 mph   ii 120 mph

Comment, with reasons, on how accurate these predictions are likely to be.

d) Comment on the use of a linear regression line in this case.

Speed (x miles per hour)	Distance (y metres)
20	7
30	13
40	24
50	35
60	48
70	64
80	81
90	102
100	123

6   Sarah says that the medals countries win at the Olympics depend on how rich they are, but Adam disagrees. He says the size of the population is more important.

The table gives data from the 2012 Olympics. Gross Domestic Product (GDP) is a measure of the wealth of each country.

Country	Population (millions)	GDP ($billions)	Medal points
Australia	22.7	981	65
China	1377.1	15034	192
France	63.5	2372	67
Germany	81.9	3501	85
Greece	11.1	282	2
Ireland	4.6	207	8
Italy	60.9	2152	53
Japan	127.5	4540	66
Russia	143.2	3449	154
Spain	46.1	1512	33
UK	63.7	2381	140
USA	313.9	16144	223

Use these data to investigate the relationships and comment on the statements given by Sarah and Adam. State clearly how you deal with any outliers.

Data from CIA World Factbook

7   The table gives the 2014–15 football results for Premiership clubs.

Premiership club	Won	Drawn	Lost	Goals for	Goals against	Goal difference	Points
Chelsea	26	9	3	73	32	41	87
Manchester City	24	7	7	83	38	45	79
Arsenal	22	9	7	71	36	35	75
Manchester United	20	10	8	62	37	25	70
Tottenham Hotspur	19	7	12	58	53	5	64
Liverpool	18	8	12	52	48	4	62
Southampton	18	6	14	54	33	21	60
Swansea City	16	8	14	46	49	−3	56
Stoke City	15	9	14	48	45	3	54
Crystal Palace	13	9	16	47	51	−4	48
Everton FC	12	11	15	48	50	−2	47
West Ham United	12	11	15	44	47	−3	47
West Bromwich Albion	11	11	16	38	51	−13	44
Leicester City	11	8	19	46	55	−9	41
Newcastle United	10	9	19	40	63	−23	39
Sunderland	7	17	14	31	53	−22	38
Aston Villa	10	8	20	31	57	−26	38
Hull City	8	11	19	33	51	−18	35
Burnley	7	12	19	28	53	−25	33
Queens Park Rangers	8	6	24	42	73	−31	30

a) Investigate how the point score is related to goal difference.
Draw a graph to illustrate your answer.

b) i   What other relationships could you investigate?
   ii Investigate one of these relationships.

Does eating chocolate make you clever?

(October 2012)
Franz Messerli of Columbia University has found that there is an 'incredibly close relationship' between the chocolate consumed in a country and the number of Nobel prizes per person that country has won.

Use the internet to find out more about this study. Look for more recent data on consumption of chocolate and Nobel prizes and investigate the correlation. Find out about other research into the benefits and disadvantages of eating chocolate.

Write a report or prepare a presentation on your findings.

Review

After working through this chapter you should:

- recognise when pairs of data are uncorrelated, correlated, strongly correlated, weakly correlated, positively correlated and negatively correlated
- appreciate that correlation does not necessarily imply causation
- be able to identify and understand outliers and make decisions about whether or not to include them when drawing a line of best fit
- be able to find a line of best fit by
  o   plotting data pairs and drawing, by eye, a line of best fit through the mean point
  o   using a calculator to find the equation of the regression line and plotting it on a scatter diagram
- be able to use interpolation with regression lines to make predictions and understand the potential problems of extrapolation
- be able to calculate the product moment correlation coefficient (pmcc) and appreciate the significance of a positive, zero or negative result
- understand that the pmcc gives the strength of correlation and that its value always lies in the range from -1 to $+1$.

Investigation

Education and careers

The Office for Standards in Education, Children's Services and Skills (Ofsted) collects a huge amount of data in order to assess progress and compare schools and colleges in England.

Here is some of the data for 12 schools.

School	Number of pupils	% who have free school meals	Income (£/pupil)	% who achieve 5A*-C at GCSE	% with special educational needs	% with English not 1st language	% absence rate	Number of teachers (full time equivalent)	Average teacher salary (£/year)
A	1445	19.6	5869	36	7.8	21.0	6.1	84.5	33010
B	1150	12.1	5548	65	7.9	19.2	4.2	78.4	36537
C	1368	7.4	5088	71	2.3	3.1	5.0	85.9	37648
D	1232	19.4	6376	32	7.5	5.3	6.3	95.0	37732
E	322	61.8	9947	12	5.3	66.1	8.3	25.1	40301
F	1297	4.9	4860	62	3.4	3.2	4.9	79.9	38475
G	1083	25.0	6958	56	4.3	24.3	4.9	74.6	38718
H	920	21.9	6752	48	5.7	20.0	5.7	64.8	38290
I	1778	11.5	5425	61	6.6	23.0	3.4	104.3	36855
J	1109	22.5	5676	42	8.9	1.6	7.6	77.2	37537
K	724	9.8	5477	58	4.3	1.0	4.1	45.2	36143
L	1286	16.3	5380	44	8.1	3.9	5.2	80.9	36592

The points score for individual students is also counted at the end of Key Stage 4 after they take GCSEs and again at the end of Key Stage 5 after A levels or other advanced qualifications. Calculating the average score for each school gives another way of comparing schools and colleges.

The tables show the 'performance points' that are used for this purpose:

GCSE grade	Points
A*	58
A	52
B	46
C	40
D	34
E	28
F	22
G	16

Grade	AS level points	A level points
A*	–	300
A	135	270
B	120	240
C	105	210
D	90	180
E	75	150

The Universities and Colleges Admission Service (UCAS) uses a different system of points (tariff points) when students apply for places at university and the universities themselves are also compared and ranked using a points system by some newspapers and other organisations.

Careers after university have different employment and pay rates depending on the subject studied, as shown in the table.

Main subject of degree	Employment rate (employed as % of all graduates)	Annual pay (£) (Median)
Medicine	95	45 600
Biological Sciences	89	28 000
Physical/Environmental Subjects	89	36 000
Maths or Computer Science	89	34 000
Engineering	89	42 000
Architecture	90	35 000
Social Sciences and Law	89	30 000
Business and Finance	90	30 000
Media and Information Studies	93	21 000
Languages	87	30 400
Humanities	84	28 000
Arts	85	21 900
Education	88	30 000

Source: Labour Force Survey

Use the methods of this chapter to investigate relationships between one of the following:

- the achievements of pupils at schools and other variables such as the absence rate or the income received per pupil
- the points achieved by students at GCSE level and those at A level
- the points awarded in university rankings, the UCAS points required for courses and the number of applicants and acceptances
- the UCAS points required for courses, the length of the courses, employment rates and average salaries for professions that follow from them.

Write a report or prepare a presentation on your findings.

8 Critical path analysis

What is logistics?

How do companies and governments manage the flow of goods efficiently?

In this chapter you will learn how to draw and use activity networks in order to plan how projects can be completed as quickly and as efficiently as possible.

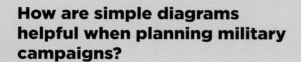

How are simple diagrams helpful when planning military campaigns?

How can the use of manpower and resources be optimised?

You will find that an excellent feature of this chapter is that it depends on very little prior mathematical knowledge beyond the ability to add and subtract numbers.

The management of the flow of goods to meet the requirements of companies and governments is called **logistics**.

The movement of containers around the world is an example of where you need to plan carefully when carrying out a large-scale project. A shipping company can earn between two thousand and five thousand dollars for transporting a full container between, say, East Asia and Europe. However, most of this trade is in one direction moving goods from east to west. Shipping companies must therefore plan how to get the containers back to East Asia without simply losing money by having to transport empty containers.

Activity 1

Think of various ways for a shipping company to earn money from the transportation of otherwise empty containers from west to east. What do you think happens in practice at the moment?

The individual activities of a project can be shown in a **network** with directed lines indicating the order in which activities must take place. This diagram shows just a small part of the process followed when importing frozen food into the UK.

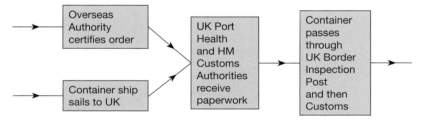

In practice, network diagrams for large projects can have many hundreds of boxes interlinking with each other.

Consider, for example, the proposed construction of London's 'Super Sewer'. This is a drainage tunnel system, proposed by Thames Water and approved by the government, as part of plans to modernise London's sewer system.

Activity 2

Think of all the many different types of activities that will need to be coordinated in the management of this project.

In this chapter, some of the methods used for analysing projects will be given as sequences of procedures. You can think of each such sequence as being simply a recipe to be followed by rote.

According to Thames Water, London's existing sewers are inadequate at present, frequently overflowing and depositing roughly 40 million tons of raw sewage into the Thames each year. The 'Super Sewer' project will be the biggest undertaken by any UK water company. Projected costs are £4.2 billion and the work is expected to take place over 7 years from 2016 to 2023.

The procedures of this chapter can easily be translated into code to enable a computer to tackle a large real-world problem or can be applied, by hand, to solve a small-scale problem.

Key point

The mathematical name for a recipe that can be translated into code is **algorithm**.

The word algorithm is derived from the Latin name of the Persian mathematician, al-Khwarizmi. The procedures or algorithms that he was interested in were those for solving equations and you will have used his procedures many times. Consider, for example, the following solution.

$$4x - 7 = 3x$$

$$4x = 3x + 7 \qquad \text{Make all quantities positive}$$

$$x = 7 \qquad \text{Collect like terms together}$$

The mathematics that you have already studied at school was full of 'algorithms' that mathematicians in the past developed for such things as multiplying and dividing numbers, finding interquartile ranges and solving equations.

For example, consider the algorithm in Activity 3.

> Don't try to draw an activity network for the 'Super Sewer' project! The analysis and even the drawing of such a diagram is done, not by hand, but by computers working through pre-defined sequences of procedures.

> Al-Khwarizmi called the first of these procedures, where you make all quantities positive, 'restoration' or 'al-jabra'. Our word algebra comes from this word. However, al-Khwarizmi himself did not use any algebraic symbols but wrote everything out in words!

Activity 3

a) Carry out the following algorithm on the set of numbers 9, 8, 3, 5, 7, 3.

       Step 1: Arrange the numbers in order of increasing size.

       Step 2: Delete the 2 outer numbers.

       Step 3: Repeat Step 2 until only 1 or 2 numbers remain.

       Step 4: Write down the number or average of the 2 numbers that remain.

b) What is this algorithm designed to find?

c) Can you see a (minor) problem with this algorithm?

Activity 4

Think about any project involving a large number of different activities. For example, in a school or college this could be putting on a play or building a new Science block.

List the separate activities in your project and estimate the time required for each activity. Draw a network to show which activities must be completed before another activity can be started.

How would you organise the start times of each activity so as to minimise the total time required to complete your project?

The first stages in the analysis of a project are to:

- break the project down into individual activities
- estimate the expected duration of each activity
- decide which activities must be completed before a given activity can be started.

The results of this analysis can usefully be represented in a **precedence table**. For example, consider the project of converting a room in a college into a new computer room. A possible precedence table is given below.

Activity	Duration (weeks)	Immediately preceding activities
A: Prepare room	2	–
B: Install cabling	1	–
C: Buy hardware	1	–
D: Install hardware	1	A, B, C
E: Buy software	1	C
F: Install software	1	D, E
G: Train staff	2	F

Activities A, B and C do not have any immediate preceding activities because none of these activities rely on anything else already having been done.

Note that activity G is preceded by *all* the other activities. However, F is its *immediate* predecessor. Likewise, G is the immediate successor of F.

Key point

- The activity X is an **immediate predecessor** of activity Y if Y depends on X and no other activity occurs between X and Y
- The activity Y is an **immediate successor** of activity X if Y depends on X and no other activity occurs between X and Y

Activity 5

Create a precedence table for the activities you listed in your chosen project in Activity 4.

The next stage in the analysis is to represent the activities in a network. The activities are drawn as boxes. The two empty boxes are for numbers which will be explained in the next section.

D	
	1	

Activity name is D
Activity duration is 1

> ## Key point
> The network should be drawn so that each activity is to the right of all its preceding activities

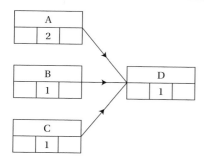

The full network diagram for the computer room project is then:

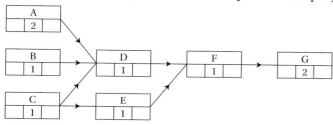

It is easy to construct activity networks for small projects. However, there are formal algorithms that a computer would follow for complex projects.

An example algorithm is given below.

> ## Network diagram algorithm
> 1 Draw a vertical line of boxes to represent all activities that have no immediate predecessors
> 2 To the right of all existing boxes draw a vertical line of boxes to represent all further activities for which all immediate predecessors are already drawn
> 3 Draw directed edges to represent all immediate predecessors
> 4 Repeat steps 2 and 3 until all activities have been drawn

Exercise 8A

1   Draw the activity network for the baking of a simple sponge cake involving the following activities.

Activity	Duration (minutes)	Immediate predecessors
A: Heat oven	6	–
B: Grease tin	0.5	–
C: Cream fat and sugar	1	–
D: Beat in eggs	0.5	C
E: Fold in flour	0.5	D
F: Put mixture in tin	0.5	B, E
G: Bake cake	12	A, F

2   The external work on a small extension has been completed. The internal work has been divided into a number of activities, as shown. Construct an activity network for the project.

Activity	Duration (days)	Immediate predecessors
A: Studding	2	–
B: Initial electrics	1	A
C: Initial plumbing	2	A
D: Plastering	3	B, C
E: Joinery	3	D
F: Final electrics	1	D
G: Final plumbing	2	D
H: Decorating	3	E, F, G

3   Some of the activities involved in bathing a baby are as follows.

Activity	Duration (minutes)	Immediate predecessors
A: Prepare and test bath water		
B: Get towel and talcum powder (if used) ready		
C: Get clean clothes ready		
D: Remove nappy		
E: Dispose of nappy		
F: Clean baby's bottom		
G: Lower baby into water and gently wash baby		
H: Play with baby in bath		
I: Place baby on towel and dry thoroughly		
J: Put new nappy and clothes on baby		

a) Estimate suitable durations for the individual activities.
b) Complete the Immediate predecessors column.

4 For the following activity network draw up a table of the activities showing their durations and immediate predecessors.

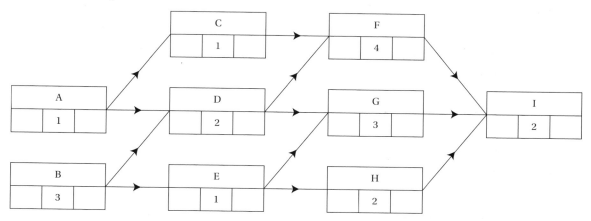

5 Draw the activity network for the baking of an apple pie involving the following activities.

Activity	Duration (minutes)	Immediate predecessors
A: Make pastry	5	–
B: Prepare apples	10	–
C: Grease tin	0.5	–
D: Chill pastry	15	A
E: Roll out pastry	2	D
F: Make pie	2	B, E
G: Bake	20	C, F

6 Draw the activity network for the following activities on a bicycle assembly line.

Activity	Duration (minutes)	Immediate predecessors
A: Prepare frame	10	–
B: Fix front wheel	5	A
C: Fix rear wheel	6	A
D: Attach chain wheel to crank	2	–
E: Fix chain wheel	2	A, D
F: Fix pedals	14	E
G: Final attachments	20	B, C, F

Early times

The next stages in the analysis of a project are to calculate the *early time* and the *late time* for each activity.

On an activity network, the early times are put into the left hand number boxes.

Example 1 Find the early times for the activities of the computer room project of Section 8.2.

You can assume that activities A, B, and C start at time 0.

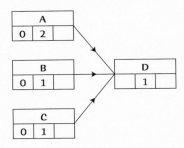

Activity D has immediate predecessors A, B, and C. Of these A is the last to finish, therefore the earliest that D can start is at time 2 (weeks).

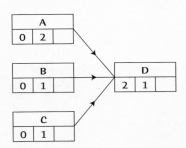

The complete diagram of early times is then as shown.

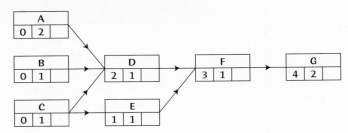

More formally, the early time for an activity X can be found from the early times of all its immediate predecessors by applying a simple algorithm.

> **Early time algorithm**
> 1. For all immediate predecessors of X calculate:
>   Early time of activity + Duration of activity
> 2. The early time of X is the maximum of all these values

Key point
Working in the direction of the directed edges in this way is termed **making a forward pass** through the network

This rule can be applied to all activities in turn, working from the left to the right of a network.

Example 2 Find the early time for activity X.

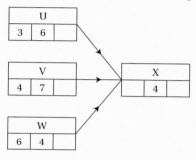

X follows U and so cannot start before time $3 + 6 = 9$

X follows V and so cannot start before time $4 + 7 = 11$

X follows W and so cannot start before time $6 + 4 = 10$

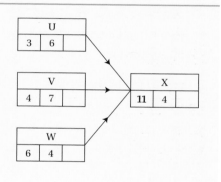

The maximum of 9, 11 and 10 is 11.

The early time of X is therefore 11.

When the early times of all activities have been calculated, the **minimum completion time** for the entire project is clearly just the largest value of

$$\text{Early time} + \text{Duration}$$

for *all* activities at the extreme right of the network.

You sometimes need to have only one activity at the extreme right of a network. This can be done for any network by simply adding an extra activity called *End* with a duration of 0. See question **2** of the consolidation exercise.

Late times

Once you have found the minimum completion time, it is then possible to calculate the late times of all other activities in such a way that the whole project is not delayed.

On an activity network, the late times are put into the right hand number boxes.

Key point
The **late time** of an activity is its latest possible finish time

Example 3 Find the late times for each vertex in this activity network.

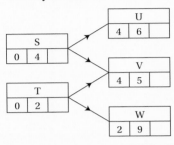

The minimum completion time is the largest value of $4 + 6 = 10$, $4 + 5 = 9$ and $2 + 9 = 11$, that is, 11.
Therefore activities U, V and W all have late time 11.

It is now easy to see that, to achieve a completion time of 11, activity T must be completed by time 2. This is just sufficient for activity W to be completed within time.

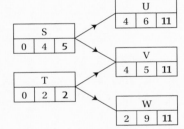

Activities U and V both depend on S. If activity S was completed by time 6, activity V could still be completed in time, but activity U would finish at time 12 which is not within time.

Similarly, the latest time by which S must be completed is 5, which is just sufficient for activity U to be completed within time.

More formally, the late time for an activity X can be found from the late times of all immediate successors by applying a simple algorithm.

Key point
Working in the opposite direction to that of the directed edges is termed **making a reverse pass** through the network

Late time algorithm

1 For all immediate successors of X calculate:
   Late time of activity – Duration of activity
2 The late time of X is the minimum of all these values

This rule can be applied to all activities in turn, working from the right to the left of a network.

Example 4 Find the late time for vertex X.

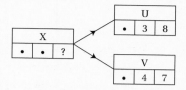

The minimum of $8 - 3 = 5$ and $7 - 4 = 3$ is 3.

The late time of X is 3.

Activity 6

Complete all the boxes on the activity network of the computer room project of Section 8.2.

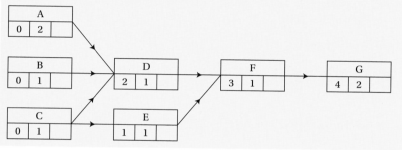

Activity 7

Find the early and late times for each activity in the task of baking a cake.

Activity	Duration (minutes)	Immediate predecessors
A: Pre-heat oven	8	–
B: Grease tin	0.25	–
C: Cream fat and sugar	2	–
D: Beat in eggs	1	C
E: Fold in flour	1	D
F: Put mixture in tin	0.5	B, E
G: Bake cake	10	A, F

You should find that the total time for baking the cake is determined by just the two activities A and G. In the next section you will see that these are called critical activities.

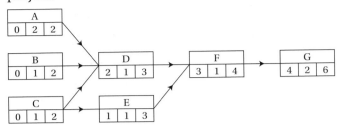

8.4 Critical activities

Consider again the activity network for the computer room project.

In particular, consider the timings for activity F.

$$\text{Late time (4)} = \text{Early time (3)} + \text{Duration (1)}$$

If the whole project is to be completed as rapidly as possible, then F must both start and finish as early as possible.

> **Key point**
>
> A **critical activity** is an activity, such as F, with no slack in its timing.

> If an activity is *not* critical, then this activity can be delayed without necessarily delaying the entire project. The delay can be of any duration up to the length of the float.

The amount of slack in an activity's timing is measured by its float.

> **Key point**
>
> The **float** of an activity is the value of
>
> $$\text{Late time} - \text{Early time} - \text{Duration}$$

The float must be greater than or equal to zero.

> **Key point**
>
> - An activity which is critical has float $= 0$
> - An activity which is not critical has float > 0

Example 5

For the activities of the computer room project

a) determine the float of each activity

b) identify which activities are not critical.

a) The floats are: A: $2 - 0 - 2 = 0$,   B: $2 - 0 - 1 = 1$,   C: $2 - 0 - 1 = 1$,
D: $3 - 2 - 1 = 0$,   E: $3 - 1 - 1 = 1$,   F: $4 - 3 - 1 = 0$,   G: $6 - 4 - 2 = 0$

b) The non-critical activities are B, C and E which all have a positive float of 1.

You may have noticed that the critical activities for the computer room project form a path through the network.

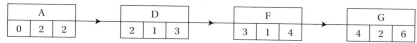

Key point

A **critical path** is a path through a network consisting only of critical activities

The same attention does not need to be paid to non-critical activities. In some cases it may even be appropriate to switch some resources from non-critical activities to critical ones.

All activity networks have at least one, and sometimes more than one, critical path.

Anyone monitoring the progress of a project must pay particular attention to these critical paths in order to ensure that the project stays on schedule.

Activity 8

Determine all the critical paths for the activity network shown below.

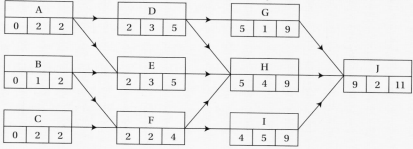

Activity 9

A project has 6 activities, as shown.

Activity	Duration (days)	Immediate predecessors	Cost of speeding up by 1 day (£)
A	3	–	1200
B	9	–	1400
C	5	A	1200
D	4	A	600
E	3	A, B	1000
F	3	B, C, D	1000

a) Find the minimum time for completion of this project if the durations are as shown.

b) Find the minimum extra cost to complete the project 1 day earlier than your answer to part **a**.

Exercise 8B

1

C				F		
2	3	5		5	1	8

State the floats of activities C and F and explain which of these activities is critical.

2 Explain to a non-mathematician what the difference is between

   a) an activity which is essential to the completion of a project

   b) an activity which is critical.

3 A small project has five activities, A, B, C, D and E. The activity network for this project is as shown, where all durations are in weeks.

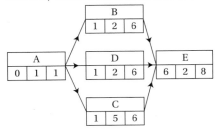

   a) State which activities are critical.

   b) Each activity can be carried out by just one person.

   i How many people are needed to complete the project in the minimum completion time?

   ii How could the work be split up to make this possible?

   c) If only one person was available for all the tasks, how long would the project take? Justify your answer.

4 A building project is to be undertaken. The table shows the activities involved.

Activity	Immediate predecessors	Duration (weeks)
A	–	2
B	–	1
C	A	3
D	A, B	2
E	B	4
F	C	1
G	C, D, E	3
H	E	5
I	F, G	2
J	H, I	3

   a) Complete an activity network for the project.

   b) Find the earliest start time for each activity.

   c) Find the latest finish time for each activity.

   d) State the minimum completion time for the building project and identify the critical paths.

(AQA, 2007)

5   The following diagram shows an activity diagram for a building project. The time needed for
    each activity is given in days.

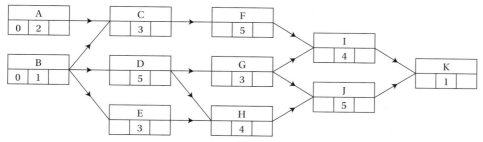

a) Complete a precedence table for the project.
b) Find the earliest start times and the latest finish times for each activity.
c) Find the critical path and state the minimum time for completion of the project.
d) Find the activity with the greatest float time and state the value of its float time.

(AQA, 2007)

6   A garage is to be converted into a granny flat. The external work on the building has been
    completed and the inside is to be fitted out. This work has been divided into a number of
    activities, as shown in the table.

Activity	Immediate predecessors	Planned duration (days)
A: Studding	–	2
B: Initial electrics	A	1
C: Initial plumbing	A	1.5
D: Insulating walls	B, C	2
E: Plastering	D	2.5
F: Artexing ceilings	E	1
G: All joinery	E	3
H: Final electrics	E	1
I: Final plumbing	E	1.5
J: Decorating	F, G	1
K: Cleaning	H, I, J	1

a) Construct an activity network for the project.
b) Find the earliest start time for each activity.
c) Find the latest finish time for each activity.
d) Write down the critical activities.
e) i   Write down the float time of activity B.
   ii State the activity with the greatest float time.
f) Both plumbing activities take twice as long as planned.
   Find the new completion time for the whole project.

(AQA, 2007)

In many large organisations, the results of a critical path analysis are often passed on to middle management in a particular diagrammatic form called a cascade diagram.

The main features of cascade diagrams are illustrated below for a project involving just five activities.

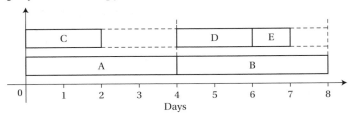

If each activity is shown on a separate row then this is also known as a Gantt chart.

Activities are often shown starting at their earliest possible time. Dashed lines show the range of times possible for an activity.

- The vertical dashed line at day 4 shows that activity C must be completed by that time. It also shows that activity D cannot start earlier than that time.
- Activity C has a float of 2 days.
- Each of activities D and E has a float of 1 day. Note that if either of these activities takes an extra day then there is no freedom in scheduling the other activity.

Key point

When drawing a cascade diagram it is usually best to start by putting all the critical activities across the bottom row(s) of the diagram.

A manager will decide on a particular timing of non-critical activities according to what factors, including time, are most important. For example:

- It may be wise to do all activities as early as possible in case of unforeseen delays later in the project.
- Some activities may involve considerable financial investment and therefore may be left as late as possible.

One particularly important aspect of the timing of non-critical activities concerns the use of the workforce or of equipment. For example, a manager may try to time activities so that there is no need to hire extra workers who will then be under-employed at other stages of the work.

Example 6 The activity network for a small building project is as shown.
All durations are in days.

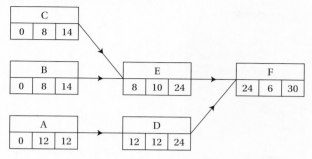

a) Construct a cascade diagram for this project.

b) Assume that each activity involves one worker. If only two workers are available, by how long will the project be delayed?

The use of early and late times to find the shortest time to complete the project assumes you are not restricted by the number of workers you have, how many activities you can pay for at the same time, the availability of machines and so on. A Gantt chart or cascade diagram enables you to see clearly which activities must run at the same time.

a)

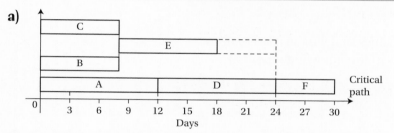

b) Put C after B. Then E is delayed by 8 days and therefore F is delayed by 2 days.

The project is therefore delayed by just 2 days.

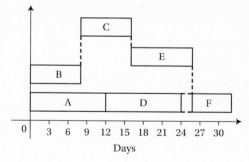

Activity 10

A project has 7 activities, A, B, C, D, E, F and G. The early time, duration and late time for each activity are as shown.

A (0, 2, 3), B (0, 1, 3), C (0, 4, 4), D (1, 3, 6), E (2, 1, 4), F (4, 1, 7), G (4, 3, 7).

Activities A, B and G each require two workers. Each of the other activities requires a single worker. Draw a cascade diagram to illustrate how the project can be completed as soon as possible using at most three workers at any time.

Cascade diagrams are easier to draw than Gantt charts. However, it can be convenient to have each activity on a separate line when allocating tasks to workers and monitoring progress.

One of the many computer packages that can be used to draw Gantt charts produces diagrams of the following type.

Task	Predecessors	Duration (days)	Responsibility	S	M	T	W	T	F	S	S	M	T	W	T	F	S	S	M	T	W	T
A		4	Marta																			
B		6	Steve																			
C	A	5	Aisha																			
D	A	7	Matt																			
E	B, C	5	Marta																			

Activity 11

★ Use your knowledge of critical path analysis to suggest the significance of:
- the activities coloured red
- the black lines.

★ What do you think is indicated by the blue and red arrows?

★ Why are the lengths of some of the activity boxes greater than the stated durations?

Project management charts were developed independently by a number of people, including Henry Gantt, 1861–1919. Notable use of Gantt charts was made in managing the construction of the Hoover Dam, some years after Henry Gantt's death.

Key point

In the AQA examination, only draw arrows or use particular colours/shading if this is specifically requested.

Example 7 A project has activities as given in the table.

Activity	Duration (weeks)	Immediate predecessors
A	7	–
B	5	–
C	4	–
D	6	A, B
E	7	B
F	8	B, C
G	5	D
H	6	D, E, F
I	8	G, H
J	9	G, H

a) Draw an activity network for this project.

b) Find the earliest start time and latest finish time for each activity and insert the values on your activity network.

c) Draw a Gantt chart to illustrate how the project can be completed in the minimum time, assuming that each activity starts as early as possible.

d) Given that each activity requires one worker and that there are only two workers available, find the minimum completion time and describe a suitable allocation of tasks.

---

a), b)

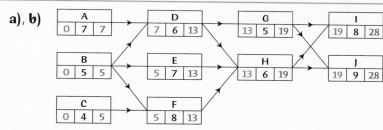

First draw the network with durations. Then complete the red numbers on a forward pass and the blue numbers on a backward pass.

c)

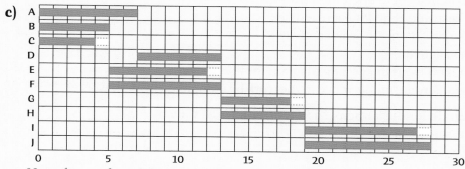

Note that each activity must start as soon as possible.

d) Activities H and J have to take 15 weeks after A, B, C, D, E and F have all been completed. These six activities have total duration 37 weeks and so (assuming no activity is split) must take at least 19 weeks. The minimum of 34 weeks can then be achieved by allocating A, C, F, G and J to one worker and B, E, D, H and I to the other worker.

Exercise 8C

1   A small building project is to be undertaken. The time needed for each activity is given in days

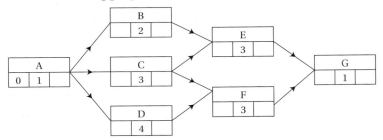

a) Find the earliest start time for each activity.
b) Find the latest finish time for each activity.
c) List the critical activities.
d) Construct a Gantt chart or cascade diagram for the project, starting each activity as early as possible.
e) Construct a Gantt chart or cascade diagram for the project, starting each activity as late as possible.
f) State one advantage for each of the following choices.
   i   Starting each activity as early as possible.
   ii Starting each activity as late as possible.

2   The table shows the main activities involved in a project.

Activity	Duration (weeks)	Immediate predecessors
A	4	–
B	1	–
C	2	–
D	1	A, B
E	6	B, C
F	1	D, E

a) Construct an activity network for the project.
b) Find the earliest start time for each activity.
c) Find the latest finish time for each activity.
d) List the critical activities.
e) Construct a Gantt chart or cascade diagram for the project.
f) If sufficient workers are employed, can the work be completed in eight weeks?

3   The cascade diagram for the launch of a new product is as shown.

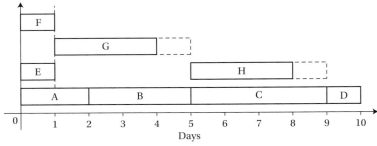

a) List the critical activities.

b) What are the two possibilities for the immediate predecessor or predecessors of activity H?

c) For one of these possibilities, construct an activity network for the project.

d) Find the earliest start time for each activity.

e) Find the latest finish time for each activity.

4   In 2012, Chestnut Liqueurs won an order to supply a chain of farm shops with a new range of products. Chestnut Liqueurs is a small family owned firm and the work had to be planned carefully to ensure that it would fit in with their existing orders. The project was divided into a number of activities as shown in the following table.

Activity	Immediate predecessors	Planned duration (weeks)
A: Trial the blending process	–	2
B: Trial the product with tasters	A	4
C: Trial the product at shows	A, F	8
D: Choose glassware	–	1
E: Purchase glassware	C	2
F: Design packaging	D	2
G: Purchase packaging	C	4
H: Bulk purchase base products	B	2
I: Carry out mass production	H	2
J: Complete and deliver the order	E, G, I	3

a) Complete an activity network for the project.

b) Find the earliest start time for each activity.

c) Find the latest finish time for each activity.

d) State the minimum completion time for the project and write down the critical path.

e) Construct a Gantt chart or cascade diagram for the project.

f) For some designs, the purchasing of packaging can be reduced to 1 week. What effect would this have on the minimum completion time?

Consolidation exercise 8

1 A consultancy company has been hired to assess the work involved in setting up an ICT system for a new school. The consultancy company has divided up the work into activities as shown in the table.

Activity	Immediate predecessors	Planned duration (weeks)
A: Decide on new system	–	1
B: Prepare ICT control room	A	2
C: Buy hardware (including delivery)	A	5
D: Buy software (including delivery)	A	2
E: Train ICT staff	B, C, D	2
F: Install cabling	C	2
G: Install hardware	E, F	1
H: Install software	G	1
I: Prepare pupil/staff data	A	5
J: Install data	H, I	1
K: Train teaching staff	H	2
L: Test system	J, K	1

a) Construct an activity network for the project.
b) Find the earliest start time for each activity.
c) Find the latest finish time for each activity.
d) List the critical activities.
e) Construct a Gantt chart or cascade diagram for the project.

(AQA, 2008)

2 A small construction project has an activity network as shown.

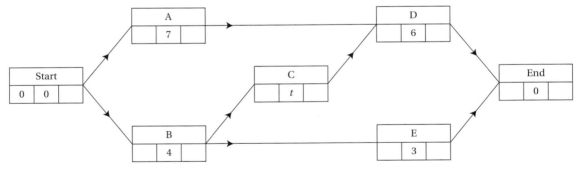

a) Complete the early and late times for the network when the duration of activity C is $t = 2$.
b) For what range of values of t is activity C critical?

(AQA Specimen Question, 2008)

3   A construction project is to be undertaken. The table shows the activities involved.

Activity	Immediate predecessors	Duration (days)
A	–	2
B	A	5
C	A	8
D	B	8
E	B	10
F	B	4
G	C, F	7
H	D, E	4
I	G, H	3

a) Construct an activity network for the project.
b) Find the earliest start time for each activity.
c) Find the latest finish time for each activity.
d) Find the critical path.
e) State the float time for each non-critical activity.
f) Draw a Gantt chart or cascade diagram for the project,
    assuming each activity starts as late as possible.

(AQA, 2006)

4   A decorating project is to be undertaken. The table shows the activities involved.

Activity	Immediate predecessors	Duration (days)
A	–	5
B	–	3
C	–	2
D	A, B	4
E	B, C	1
F	D	2
G	E	9
H	F, G	1
I	H	6
J	H	5
K	I, J	2

a) Construct an activity network for the project.
b) i   Find the earliest start time for each activity.
    ii   Find the latest finish time for each activity.
c) State the minimum completion time for the decorating
    project and identify the critical path.
d) Activity F takes 4 days longer than first expected.
    i   Determine the new earliest start time for activities H and I.
    ii   State the minimum delay in completing the project.

(AQA, 2009)

5 The activity network for a project is shown below.

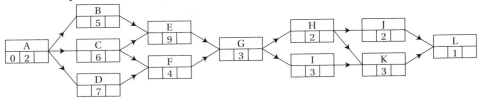

a) Find the earliest start time and the latest finish time for each activity.

b) Hence find:
  i the critical path
  ii the float time for activity D.

(AQA, 2008)

6 A decorating project is to be undertaken.
The time needed for each activity is given in days.

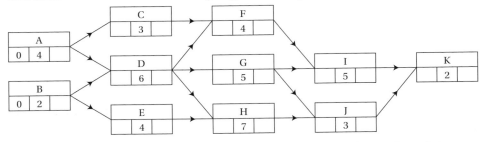

a) Find the earliest start time and the latest finish time for each activity.

b) Find the critical paths and state the minimum time for completion.

c) Draw a Gantt chart or cascade diagram for the project, assuming each activity starts as early as possible.

d) Activity C takes 5 days longer than first expected. Determine the effect on the earliest start time for other activities and the minimum completion time for the decorating project.

(AQA, 2011)

7 Some of the activities involved in a loft conversion are as follows.

Activity	Duration (days)	Immediate predecessors
A: Draw up and approve plans	8	–
B: Get planning permission	42	A
C: Clear site	14	–
D: Order stairs, windows and door	14	B
E: Build new entrance	3	C, D
F: Electrical work	4	B, C
G: Fit flooring	3	E, F
H: Fit windows	2	E
I: Plaster walls	3	G, H
J: Decorate	3	I

a) Draw an activity network.

b) Find the earliest start time and latest finish time for each activity.

c) State the minimum time for completion.
d) A way of speeding up the project would be to start activity D before completing A and B.
   i   By how much would this reduce the minimum time for completion?
   ii What is a major drawback of this method of speeding up the project?

8 Thimbleby Technology won a contract to install a custom-designed accounting system.
   An activity network for the project was as shown.

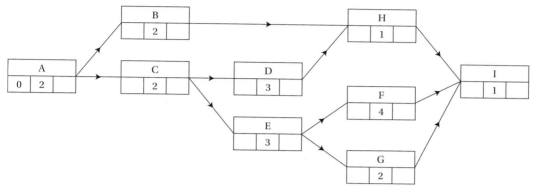

a) Find the earliest start time for each activity.
b) Find the latest finish time for each activity.
c) State the minimum completion time for the project and write down the critical path.
d) State the floats on activities B and D.
e) The total expected programming time for activities D and F was 7 weeks. Due to staff
   illness, this total programming time was doubled. How did the Senior Analyst at
   Thimbleby Technology reorganise the work on activities D and F so as to complete the
   project in the shortest time?
   Explain your answer.

Review

You should now know how to:

- construct and understand activity networks
- calculate early and late times by performing forward and reverse passes
- identify critical activities and paths
- construct and understand Gantt charts (cascade diagrams).

Investigation

Critical path analysis

An army marches on its stomach.

Napoleon Bonaparte

Logistical support, or the lack of it, has been crucial in many military campaigns. Its importance was shown in the First Gulf War (1990–91) where months of planning resulted in the transportation of half a million troops and seven million tonnes of supplies half the way around the world. This prepared the way for what was effectively just four days of ground operations.

Critical Path Analysis has a very important part to play when planning complicated projects. You can use these investigations to practise analysing and planning projects in this way.

Critical Path Analysis is the first chapter in this book about a topic where the initial development was not from mathematical or scientific interest but directly because of management and military needs.

Investigation 1

Find out more about the early development of Critical Path Analysis by the Du Pont Corporation and of PERT analysis by the US Navy. What is the most important difference between PERT analysis and the techniques you have studied in this chapter?

The Du Pont Corporation is a global science company that has had a major role in scientific advances in the USA. Initially important for the production of explosives, its more recent inventions have included Nylon and Teflon. If you are interested in Accounting and Investment you could research *Du Pont Analysis* and the formula for Return on Equity (ROE) given by Profit Margin × Asset Turnover × Equity Mutiplier.

Investigation 2

A wedding planner will charge perhaps £2000 for a full service including:

- preparing a budget
- setting a wedding schedule
- finding a venue
- organising guest lists
- coordinating on the day.

> Could you be a wedding planner and deliver the perfect wedding down to the last detail? Set up an activity network to cover all the main aspects of a wedding and suggest a suitable timetable for these activities.

Investigation 3

Passwell Academy Staff Timetable 2016–17

Period Staff	Mon 1	Mon 2	Mon 3	Mon 4	Mon 5	Tue 1
B.T.A	7A Maths	10C Maths	11B Maths	11 Games	11 Games	9C Maths
C.A	9C English	12B English	8A English	Free	13B English	10A English
F.A.A	8C Science	8D Physics	9B Maths	7C Science	9D Physics	13A Physics
J.A	10A Science	9B Biology	Free	8A Science	12A Biology	Free
G.B.B	7B Maths	10C Maths	11B Maths	12A Maths	13B Maths	8E Maths
K.B	10B French	Free	7A French	13A French	10C French	8B French

The production of a timetable for a school or college is a complex task that has to be completed on schedule each year. It depends, for example, on expected student numbers, student choices, the appointment of teaching staff and the availability of suitable teaching rooms and laboratories. It may also be affected by the need to share teaching or facilities with nearby colleges.

> Discuss your school or college's timetabling with the members of staff responsible. Draw up and analyse a network diagram for the production of such a timetable.

Investigate what parts of the timetabling process are performed by computer. Think of a few ways the overall procedure might be improved and check with the staff concerned if there are issues you have not considered which would make your ideas difficult to implement.

9 Expectation

Should a tennis player risk more double faults by using a more aggressive second serve?

Which of your team's players should take a penalty?

The chance of a single person born today dying in an asteroid strike is around 1 in 200 000.

In this chapter, you will learn how to use and in some cases calculate the probabilities of events such as those listed on this page. You will also meet the important idea of expected value and, in particular, expected financial gain.

What is the probability that you will pass the theory test but fail the practical driving test?

What is the mathematical connection between obesity and diabetes?

You should know how to:

- list all the possible outcomes of an event 1199
- use percentages and simple ratios 1961
- use a calculator for addition and multiplication of fractions and decimals. 1933

Diabetes and obesity rates soar in UK

- 64% of people are overweight or obese.
- 80% of people with Type 2 diabetes are overweight or obese.
- About 1 in 20 people are now being treated for diabetes.

These latest figures are extremely worrying. If we don't stop the rising tides of obesity and diabetes, millions will face a future of ill-health and will put an ever-growing strain on NHS resources.

Type 2 diabetes occurs when the body does not produce enough insulin to function properly. It is strongly linked to obesity, diet, lack of exercise and ageing. The first three of these factors are avoidable!

Activity 1 🗨

If you were to critically analyse the article above, what aspects of the information would you need to clarify?

Once you are happy with the information in an article it can be very helpful to put the data into an easy-to-understand form. A **Venn diagram** is ideal for this. ⊞ 1921, 1922

The rectangle represents all the people in the UK and the two circles represent the people who are diabetic and overweight, respectively.

For simplicity, ignore any differences between types of diabetes and between overweight and obese.

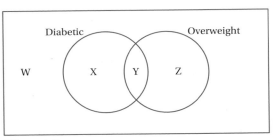

UK population

W People who are not overweight and do not have diabetes.
X People who are not overweight but do have diabetes.
Y People who are over-weight and diabetic.
Z People who are overweight but do not have diabetes.

Filling in the percentages for these categories gives:

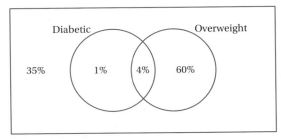

UK population

Activity 2

Explain how each of the three bullet points of the newspaper article are represented in this diagram.

You can easily see from the Venn diagram that 35% of people are neither overweight nor diabetic, but it is not so easy to see this from the original article.

Paper 2B

Example 1 Comment on the statement that 'You are twice as likely to acquire Type 2 diabetes if you are overweight.'

Look at the information in the Venn diagram and work out the chance of an overweight person being diabetic.

64% are overweight.

4% are overweight and diabetic.

So, overweight people have a 1 in 16 chance of being diabetic.

Now work out the chance of a person who is not overweight being diabetic.

36% are not overweight.

1% are not overweight but are diabetic.

So, people who are not overweight have a 1 in 36 chance of being diabetic.

These calculations support the statement since the odds are more than doubled.

> This example shows you how useful a Venn diagram can be when you need to combine the information from different parts of the diagram.

> $\frac{4}{64} = \frac{1}{16}$ so a 4 in 64 chance is equivalent to a 1 in 16 chance. ⊕ 1042

> $\frac{1}{16}$ is more than twice as big as $\frac{1}{36}$ so a 1 in 16 chance is more than twice as likely as a 1 in 36 chance.

Exercise 9A

1   In a group of 100 tennis players, 30 said they needed to improve their forehand and 50 said they needed to improve their backhand. Given that 8 players said they needed to improve both strokes, draw a Venn diagram to represent these data.

   What proportion of the players who were happy with their forehand said they needed to improve their backhand?

2   1000 people took part in a trial for a new test for a disease. The Venn diagram illustrates the results showing those who were indicated as having the disease and those who actually had the disease.

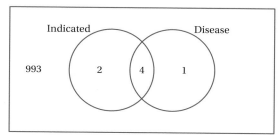
People tested

a)   Correct the statement, '5% of the people have the disease.'

b)   A statement in support of the new test claims:
   'This test has an accuracy of 99.7%.'
   Is this statement correct?

c)   Use the figures on the diagram to make two negative statements about this test.

d)   For this sample of people, what would be a very simple way of achieving 99.5% accuracy?

> Q This example shows how tests for rare diseases have to be extremely accurate as 'false positives' can be unnecessarily worrying for healthy patients. According to the *Journal of Rare Disorders*, July 2014, patients with rare diseases visit an average of 7.3 physicians before receiving an accurate diagnosis.

You call a coin **unbiased** if it is just as likely to come down heads as it is to come down tails. Similarly, when you roll a fair dice, you are equally likely to roll a 1, 2, 3, 4, 5 or 6. This idea was essential when you interpreted the Venn diagrams in Section 9.1.

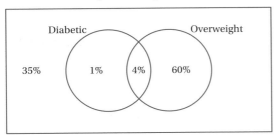

UK population

'An overweight person has a 1 in 16 chance of being diabetic.'

The statement about the chance of being diabetic is based upon the simple idea that a given overweight person is equally likely to be any one of the people in the 'overweight' circle.

Out of every 64 people in this circle, you expect 4 to be diabetic and so the odds are 4 in 64 or 1 in 16.

Although this is a very simple idea, you must be careful not to assume events are equally likely when they are not.

Imagine an experiment where you have three coins. The first is a normal coin, the second coin has heads on both sides and on the third coin both sides are tails. One of these coins is chosen at random and spun on a table. It comes to rest showing heads.

> Whenever a choice is said to be 'at random' it means that each of the possible choices has an equal chance of being chosen.

Activity 3

Someone claims that 'the face-down side of the coin is either heads or tails so the coin has a 1 in 2 chance of being double-headed.' Are they correct? (Think hard about this question; it is not as easy as you might think!)

The next example is a more straightforward situation involving equally likely events.

Example 2 Two dice are thrown. What is the chance of the total score being 8?

Start by working out how many different combinations there are. 1199

The possible results can be written as: (score on first dice, score on second dice).

You could roll (1, 1), (1, 2), (1, 3), (2, 1), (2, 2) and so on.

You have a total of 36 equally likely events.

> For each of the 6 possibilities for the first dice, there are 6 possibilities for the second dice. Then 6 × 6 = 36.

Work out how many ways there are to get a total of 8.

The pairs that have a total of 8 are then (2, 6), (3, 5), (4, 4), (5, 3) and (6, 2).

There are 5 ways to get a total of 8 so the odds are 5 in 36.

Exercise 9B

1 What are the odds of
   a) dealing a red card from a pack of 52 cards
   b) dealing a king from a pack of cards
   c) dealing a diamond from a pack of cards
   d) obtaining a multiple of 3 when throwing a dice?

2 Two dice are thrown. What is the chance of the total score being
   a) 1   b) 2   c) 7?

3 A friend has read the solution to Activity 3 but still feels that the answer has to be 1 in 2. What simple experiment could you do to convince them about the correct odds?

4 Which of the following have equally likely outcomes? Justify your answers.
   a) A tennis match between two top players.
   b) Dealing a playing card from a shuffled pack.
   c) Dropping a drawing pin and observing whether the point is up or down.

5 The Venn diagram represents a group of 30 students.

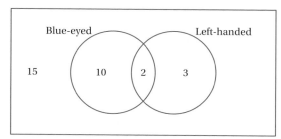

Students

a) If you meet one of these students, what are the odds that they are
   i blue-eyed
   ii left-handed
   iii blue-eyed and left-handed?
b) If you meet one of the students who is left-handed, what are the odds that they are also blue-eyed?

6 A sample of Scottish voters were asked if they had voted SNP in the 2015 General Election and also whether they wanted another vote on Scottish Independence. The results were as shown.

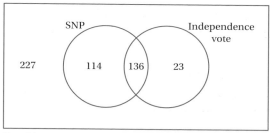

Scottish voters

a) What percentage voted SNP?
b) What percentage of SNP voters wanted another vote on independence?
c) Assuming this sample is representative of Scottish voters, what are the odds that a Scottish voter wants another vote on independence?

In many sporting contests, as well as the prestige of the players and teams, huge sums of money can be at stake. There can be controversy when matches are played to completion and even worse problems when matches have to be postponed or abandoned.

Activity 4 Q

For a variety of sports or games think of reasons why matches have been abandoned. Are there special rules for what then happens?

Different sports have different rules about abandoning matches and what is 'fair' is not always obvious. In the 17th century, the gambler Chevalier de Méré asked the mathematicians of the day to come up with a solution for a type of problem illustrated by this example.

Blaise and Pierre are evenly matched and playing a series of games. The first to win a total of 4 games will take the prize money. When Pierre is 2–1 ahead, the match has to be abandoned and cannot be resumed. Some opinions on what should happen next are given below.

> **Pierre**
> I was ahead when the game had to be abandoned and should therefore receive the prize money.

> **Blaise**
> The game was not finished so we should simply split the money 50:50.

> **Onlooker 1**
> Pierre was 2–1 ahead so a compromise would be to divide the money in that ratio.

> **Onlooker 2**
> Pierre needed 2 more wins and Blaise needed 3, so the best compromise is for the money to be split in the ratio 3:2 in favour of Pierre.

Activity 5 💬

Which one (if any) of the proposals above do you think is fair?

Can you devise an alternative scheme and justify why it is fair to both players?

Fermat came up with a general solution which could be applied to this problem.

Suppose the players had played on for four more games. There would have been 16 equally likely events for these games. (You can assume these 16 events are equally likely because you have already been told the players are evenly matched.)

Nowadays, you might think that the rules of what should happen would all be clearly laid down. However, on 4th April 2015, England's Under-19 women's football team controversially qualified for the European Championships after replaying the final 19 seconds of a match against Norway from a few days earlier!

In the original match, the referee had incorrectly stopped the match without allowing a penalty to be retaken.

PPPP PBPP BPPP BBPP
PPPB PBPB BPPB BBPB
PPBP PBBP BPBP BBBP
PPBB PBBB BPBB BBBB

PPBP means Pierre wins games 1, 2 and 4 and Blaise wins game 3.

Activity 6

a) In which of the 16 equally likely events would Pierre have been the winner?

b) In the four sequences of the form PP…, Pierre has already won after the first two games. What was Fermat's clever reason for listing all four sequences separately and not just having PP in the list?

c) Can you think why an extra *four* games were considered rather than, say, three?

Since Pierre would have won in 11 out of the 16 equally likely events, Fermat declared that Pierre should receive $\frac{11}{16}$ of the money and Blaise $\frac{5}{16}$ of the money.

Key point

Fermat's general procedure can be described as follows.

- List the equally likely events of the game.

- For each player, count those events that would mean they were the overall winner.

- Divide the money accordingly.

Exercise 9C

1 An unbiased coin is thrown repeatedly. After five throws, there have been three heads and two tails. Apply Fermat's method to find the likelihood that a total of four tails will be obtained before a total of four heads.

2 On a game show, you are in the fortunate position of having the option of opening one of two identical envelopes.

In one envelope there is a cheque for £100 000, but in the other the cheque is for just £100. Before you can open the envelope of your choice, you are offered £30 000 to not open an envelope.

a) Is this a fair offer? If not, what amount would be fair and why?

b) Why might someone (sensibly) decide to accept an offer for an amount less than what is 'fair'?

3 In the 1984 World Chess Championship, the match was supposed to be played until one player had reached 6 wins, but it was ended controversially, without result, to 'safeguard the health of the players'.

When the match was stopped, 48 games had been played and the score was Karpov 5 – Kasparov 3 (the other 40 games had been drawn).

A new match was played in 1985 and won by Kasparov.

Assume that the 1984 match had been played to completion and that the players were equally likely to win any game.

a) List the possibilities for the next three games. (Ignore any drawn games.)

b) What do you estimate as Kasparov's chance of winning the match?

The word that is used in mathematics when referring to 'odds' and 'chance' is **probability**. The probability of an event is how likely the event is to occur and you represent it using a number between 0 and 1. Probability is usually given as a decimal or, for simple numbers, as a fraction.

> **Key point**
> An event with a probability of 0.2 or $\frac{1}{5}$ means there is a 1 in 5 chance of it occurring.

The Venn diagram you drew in Exercise 9A for tennis players who need to improve their strokes can also be drawn using probabilities.

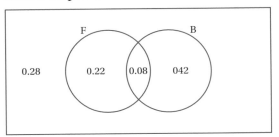

> Note that
> $0.28 + 0.22 + 0.08 + 0.42 = 1$

> **Key point**
> - The probabilities of all possible outcomes add up to 1. Such outcomes are called exhaustive. ⊕ 1262
> - The notation for the probability of an event can be written as P(Event). So in the tennis example P(F) = 0.3.
> - You need to be familiar with three ways of combining events:
>   - ▶ A′   the event that A does *not* occur
>   - ▶ A ∪ B   the event that A or B or both occur
>   - ▶ A ∩ B   the event that both A and B occur. ⊕ 1921

> $P(A') = 1 - P(A)$

> ∪ is called the union and ∩ is called the intersection.

Activity 7

★ For F and B as in the Venn diagram for tennis players, work out

     P(F′)   P(F ∩ B)   P(F ∪ B).

★ Can you think how to represent the event of neither F nor B, using this notation?

There are three ways of finding or calculating the probability of an event.

Method 1

When the possible outcomes are all equally likely, the probability of event A is given by

$$P(A) = \frac{\text{Number of outcomes resulting in A}}{\text{Total number of equally likely outcomes}}$$

> So, as you have already seen, the probability of a total score of 8 when two dice are thrown is
> $$P(8) = \frac{5}{36}$$

Paper 2B

Method 2

You can estimate probabilities based upon experience (for example the probability it will rain on April Fool's Day) or a large sample (such as the sample of Scottish voters). This is known as using relative frequencies. ⊕ 1211, 1264

Method 3

You can use mathematical ideas to calculate probabilities of some events using the probabilities of simpler events. This will be covered in later sections of this chapter.

> In the case of April Fool's Day, historical records for London show that it has rained in roughly 44% of the years on record. Therefore, if you are planning a wedding or a sports event for next year in London, you could assume that the probability of rain on April 1st will be 0.44.

Example 3 Here is a Venn diagram for two events.

a) Find the missing probability.

b) Given that event A occurs, what is the probability that event B also occurs?

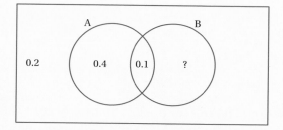

a) You can use the fact that the probabilities of the outcomes add up to 1 to work out the missing probability.

$$1 - (0.2 + 0.4 + 0.1) = 0.3$$

b) You might find it easier to think in terms of actual occurrences rather than probabilities. You could replace the numbers 0.2, 0.4, 0.1 and 0.3 by 2, 4, 1 and 3 for 10 possible outcomes.

In every $4 + 1 = 5$ occurrences of A, event B also occurs just once. Therefore the probability of B given A is

$$\frac{1}{4 + 1} = \frac{1}{5} \text{ or } 0.2$$

A quicker way of carrying out this calculation is to use the original probabilities:

$$\frac{0.1}{0.4 + 0.1} = \frac{1}{5}$$

Exercise 9D

1   A Venn diagram for three events A, B and C is as shown.

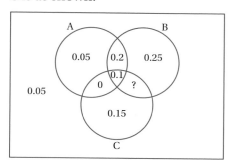

a)   Find the missing probability.

b)   What are
   i   P(A)   ii   P(A′)
   iii P(A ∩ C)   iv   P(B ∪ C)?

c)   Find the probability of neither A nor B.

d)   Given that event A occurs, what is the probability that event B also occurs?

e)   Given that both A and C occur, what is the probability that B also occurs?

2   David and Siska are expecting a baby.
   a)   David says that the probability of the baby being a girl is 0.5. What is he assuming?

   b)   Siska says that across the world about 105 boys are born for every 100 girls. Using these relative frequencies, what does Siska think is the probability of the baby being a girl?

A tree diagram is a good way to show all the possible outcomes of a series of events. For example, this tree diagram shows the outcomes of taking the two-part driving test.

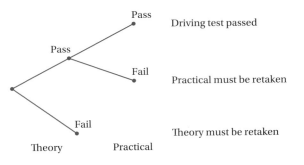

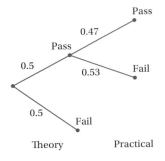

National figures for the last quarter of 2014 showed a pass rate of 50% for the theory test and 47% for the practical test. These figures can be added to the diagram and can be thought of as the probabilities for a 'typical' learner driver.

Key point
- The natural sequence of events on a tree diagram is from left to right.
- The probabilities on all branches emerging on the right of a point must sum to 1.

Note how the branches to the right of the theory test pass have probabilities 0.53 and 0.47, where $0.53 + 0.47 = 1$.

The most important use of a tree diagram is to find the probability of combined events.

Activity 8
The figures given for the pass rates were based upon the results of roughly half a million people.

★ Assuming precisely half a million learner drivers attempted the driving test, how many of these would you expect to pass the theory test but fail the practical?

★ What is the probability of this happening for a typical learner driver?

Key point
- To find the probability of a particular route from left to right of a tree diagram, multiply along the branches.
- To find the combined probability of several routes, add the probabilities of each route. ⊕ 1208

Fermi estimates for passing the driving test are very simple because the probabilities of passing the two parts of the test are both roughly $\frac{1}{2}$. So, half of the learner drivers will pass the theory test and then half of these will pass the practical. Overall, the proportion who pass will be $\frac{1}{2} \times \frac{1}{2} = \frac{1}{4}$. More precisely, the probability of passing the driving test is $0.5 \times 0.47 = 0.235$ as shown in Example 4.

Example 4 For a 'typical' learner driver, taking the test in the last quarter of 2014, what was the probability that they

a) passed the driving test

b) passed the theory test but failed the practical

c) failed the driving test?

---

a) Passing the driving test means passing both the theory and the practical. Multiply the probabilities as you move along the two 'pass' branches.

$0.5 \times 0.47 = 0.235$

b) To find the probability of passing the theory test but failing the practical, multiply the probabilities on these branches.

$0.5 \times 0.53 = 0.265$

c) A good way of finding this probability is to say it is 1 – Probability of passing.

$$P(\text{failing the driving test}) = 1 - 0.235$$
$$= 0.765$$

Alternatively, you could add the probabilities for the ways of failing.

$$P(\text{failing the driving test}) = P(\text{fail theory}) + P(\text{pass theory, but fail practical})$$
$$= 0.5 + 0.265 = 0.765$$

The event A∪B is the event that A or B or both occur. This is connected with adding different branches of a tree diagram.

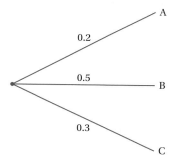

The event A ∪ B has probability

$0.2 + 0.5 = 0.7$

Similarly,

$P(A \cup C) = 0.2 + 0.3 = 0.5$

$P(A \cup B \cup C) = 0.2 + 0.5 + 0.3$
$\qquad\qquad\qquad = 1$

The event A∩B is the event that both A and B occur. This is connected with multiplying along a branch of a tree diagram.

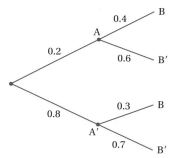

The event A ∩ B has probability

$0.2 \times 0.4 = 0.08$

Similarly,

$P(A \cap B') = 0.2 \times 0.6 = 0.12$

$P(A' \cap B) = 0.8 \times 0.3 = 0.24$

$P(A' \cap B') = 0.8 \times 0.7 = 0.56$

Exercise 9E

1   A tree diagram has outcomes X, Y and Z as shown.

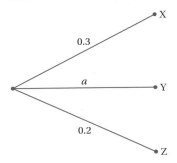

a)   Find probability a.
b)   Find the probability of
    i $X \cup Y$ ii $Y \cup Z$ iii $X \cup Y \cup Z$.

2   This tree diagram has four possible outcomes W, X, Y and Z.

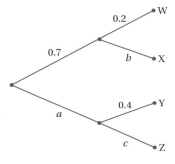

a)   Find a, b and c.
b)   Find the probabilities of
    i X ii $X \cup Y$ iii $X \cap Y$.

3   A hand of 13 playing cards contains 8 red cards and 5 black cards. Two cards are selected at random and discarded. The probability tree diagram represents the possibilities for the discarded cards.

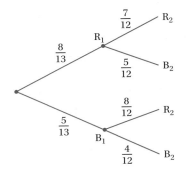

a)   State the value of $P(R_1)$.
b)   Calculate the value of $P(R_2)$.
c)   Find the probability that both discarded cards are red.
d)   Find the probability that the discarded cards have different colours.
e)   What event has a probability which when added to the answers to parts c) and d) will equal 1?

4   Another hand of 13 playing cards contains 6 red cards and 7 black cards. Three cards are selected at random and discarded. What is the probability that there will be equal numbers of red and black cards remaining?

5   Sally is taking a test to work in an insurance office. The test consists of three parts. The probabilities of Sally passing each part are shown in the probability tree diagram.

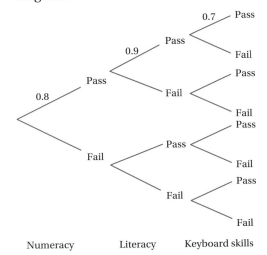

Numeracy   Literacy   Keyboard skills

a) Stating any necessary assumptions, copy and complete the tree diagram.

b) What is the probability Sally will pass all the tests?

c) What is the probability that she passes exactly two tests?

d) Providing she only fails one test, she is allowed to retake that test. What is the probability that she has to retake a test and passes it the second time?

6 In a play-off to decide a chess tournament, Anand and Carlsen play a game with Carlsen playing the white pieces. If the game is drawn, then a second game is played with Anand as white. The probabilities are as shown on the probability tree diagram.

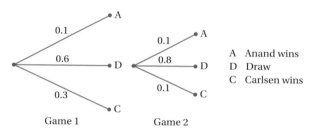

A	Anand wins
D	Draw
C	Carlsen wins

Game 1   Game 2

a) What is the probability that
  i Anand wins
  ii Carlsen wins?

b) If both games are drawn, then the whole process is repeated with the same probabilities as in the tree diagram. What is the probability that the tournament will be undecided after 8 play-off games?

7 Three boxes contain lamp bulbs. The proportions of defective bulbs in boxes A, B and C are $\frac{1}{4}$, $\frac{1}{8}$ and $\frac{1}{32}$, respectively. One box is selected at random and a bulb is selected from that box.

a) Draw a tree diagram to represent this situation.

b) What is the probability that a defective bulb from box A has been chosen?

c) What is the probability that a good bulb from box A has been chosen?

d) Explain why the sum of the answers to parts **b)** and **c)** is $\frac{1}{3}$.

e) Find the probability that the selected bulb will be defective.

8 A model for the serve of a tennis player is as shown.

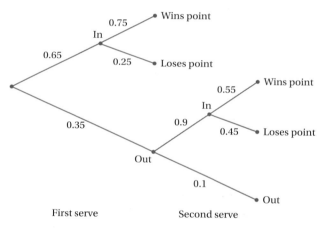

First serve   Second serve

a) What is the probability of a double fault?

b) What is the probability the server wins the point on the first serve?

c) What is the total probability of the server winning the point?

If a tennis player's first serve is out, they then have a second serve. A double fault occurs when a player's first and second serves are both out.

When a penalty kick is taken, a goalkeeper is only able to save the penalty in a 'reachable' zone which covers roughly 75% of the goal. Outside this zone the penalty taker will always score unless they miss the goal entirely. Success rates differ across different leagues and competitions but the probabilities are roughly as shown.

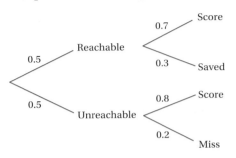

Activity 9

★ Use this probability tree to find the probabilities that a penalty is

missed   saved   scored.

★ One probability on the tree diagram is 0.3. Of what is this the probability?

Conditional probabilities refer to the numbers in the second line of probabilities. In the probability tree above these are 0.7, 0.3, 0.8 and 0.2.

The probability 0.7 is the probability that the penalty is scored *given that* (or based on the condition that) it was in the keeper's reachable zone.

> Note that 0.2 is *not* the probability the penalty was missed.

The probability 0.2 is the probability that the penalty is missed *given that* it was placed outside the keeper's reachable zone.

Key point

The notation for the probability of event A given that B has occurred is P(A | B).

A typical tree diagram looks like this:

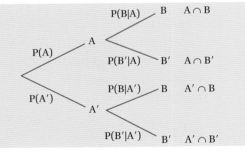

Applying the standard rule on probability trees gives the following extremely important law of probabilities.

> Remember the standard idea of multiplying along branches.

Activity 10

★ Think of examples of events A and B such that $P(B|A) = P(B)$.

What does this imply about the relationship between A and B?

What simple form does the multiplication law now take?

★ Think of examples of events A and B such that $P(B|A) \cdot P(B)$.

Example 5

a) For the penalty kick tree diagram, what is the probability that the penalty kick is reachable but a goal is still scored?

b) Is scoring independent of the kick being reachable?

a) Multiply along the branches to find the probability.
P(Reachable and Scores) = 0.5 × 0.7 = 0.35

b) Compare P(Score | Reachable) with P(Score). If they are equal the events are independent.
P(Score | Reachable) = 0.7 whereas P(Score) = 0.75. So scoring is **not** independent of reachability.

Exercise 9F

1 Two events A and B are such that $P(A) = \frac{1}{3}$, $P(B) = \frac{1}{4}$, $P(B|A) = \frac{1}{2}$ and $P(B|A') = x$.

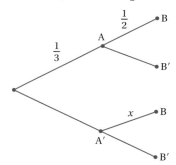

a) Copy and complete the probability tree diagram, giving probabilities in terms of x if necessary.

b) Write down an expression for P(B) in terms of x and hence find the value of x.

> Hint: remember to include all the ways that B can occur in your calculation.

c) Given that A′ has occurred, what is the probability of B′?

2 When dealing a card from a shuffled pack of playing cards, let A, B and C represent events as shown.

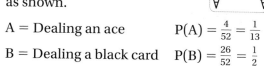

A = Dealing an ace $P(A) = \frac{4}{52} = \frac{1}{13}$

B = Dealing a black card $P(B) = \frac{26}{52} = \frac{1}{2}$

C = Dealing a club $P(C) = \frac{13}{52} = \frac{1}{4}$

a) Find $P(A|C)$ and $P(B|C)$.

b) Are A and C independent? Justify your answer.

c) Are B and C independent? Justify your answer.

In Section 9.3, you saw how Fermat managed to solve a long-standing problem about fairness connected with gambling. Fermat's idea is useful not just in gambling but also in investment, insurance, business planning and project management.

Future events are almost always uncertain and so you use Fermat's idea in order to analyse these events mathematically. The things you have studied so far in this chapter on probability theory put you in a good position to understand how another mathematician called Pascal extended Fermat's idea.

The following scenario shows how the idea can be extended.

Imagine that one afternoon, you intend to take a bus to the railway station and then catch a train to complete your journey. Your bus ticket costs £2 and your train ticket should cost £25.

However, the bus has a probability of 0.2 of being late which means you will reach the train station during rush hour and the cost of the ticket will increase to £35. What will the cost of your journey be on average?

Start by imagining that you made this trip repeatedly. The bus has a probability of 0.2 of being late which means if you made the trip 10 times you would expect the bus to be late twice and on time for the other eight trips.

For the 8 out of 10 occasions when the bus is on time the cost is £2 + £25 = £27.

On the other 2 out of 10 occasions the cost is £2 + £35 = £37.

So, the expected cost is $\frac{8 \times £27 + 2 \times £37}{10} = \frac{£290}{10} = £29$.

Pascal realised that there was no need to 'imagine' these 10 events. All you need do is calculate $0.8 \times £27 + 0.2 \times £37 = £29$.

> **Pascal** avoided the need to always use equally likely events and extended Fermat's method to any events for which the probabilities could be calculated.

> Pascal (1623 – 1662) was a French mathematician and philosopher. He worked in several areas of mathematics and he built a number of early mechanical calculators.
>
> You may know of Pascal from Pascal's triangle:
> ```
>          1
>        1   1
>      1   2   1
>    1   3   3   1
>  1   4   6   4   1
>  . . . . . .
> ```

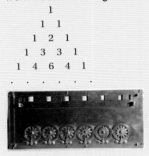

> This average cost is called the **expected cost**.

Key point

Suppose that precisely one of n events must occur and that the events have probabilities $p_1, p_2, ..., p_n$.

Suppose that each event has a 'value', $v_1, v_2, ..., v_n$, associated with it.

The **expected value** is $p_1 v_1 + \cdots + p_n v_n$. ⊕ 1264

> In this scenario:
> • the events are bus on time and bus late
> • the probabilities are 0.8 and 0.2
> • the 'values' are £27 and £37.

'Value' might be a profit or a cost or a time. It can be anything which has a numerical value.

Example 6 Based upon an analysis of the results of premiership soccer matches, the probabilities for the goal difference at the end of a match are estimated to be as follows.

Goal difference	0	1	2	3	4	5
Probability	0.25	0.42	0.23	0.08	0.01	0.01

After a drawn game, each team receives 1 point. Otherwise, the winning team receives 3 points and the losing team receives 0 points.

a) What is the expected goal difference for a match?

b) What is the expected number of points awarded in a premiership match?

a) Expected goal difference

$= 0.25 \times 0 + 0.42 \times 1 + 0.23 \times 2 + 0.08 \times 3 + 0.01 \times 4 + 0.01 \times 5$

$= 1.21$

b) The probability of a draw is 0.25 and the probability of a team winning is 0.75.

Expected points awarded $= 0.25 \times 2 + 0.75 \times 3$

$= 2.75$

Activity 11

A garage is examining a car with a fault. From the manufacturer's manual it is known that the fault will be the result of the failure of one of two equally expensive parts, A or B. The only remedy is to replace one of the parts and, if the fault is not solved, then replace the other part.

In such cases, the garage has the common-sense policy of always replacing first the part which is most likely to have failed. In this case, from experience it is known that the fault will be the result of part A failing roughly $\frac{3}{5}$ of the time.

★ Replacing part A takes 2 hours and replacing part B takes 1 hour. Find the expected time that the repair will take.

★ Comment fully on the garage's policy.

Exercise 9G

1 A small project has four possible outcomes for the profit it generates, as shown in the probability tree diagram.

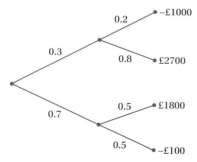

What is the expected profit of this project?

2 In a charity raffle, 2000 tickets have been sold. Each ticket costs £1 and the prize is £1000.

A philanthropist has bought 100 tickets. What is her expected loss?

3 A garage is repairing your car. The only remedy is to replace one of two parts, A and B, and if the fault is not cured, to then replace the other part. From experience, it is known that the fault will be the result of part A failing roughly $\frac{3}{5}$ of the time.

a) You are told that replacing part A will cost £160 and replacing part B will cost £120. Which part should you ask the garage to replace first? Show calculations to justify your choice.

b) What is the expected difference in cost between the two options that you have?

Consolidation exercise 9

1   For each of the following pairs of events, decide whether they are likely to be dependent or independent and explain you reasoning.

   a)   Mrs Dickens is a churchgoer; Mr Dickens is a churchgoer.

   b)   Mrs Dickens has toothache; Mr Dickens has toothache.

   c)   The weather is fine; I walk to work.

2   A letter is picked at random from the alphabet. Find the probability that the letter is

   a)   a vowel

   b)   in the word PROBABILITY.

3   Kat has probability 0.1 of being late for work and, independently, Jules has probability 0.2 of being late.

   a)   Draw a tree diagram to represent the possibilities.

   b)   What is the probability that both Jules and Kat are late?

   c)   What is the probability that just one of them is late?

   d)   Of these two people, what is the expected number who will be late?

4   A tetrahedral dice has its faces numbered 1, 2, 3 and 4. The score when the dice is thrown is the number on its bottom face. Stating any necessary assumption, find the expected score when the dice is thrown.

5   A bowler delivers off-breaks, leg breaks and straight balls in the ratio $1:1:2$. Her chances of getting a wicket with one of these deliveries are $\frac{1}{20}, \frac{1}{40}$ and $\frac{1}{40}$ respectively.

   a)   What is her chance of getting a wicket with a randomly chosen delivery?

   b)   How many wickets is she expected to get in 6 deliveries?

6   50 white balls and 50 black balls are divided between two pots. A pot is chosen at random and then a ball is chosen at random from this pot.

   a)   Can the balls be divided between the pots in such a way as to make the chance of choosing a white ball greater than 50%?

   b)   What is the maximum possible chance of choosing a white ball?

> This question shows how important your sampling method is when you want to make sure that your sample is not biased.

7   20% of a population are vaccinated against a disease. In an epidemic, the chance of a vaccinated person being infected is 5%. Otherwise, the chance is 25%.

   a)   What proportion of the population will be infected in the epidemic?

   b)   Represent this situation with a Venn diagram.

   c)   What is the probability that a person is vaccinated given that they are infected?

8   The chance of tomorrow being fine is $\frac{3}{4}$. If it is fine, the probability of the favourite winning a particular race is $\frac{4}{5}$. Otherwise, the probability of the favourite winning is only $\frac{1}{2}$.

   a)   Draw a tree diagram to represent this situation.

   b)   What is the probability that the favourite will win tomorrow?

9   In a fairground game, a player rolls a ball towards five channels as shown.

The numbers are the score for that channel.

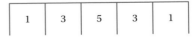

1	3	5	3	1

The ball has a probability of $\frac{2}{5}$ of going into the middle channel, $\frac{1}{5}$ of going into either adjacent channel and $\frac{1}{10}$ of going into either outside channel. What is the player's expected score?

10 30% of people in a city have been inoculated against influenza. In an outbreak of flu the chance of infection is $\frac{1}{10}$ if inoculated, but $\frac{1}{2}$ otherwise.

a) What proportion of the inhabitants will be infected?

b) What is the probability that an infected person will have been inoculated?

11 The probability that it is fine on the day of a local charity fête is $\frac{3}{4}$. The profit is likely to be £2500 if it is fine but otherwise it is only £1500. What is the expected profit for the charity?

12 The prize money in the 2015 Snooker World Championship included these prizes for the last remaining 8 players.

£300 000	Winner
£125 000	Losing finalist
£60 000	Losing semi-finalist
£30 000	Losing quarter-finalist

When the 10th seed, Stuart Bingham, reached the quarter-finals he had to play the 2nd seed, Ronnie O'Sullivan. Although Stuart Bingham was 'expected' to lose this match, his expected winnings were, of course, still greater than the £30 000 that he was guaranteed.

Based upon data from the Championship, a simple model of the probabilities was that Stuart Bingham had a probability $\frac{1}{6}$ of winning the quarter-final with a similar probability for further matches, if he made it through to any.

Before the quarter-final was played, calculate

a) the probability that he wins the championship

b) his expected winnings.

13 A contractor is undertaking a building project. There is a probability of 0.2 of bad weather and a probability of 0.1 that a sub-contractor will let the contractor down. If either or both of these problems occur then the contractor will be subject to a £25 000 penalty.

Stating any necessary assumptions, find the expected penalty.

Review

After working through this chapter you should:

- know how to use Venn diagrams and probability tree diagrams to represent events
- understand the importance of equally likely events and of other methods for determining probabilities
- know that the probabilities of all possible outcomes sum to 1
- be familiar with the notations P(A), A', A ∪ B, A ∩ B and A|B
- be able to calculate the probability of combined events such as A ∩ B for both dependent and independent events
- be able to calculate expected values such as financial loss and gain.

Investigation

Sporting strategies

In top-class tennis matches, almost all players use the same overall strategy when they serve. The first serve is extremely fast, whereas the second serve is slower and usually played with more spin. You have already seen a typical probability tree for a single point in tennis.

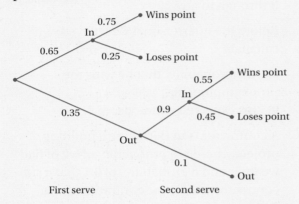

For men at the US Open, typical speeds of first serves have been about 110 mph whereas typical speeds of second serves have been about 90 mph.

Unofficial record speeds for first serves are, for men, 163 mph by Samuel Groth and, for women, 131 mph by Sabine Lisicki.

For this particular match, the probability that the point will be won by the server is

$$0.65 \times 0.75 + 0.35 \times 0.9 \times 0.55 = 0.66$$

Investigate what this probability would be if the server adopted different strategies.

1 The 'play-safe' strategy of using the 'second serve' for both serves.

2 The 'aggressive' strategy of using the 'first serve' for both serves.

Tennis players have variety in the types of first and second serve they use. For the tree diagram used in this investigation, the probability of winning a point on the first serve is

$$0.65 \times 0.75 = 0.49$$

Suppose that the player could play a really powerful first serve which has a probability of only 0.5 of being in. If it goes in, it is almost unreturnable and the probability that the server wins the point is 0.98.

How would using this serve affect the probability of winning the point with either strategy 1 or 2? Does it matter that the player's chance of serving a double fault would increase?

- Write a brief report on what your analysis has indicated so far.
- Research data for actual players and tournaments and repeat the analysis for real data.
- Can you find players for whom strategy 2 would appear to be the best strategy? Are there any players who do use this strategy sometimes?

Other sports and games

The tennis serve is good area to investigate using probabilities because a player has a clear cut decision over what strategy to use and because there are plenty of data available to study. However, this also applies to many other sports and games. You will be able to think of examples in sports and games in which you are particularly interested, but here are a few ideas of occasions where strategic decisions need to be made.

Soccer

Committing a professional foul may be morally wrong but it is called a 'professional' foul because it is believed to be right for the team's chances of winning.

Cricket

- Should the follow-on always be enforced?
- How does the order of your side's batsmen affect the chance of winning?

Analyse some circumstances in which professional fouls occur and model the associated probabilities.

Rugby

When should a penalty kicker go for goal and when for position?

Formula 1

Race strategy depends on a number of factors, including the number of pit stops, the type of tyres fitted, the probability of rain and the chances of an accident requiring the use of the safety car.

Can you model some or all of these factors with appropriate probabilities?

An infamous example of a professional foul was Thierry Henry handling the ball in a Northern Ireland v France match. This led to an equalising goal for France and, as a result, Northern Ireland were eliminated from the 2010 FIFA World Cup.

The financial significance that a professional foul can have was underlined when, in June 2015, it was revealed that FIFA had made a 5 million euro payment to the Football Association of Ireland in return for no legal action being taken against FIFA over this matter.

10 Cost–benefit analysis

Is it worth going to university?

Are wind farms 'expensive and deeply inefficient'?

In this chapter you will study a method of decision making which has widespread uses even when outcomes are uncertain. This method is called cost–benefit analysis. It involves calculating the benefit you expect to receive from carrying out a project and comparing it with the expected cost.

Why spend 1.4 billion euros landing on a comet?

Is it worth building the HS2 line?

You should know how to:

- calculate with fractions, decimals and percentages 1933
- calculate expected values (Chapter 9)
- interpret network diagrams (Chapter 8).

On the opening pages of this chapter you read some questions where good decision making is (or should be) essential. Being able to make good decisions is very important in your personal life just as it is in business and politics. In this chapter you will learn about some of the tools you can use to make good decisions.

This principle may seem like simple common sense so let's look at its implications for a genuine (but simplified) decision. This type of decision is central to running a business.

Suppose you are the Managing Director of a contracting company with expected annual revenue of around £1.3 m. The direct costs of taking on this work are £1 m and you have annual overheads of £200k.

You have the opportunity to take on either or both of two new contracts. From your Finance Director's very detailed report you can see that one contract is more complicated to manage and therefore has much higher overheads, but will bring in more revenue. Otherwise the contracts are similar.

You extract the following crucial information.

	Extra revenue	Extra direct costs	Extra overheads
Contract A	£110k	£90k	£25k
Contract B	£100k	£85k	£5k

> **Key point**
> **The cost–benefit principle**
> Take an action if, and only if, the extra benefit from taking it is greater than the extra cost.

> A company's revenue is the total (gross) income received by the company. This is also sometimes referred to as the company's 'turnover'.

> Costs that are not easily attributed to individual projects are called 'overheads' or 'indirect costs'. These often include such things as the costs of premises and of administration and personnel departments.

Activity 1

Before reading on, decide whether you should take on either or both of these contracts.

Paper 2B

What you haven't been told until now is that the Finance Director's report recommends that you should give the green light to Contract A but not to Contract B.

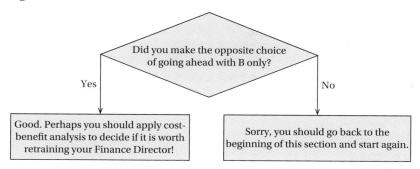

Did you make the opposite choice of going ahead with B only?

Yes

No

Good. Perhaps you should apply cost-benefit analysis to decide if it is worth retraining your Finance Director!

Sorry, you should go back to the beginning of this section and start again.

The Finance Director made his decision by assigning overheads throughout the company's operations as a fixed proportion of direct costs. For 'your' company, with annual expectations of £1m of direct costs and £200k of overheads, this proportion is:

$$\frac{200k}{1m} \times 100 = 20\%$$

He then assessed each of the company's separate operations by working out the direct costs and adding 20%. So for the two new contracts:

Contract A.   Total cost is £90 000 × 1.2 = £108 000

This is less than the £110 000 extra revenue.

Contract B.   Total cost is £85 000 + £17 000 = £102 000

This is greater than the £100 000 extra revenue.

These calculations make it appear sensible to go ahead with A but not with B, but this is definitely the wrong choice.

This example highlights the importance of the word *extra* that appears in the cost–benefit principle. This word makes your analysis much easier than it would otherwise be.

How can something so important be so simple?

You may have thought of a few problems that might arise when applying cost–benefit analysis to decisions. Critically evaluating the idea now will make it much easier for you to apply it correctly later.

Activity 2

Before reading on, make a list of possible shortcomings of cost–benefit analysis.

Some further issues may occur to you when tackling the questions in Exercise 10A. Some important issues will be summarised after the exercise.

Assuming that you did answer yes, you might be feeling very surprised that reading a single page of a text book and applying common sense has allowed you to make a better decision than a trained professional.

You can add 20% either by multiplying the original amount by 1.2 or by finding 20% of the original quantity and adding it to the original amount.   1060

This way of assigning overheads is quite a common practice in accounting. You must hope that your competitors continue to apply this mistaken methodology and win the contracts for jobs that will require a lot of administrative overheads while you apply the cost–benefit principle and take the easily managed contracts.

Key point
- Calculate the *extra* benefits, B, of taking an action.
- Calculate the *extra* costs, C, of taking the action.
- Check which of B or C is larger.

Exercise 10A

1   Some batches of a product produced by a major international company have a fault. The total cost of recalling these batches will be $137.5 billion.

The only alternative is to simply pay compensation to those customers who find they have received faulty goods and who subsequently claim compensation. An analysis of this alternative has shown that the claims will cost the company approximately $49.5 billion.

a) Assuming that no other factors need be considered, what would you recommend?

b) This question is based upon a real example. Unfortunately, the company had overlooked a major consideration. Can you think of what other factor they should have considered?

2   A design company has annual revenue of £100 000. The company has to outsource approximately 60 hours of this work each month at a cost of £40 per hour. The company could expand so that all this work could be done in-house. The only costs involved in this expansion would be:

- additional salary and associated costs of £4000 per month
- leasing another $20 \, \text{m}^2$ of office space at an annual cost of £90 per m^2
- other additional overheads of £300 per month.

This expansion would also be expected to increase the company's revenue by 20%.

a) Would expanding be worthwhile? Fully justify your answer with numerical calculations and state any assumptions that you have made.

b) What is the smallest percentage increase in revenue needed to justify the expansion?

'Outsourcing' occurs when one company buys goods or labour from another company (or from freelance workers) rather than using in-house resources. This can be a cost-saving strategy, especially if it allows a company to deal with periods of high activity whilst not having staff or machines relatively idle at other times.

3   You have decided that the value to you of going to a particular concert just about outweighs the cost.

You buy a ticket but unfortunately lose it before the concert. You can however buy another ticket. Carefully stating any assumptions that you need to make, decide what you should do.

4   In 1981, a project to improve the waterway infrastructure in the USA had already cost $1.1 billion. It was then proposed that the project should be stopped on the grounds that the total value of the project to the country was even less than the additional $0.9 billion required to complete the project.

Assuming that the valuations given above are accurate, comment on the following argument made by a supporter of completing the project.

To terminate a project in which $1.1 billion has already been invested represents an unconscionable mishandling of taxpayers' money.

Senator Jeremiah Dalton

Shortcomings of cost–benefit analysis

In answer to Activity 2, you may have thought of some or all of the following issues connected with the application of the cost–benefit principle.

> **There is considerable uncertainty in the costs and benefits of many actions.**
>
> This is a very important point which often has to be considered when carrying out a cost–benefit analysis. The technique which has to be used will be introduced in Section 10.2 and depends on using the idea of expected value that you met in Chapter 9.

> **Not all costs and benefits have monetary values.**
>
> Cost–benefit analysis is sometimes carried out using non-monetary scoring systems. However, it is much more common to attempt to give a monetary value to all relevant costs and benefits, no matter how difficult this might be.
>
> An extreme example of this is that many people feel uncomfortable about putting monetary values on things such as a human life but it is done all the time in insurance tables, compensation claims and decisions on health spending.

> **Cost–benefit analysis ignores ethical and legal issues.**
>
> This is a very contentious area. Decisions based upon cost–benefit analysis have been successfully defended in legal cases on ethical grounds that if money had been spent on a project with limited monetary benefit then the opportunity to improve human welfare through other projects would have been lost.
>
> Q For example, it sounds unethical for a medical treatment to be unavailable simply because it is expensive. You can understand why a patient's family would be deeply upset by this. However, decisions have to be made as to where money and effort are most effectively spent.

Some people make a very different type of comment about cost-benefit analysis and say that people use this method naturally and that it doesn't need to be taught. However, this isn't entirely true!

People do things for many reasons: what is easy, what they've done in the past, what everyone else does, what their parents don't do, what immediately pleases them and so on. It is true that there are strong human tendencies to behave according to a type of cost-benefit principle but it is rarely applied in a totally rational manner.

Overall, cost-benefit analysis is an extremely important technique for making good decisions. It may be difficult to quantify all costs and benefits but at least attempting to do so is more sensible than making decisions on a whim. When a cost-benefit analysis goes wrong it is usually because an important cost, such as damage to customer relations, has been overlooked rather than because of any shortcomings of cost-benefit analysis.

All actions have risks of various types. In business there are always financial risks but there are also risks such as those to the health of employees.

By planning ahead, you can try to manage these risks and the steps that you then take are called **control measures**. The following extract shows that employers have legal duties to carry out some control measures.

Advice for employers: prevention measures

Where the risks to workers are not prevented, control measures should be applied to remove or reduce the risks to workers' health. The following control hierarchy should be followed:

1. Design work processes and controls, and use adequate equipment and materials to reduce the release of dangerous substances.

2. Apply collective protection measures at the source of the risk, such as ventilation and appropriate organisational measures.

3. Apply individual protection measures including using personal protective equipment (PPE). By law this is the last resort, and should only occur where exposure cannot be adequately controlled by other means. Where PPE is given to workers, they must be trained in its use.

European Agency for Safety and Health at Work

Whether it is legally required or not, each control measure will have a cost and this has to be taken into account when performing a cost–benefit analysis.

One important way of managing a risk is to insure against that risk.

- You pay an insurance company a sum of money called the premium.
- If what you are insured against occurs then the insurance company pays you a sum of money.

As with control measures for employee safety, there are legal obligations to take out some forms of insurance. For example, employers' liability insurance is compulsory and all car owners are obliged to have at least third-party car insurance.

Any insurance you take out above the minimum requirement is an action and you should apply cost–benefit analysis to it. This will be considered more fully later in Section 10.3.

In 2009, Real Madrid was reported to have insured Christiano Ronaldo's legs for £90 million. Being insured does not stop an accident happening but, if one does occur, then at least you have compensation.

Paper 2B

Deciding whether or not to take out insurance is something that that few people have the knowledge to approach rationally. The consequence of this is that people have been mis-sold insurance on a massive scale.

> *Payment Protection Insurance (PPI) was set up to cover your debt repayments if, for example due to redundancy or illness, you were unable to work.*
>
> *But our research shows that at least two million people may have a PPI policy they couldn't ever claim on.*
>
> Which? Consumer Rights website

Thinking rationally about insurance requires some understanding of probability.

Example 1 Suppose that the chances of Real Madrid qualifying for a £90 million payout on Ronaldo's legs in 1 year were 1%. What would a fair premium for the insurance be?

A simple model for this calculation would be to use the expected value of the payout.

This would be

$$0.99 \times £0 + 0.01 \times £90 \text{ million} = £900\,000.$$

The mis-selling included policies being sold to people who were unemployed or retired and who would therefore not be eligible to claim on the policy. It is relatively straightforward to claim compensation for a mis-sold PPI policy. There is certainly no need to use one of the many 'claims management' companies that will charge commission on any successful claim. One firm charges 30% commission and so on a typical £3000 claim you would lose £900 unnecessarily.

Remember:
Expected value $= p_1 v_1 + p_2 v_2 + \cdots + p_n v_n$

The answer of £900 000 is a sensible figure but it does ignore a number of factors. For example, the insurers will expect (on average) to make a profit from the policies they sell and so would charge a higher premium than this.

As with all insurance, the person requesting the insurance could have special knowledge. For example, they might know that a particular player has an undisclosed medical problem. That is why, when you take out insurance, you are obliged to disclose all relevant information. Not doing this will make the policy invalid.

One further factor is that the insurers take money in at the beginning of the period of insurance but only pay out later (if at all). Having use of this money is a benefit for the insurer which has to be taken into account in any thorough analysis.

As you have seen, there can be considerable uncertainty in decision making and this is just as true of decisions in a person's private life as it is of their business decisions.

You can clearly see and understand some of the decision-making principles in games of chance.

This factor is just as relevant to almost all cost–benefit analyses and there are standard methods of dealing with this 'time value of money' which are not included in this course.

Backgammon

You can win a game of backgammon simply by being lucky.

However, professional backgammon players rely on steadily winning money, on average, from less skilful players.

The following is a simple but realistic choice of a type that you might often face during a series of games.

Example 2 Suppose that you are playing a game of backgammon for a stake of £1. During the game your opponent has the right to ask for the stake to be doubled. When this happens you can choose either to play on or to simply lose the original stake.

> The stake is the amount of money that the loser owes the winner at the end of the game.

In the middle of a game, suppose you calculate your chance of winning to be $\frac{3}{10}$. This means, for example, that if you played on another 100 times from this position you would expect to win on about 30 occasions and lose the other 70. If your opponent now asks to double the stake what should you do?

Work out what you would expect to lose if you played on and compare this with losing the original stake.

If you play on, your expected loss will be
$$\frac{7}{10} \times £2 - \frac{3}{10} \times £2 = £0.80$$
If you don't play on you lose £1.
You should therefore play on.

> $\frac{7}{10}$ of the time you will lose £2 and $\frac{3}{10}$ of the time you will win £2 (writing this as a loss is −£2).

The key point shows how the cost–benefit principle can be modified to deal with uncertainty.

Whilst an experienced backgammon player might find this type of cost–benefit analysis to be straightforward, most people find it only too easy to choose incorrectly when there is uncertainty. Instead of calculating (at least approximately), people are sometimes overly optimistic when they should be cautious and vice versa.

Key point
The cost–benefit principle

Take an action if, and only if, the *expected* extra benefit from taking it is greater than the *expected* extra cost.

Now consider a game played for much higher stakes.

Settle for less?

During part of a TV gameshow, the participant can choose one of four envelopes. The envelopes contain the amounts £1, £10, £1000 and £20 000.

The choice of envelope, no matter how this choice might be made to appear, is entirely one of chance.

However, the game is made more interesting by the participant being offered a sum of money, say £3000, to give up the opportunity of opening an envelope.

What should they do?

Activity 3 ●

★ Decide what you would do in this situation and justify your choice.

★ In practice, people in the type of situation described here often accept relatively low offers. Why do you think this is the case?

★ What factors other than simply reducing the expected cost will be important for the TV company when making the offer to settle?

★ What factors other than simply increasing their expected financial gain will be important for the participant when considering the offer to settle?

★ In what ways, if any, does the choice in this game differ from that in the backgammon game of Example 2?

The type of analysis used in Activity 3 is crucial when deciding how much to spend on control measures.

Activity 4 ●

The contract for a building project has a £20 000 penalty clause if the completion of the project is delayed.

The project will be delayed if either or both of the critical activities x and y are delayed.

The probability of x being delayed is 0.2. A control measure to prevent x being delayed costs £4500.

The probability of y being delayed is 0.4. A control measure to prevent y being delayed costs £5500.

★ Work out what the expected loss would be if no control measures are taken. State any assumptions that you have made.

If only x is controlled, then this costs £4500. There is then a probability 0.4 of delay and the expected penalty is therefore £8000. The total expected loss is therefore £12 500.

★ Carry out similar analyses for controlling only y and for controlling both x and y.

★ Draw up a table of results that would be useful in presenting your analysis to the managers of this project.

★ Which control measures, if any, should be taken to minimise the expected loss?

★ Suppose that, instead of minimising the *expected loss*, the contractor decided to minimise the *maximum possible loss*. What would you then recommend?

Considering the maximum possible loss in a situation is often described as 'considering the worst-case scenario'.

You may well feel that you have no intention of becoming a professional gambler and are unlikely to find yourself on a TV game show. In that case, this is the important section for you to read and understand since the benefits and costs of insurance affect everyone.

A common trick of people trying to sell insurance policies is to imply that only a totally reckless fool would *not* have the insurance. If this happens to you, then remember the mis-selling of PPI to millions of people that was mentioned earlier. It would be nearer the truth to say that only a reckless fool would have insurance if they had not thought carefully about its costs and benefits.

What you can rely on is that insurance companies will have used highly qualified analysts, considerable computer time and well-researched actuarial tables to carry out a cost–benefit analysis for the products they have on offer. This analysis will have proved that offering the products has an overall financial benefit for the insurance company.

This table shows the annual results of offering a particular policy. It is the type of situation the insurance company would be aiming for.

Premiums	Payouts	Overheads[1]	Profit
£100 000	£75 000	£20 000	£5000

[1] Consider all those sales staff, computers and imposing buildings.

This means that all the policy holders *together* will have paid out £100 000 and received £75 000 back. This difference means that you should not pay for an insurance policy unless you are getting some benefit not included in a simple expected value assessment of the risk.

You might like to think of as many of these benefits as you can, but a simple rule will cover most people's needs.

> **Key point**
> You should buy insurance under the following circumstances.
>
> - The insurance is a legal necessity.
> - You cannot afford to take the risk.

For most people this means, for example, that they *should* insure their house but *not* their mobile phone. The next example will illustrate this for a lifetime of paying insurance premiums and receiving the occasional payout.

Actuaries are employed to manage risk for businesses and governments. They evaluate the likelihood of future events and prepare tables of probabilities for insurance companies.

The table shows that the insurance company has calculated that on average they will have to pay out 75% of each premium. These payouts will, for example, cover the costs of replacing or repairing insured items.

The difference between these numbers should not be a surprise since the insurance companies have to cover their costs and can legitimately expect to make a profit from offering their products.

Personal circumstances may affect the risk you are willing to take. For example, while you're a student you may want to insure your mobile phone. However, when you're earning a salary you may consider the risk to be affordable.

Example 3 Over the years, the Smith family have paid out £5000 in insurance premiums for policies on relatively small items. The insurance company worked out the costings of these policies as shown in the previous table.

How much better off would the family be if they had invested the premiums and simply paid for replacements themselves as accidents had occurred?

The expected cost of replacements would be £5000 × 0.75 = £3750.

This means that if the family had invested the premiums they would expect to be better off by about £1250 plus any returns on their investments.

The premiums on many insurance policies can be reduced by accepting that any payout will be reduced by a set amount, called a **voluntary excess**. Since most people *will* be able to afford the loss of the set amount from any claim, these are generally worth taking.

Example 4 A compulsory insurance product has the option of a voluntary excess. For a reduction in the premium of £100 per annum, you would have to pay the first £500 of any claim. You estimate that the probability of a payout in 1 year is 0.1. Is it worth taking this option?

Compare the expected extra benefit with the expected extra cost of taking the voluntary excess.

Over 1 year, taking the voluntary excess has an extra benefit of £100.

The expected extra cost is 0.1 × £500 = £50.

Assuming that you can afford to pay the £500 should that eventuality arise, the voluntary excess should be taken because your expected savings are £50 per year.

Exercise 10B

1   The probability of bad weather seriously affecting a building project is estimated to be 0.1. The delays would cost approximately £100 000. The work could be set up differently to avoid these problems at an extra cost of £15 000.

What does a cost–benefit analysis indicate that the building company should do? Why might the company consider not following the conclusions of this analysis?

2   There is a £30 000 penalty if a contract is not carried out on time. To reduce the probability of this happening either one or both of two control measures, X and Y can be taken.

Control measure	Cost of the measure (£)	Probability of delay
X only	4000	0.3
Y only	8000	0.1
Both X and Y	12 000	0

What would you advise? Justify your answer.

3   To include a ring worth £150 on a contents policy costs an extra £5 per year. If you assume that you have a probability of 0.01 of losing the ring in the year, should you pay the extra premium? Justify your answer with a numerical calculation.

Consolidation exercise 10

1 You have to choose one of three possible options. The consequences of these options are shown below.

Option	Expected gain (£)	Maximum possible gain (£)	Maximum possible loss (£)
A	10 000	50 000	20 000
B	5000	100 000	50 000
C	2000	20 000	10 000

a) For each option describe why a person might prefer that option to either of the others.

b) List a few examples where you or other people are persuaded away from considering expected gains in order to increase the maximum possible gain.

c) List a few examples where you or other people are persuaded away from considering expected gains in order to reduce the maximum possible loss.

2 In backgammon it frequently happens that a relatively weak player makes a bad decision but then wins by a lucky roll of the dice.

Why, when one expert player had been beaten in this way, did he comment 'I want to get into similar positions as that as often as possible'? If he did get into the same position again why would he play in exactly the same way as had led to his loss?

3 The main stages in a building project are as shown, where the times are in weeks. There is a £9000 penalty if there is a delay.

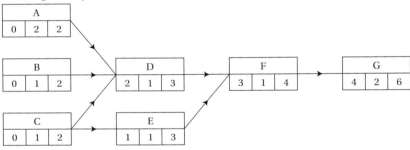

Activities A, E and G could be subject to delays of 1 week each. The probabilities of these delays and the cost of control measures to prevent the delays are shown in the table.

Activity	Probability of delay	Cost of control measure (£)
A	0.1	1000
E	0.1	500
G	0.2	500

a) Why would you not consider trying to control a delay in activity E?

b) If no control measures are taken, show that the probability of there being a delay in completion is 0.28. State any assumption that you have had to make.

c) What is the expected penalty if no control measures are taken?

d) What would you recommend on the basis of a cost–benefit analysis?

4 In the game of *Settle for less?* suppose the envelopes contain the amounts of £1, £10, £5000 and £20 000.

The participant is now allowed to discard one of the envelopes. The amount in this envelope is then shown to the player, who is usually delighted if it is one of the two smaller amounts and upset if it is one of the two larger amounts.

Suppose that the discarded envelope contains £5000. By what amount is the expected value of this game to the participant changed by discarding this envelope?

5 For a small freezer, an advertisement for cover to replace the contents is as shown.

a) What benefit is being suggested by the word 'relax'?

b) Why would a company charge £10 less for payment by Direct Debit when this delays their receipt of your money?

c) Give an example of the value for the food in a freezer and the probability of freezer failure which would make it sensible to have this insurance. State any assumptions you make.

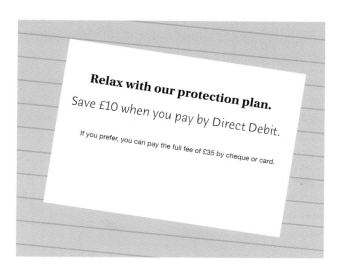

Relax with our protection plan.

Save £10 when you pay by Direct Debit.

If you prefer, you can pay the full fee of £35 by cheque or card.

Review

After working through this chapter you should:

- be familiar with the basic idea of cost–benefit analysis
- be able to calculate the expected extra benefit from taking an action
- be able to calculate the expected extra cost of taking the action
- know and understand the cost–benefit principle: take an action if, and only if, the expected extra benefit from taking it is greater than the expected extra cost
- know that selecting an action according to the cost–benefit principle is the way to optimise your overall expected gain
- understand that for a one-off decision, a person might be more swayed by other factors, but that the advantages of applying the cost–benefit principle are clear when making a long series of decisions.

Investigation

Costs and benefits of major projects
Rating university education

> *Everybody's gettin' so goddam educated in this country there'll be nobody to take away the garbage.*
>
> 'All My Sons' Arthur Miller

University education is extremely expensive (both for the state and the student) and value for money is hard to measure. Investigating the costs and benefits is an interesting activity to carry out and can involve mathematical modelling of factors such as future earnings and making Fermi estimates of other amounts.

One difficulty you may find is that many measures of how well a university is performing are based not on its teaching but on its research. On these measures, USA and UK universities do extremely well. For example, using the Shanghai ranking scheme, the USA has over half the universities in the top 100.

However, in studies of numeracy and literacy skills of graduates, the USA and UK do relatively poorly and are below the OECD average.

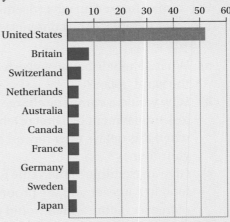

Data from www.ShanghaiRanking.com

A second potential difficulty concerns the causal connection between having a degree and having a good salary. Most studies support the idea that there is a significant financial benefit to having a degree. However, is this because the quality of the graduate's education has increased their value to employers?

> OECD is the Organisation for Economic Co-operation and Development.

It might simply be, for example, that potential employers are using 2.1 degrees as a hurdle for well-paid jobs and not that the university has educated the graduates to a standard that makes them appropriately qualified for the job.

You could investigate how universities might be ranked and costed to find out, for example, which countries have universities giving the best value to the state.

Student loans

In the USA, there are grants for the less well off and, for everyone else, loans which have to be repaid irrespective of earnings. You could research the effect of this on access to higher education. Other countries, such as Australia and now the UK, have 'income-contingent loans'. These have the benefit of not discouraging the less well off, but run the risk of high costs for the state if a large number of loans are not repaid. You could research and investigate the expected cost to the state for student loans in the UK.

Financial benefit of a degree to the student

This area is perhaps one where the costs and benefits are relatively easy to research. However, when researching figures on the internet you should be aware that financial benefits differ widely from country to country and from one academic subject to another. Your investigation should include factors such as the costs while at university, the loss of earnings during the course and the expected benefits of the education in terms of improved earnings. You will need to decide whether or not to include the effect of interest on the student loan and the effect of inflation during the long period of repaying the loan from your higher earnings.

You can also use cost–benefit analysis to explore the questions that were asked at the start of this chapter. You could also think about any local schemes to improve transport or housing and make a cost–benefit analysis.

Wind turbines

On 7th October 2014, a report by the Adam Smith Institute and the Scientific Alliance claimed that wind farms were expensive and deeply inefficient and also reduced the value of nearby housing. You could investigate one or more of these claims.

You could for example compare the expected cost of a new nuclear power station with the cost of all the wind turbines that would be needed to produce an equivalent energy output. Don't forget to include running costs and decommissioning costs.

Alternatively, imagine you have the space and planning permission for a small wind turbine. Make a cost–benefit analysis of such a project. How sensitive would any profit be to changes in government policy?

HS2

HS2 is a proposed high-speed rail line from London to at least as far north as Birmingham. What are the advantages and disadvantages of the plans to build the HS2 line?

As a start to your analysis of this project you could read the analysis carried out by the Centre for Economic Performance in its online magazine *Centre Piece Winter 2011/12*. This should be contrasted with the many statements made about the project by government spokespeople.

Space exploration

Is it possible to make a cost–benefit analysis of projects such as the Apollo program, landing on a comet or establishing a colony on Mars?

The Rosetta mission successfully landed a probe on the comet 67P. This was paid for by subscriptions to the European Space Agency. The UK contributed £240 million to the total budget.

You could research the various possible benefits of space missions such as the Rosetta mission and consider whether or not these can be given monetary values. If not, how can governments decide what projects of this type should be supported?

11 Graphical methods

What selling price for an item maximises the annual profit?

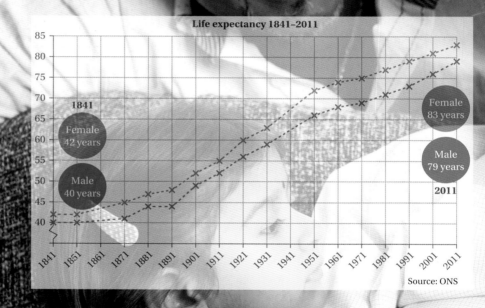

Life expectancy 1841–2011

1841
Female 42 years
Male 40 years

Female 83 years
Male 79 years
2011

Source: ONS

Will life expectancy continue to increase throughout the 21st century?

In this chapter you will study how graphs are used to show the connection between two (or more) quantities.

How can graphs help you compare two mortgages?

Do house prices increase faster than inflation?

What dimensions give the maximum volume for a box made by folding a piece of card?

You should know how to:

- write algebraic terms as simply as possible ⊕ 1178, 1179
- plot a graph from a table of values ⊕ 1396
- read information from graphs
- use the formula speed $= \dfrac{\text{distance}}{\text{time}}$ ⊕ 1121
- rearrange and solve equations. ⊕ 1154, 1925

11.1 Graphs tell a story

Suppose you tested people of different ages on the Highway Code and calculated the mean test score for each age. You could plot the mean test scores against age and join the points with a curve to give a graph like this.

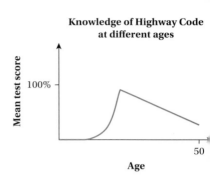

Knowledge of Highway Code at different ages

Activity 1

★ What is the story of typical learner driver's knowledge of the Highway Code according to this graph?

★ How would you expect a real graph of knowledge of the Highway Code against age to differ from the one shown above?

Key point

A graph illustrates the relationship between two quantities. Each quantity is measured on its own axis.

Each coordinate pair:

(Value of one quantity, Value of other quantity)

identifies a point on the graph.

The UK has what is called a 'progressive' tax system. This means that there is not one single tax rate. Instead, the rate increases as your income increases. The graph shows this clearly.

> The tax year 2015–16 was from 6 April 2015 to 5 April 2016. Tax rates usually change every year.

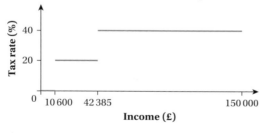

Income tax 2015–16 (For income up to £150 000)

Activity 2

★ For the income tax graph, explain the significance of the points:

(10 600, 0)

(42 386, 40)

★ What is the tax rate for incomes over £150 000?

To answer the questions in the activity, you needed to read the axis labels and scales.

Key point

A graph must have:

- a clear title and clearly labelled axes
- a clearly marked scale on each axis, increasing in equal-sized steps.

The individual points on a graph are usually joined up to form a single line or curve. However, in the income tax graph there are sudden jumps (called **discontinuities**) where the tax rate 'jumps' to another level. This graph also illustrates how extending a graph beyond the points plotted from known values (called **extrapolation**) can be very unreliable.

> Could you accurately predict the tax rates for incomes over £150 000 from the graph?

The income tax graph is a graph of 'tax rates against income'. This description tells you which quantity is on which axis. 'Tax rates' are plotted on the vertical axis (*y*-axis), against 'income' on the horizontal axis (*x*-axis).

Activity 3

In 1894, The Times newspaper predicted that

> In 50 years, every street in London will be buried under 9 feet of manure.

★ Sketch a graph of 'depth of manure against time' to illustrate this prediction. Label your axes clearly.

★ Why was extrapolation so inappropriate in this case?

The table shows share prices in some FTSE 100 companies at the end of eight successive trading periods in May 2015.

> The FTSE 100 is an average of the share price of the 100 companies with the highest market value.

FTSE 100 Index	6950	6900	6875	6925	6950	6955	6975	6930
Trading period	1	2	3	4	5	6	7	8

Activity 4

★ When plotting a graph of the FTSE 100 against trading period what is the disadvantage of using a uniform scale (going up in equal sized steps) from 0 to 7000 on the *y*-axis?

★ How would using each of these scales on the *y*-axis change the overall appearance of the graph and what it appears to show?

> You could try plotting the graphs to compare them.

0 to 7000   6000 to 7000   6800 to 7000

If you are not using a uniform scale from 0 on an axis, then you should indicate this clearly, as shown on this graph.

The zig zag line shows that the values from 0 to 6874 are not shown on the axis.

Key point

A zig zag line like this on an axis shows there is a break or gap in the values shown.

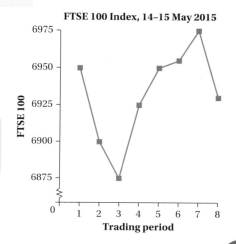

FTSE 100 Index, 14–15 May 2015

This graph shows how house prices have changed over time. To allow a fair comparison, the graph uses the 2011 value of money for all prices.

> A trend line shows the general pattern that the points on the graph seem to be following.

The graph shows that although house prices have fluctuated quite wildly, the general trend has been upwards. This tells you that house prices from 1975 to 2011 have risen faster than general inflation.

House prices in real terms

Example 1 For the house price graph, find the average yearly increase in house prices and describe the trend line.

Read the prices from the ends of the trend line.

On the trend line, the house prices changed from £65 000 in 1975 to £170 000 in 2011, 36 years later.

This is a change of £170 000 − £65 000 = £105 000

Calculate the mean change.

This is an average change of $\frac{£105\,000}{36} \approx$ £2900 each year.

The trend line shows that house prices increased, on average, by approximately £2900 every year until 2011.

> The average change each year is the gradient of the graph. The gradient of a line tells you how much the y-value increases for every 1 unit increase in x.

Key point

Linear graphs are straight-line graphs. The equation of a linear graph is $y = mx + c$.

- m is the gradient.
- c is the intercept with the y-axis. 1153

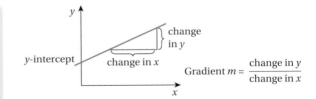

For the house price graph, you can use $y = mx + c$ to find the equation of the trend line.

Example 2 For the house price graph, y is the average house price x years after 1975.

Find the equation of the trend line. ⊕ 1957

Find the coordinates of two points on the line.

The line passes through $(0, 65\,000)$ and $(36, 170\,000)$.

Calculate the gradient of the line.

$$\text{Gradient} = \frac{\text{change in } y}{\text{change in } x} = \frac{170\,000 - 65\,000}{36} \approx 2900$$

Find the value of c, where the graph meets the y-axis.

$c = 65\,000$

Substitute your values for m and c into $y = mx + c$.

The equation of the trend line is $y = 2900x + 65\,000$.

This equation of the trend line is a **mathematical model** for house prices during this period. The gradient, 2900, tells you the average house price increase every year.

Example 3 A linear graph passes through the points $(0, 10)$ and $(5, 25)$. Find:

a) the y-intercept

b) the gradient

c) the equation of the graph

d) the value of y such that $(12, y)$ is a point on the linear graph.

Sketch a pair of axes, mark on the two points and join them with a straight line.

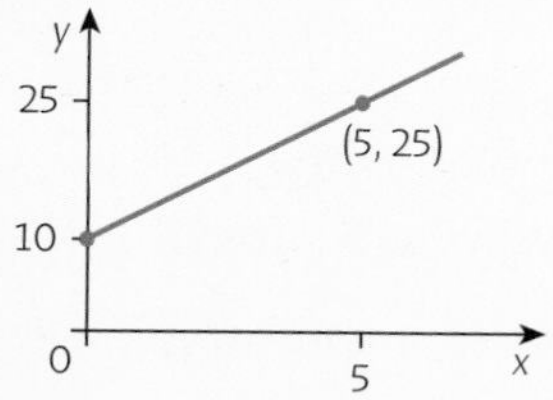

The y-intercept c is where the graph crosses the y-axis.

a) y-intercept $= 10$

b) Gradient $m = \dfrac{\text{change in } y}{\text{change in } x} = \dfrac{25 - 10}{5 - 0} = 3$

c) $y = 3x + 10$

d) Substitute $x = 12$ into the equation of the line.

$y = 3 \times 12 + 10 = 46$

Example 4 At the start of a drought, a reservoir is full and contains 200 000 gallons of water. After one week, the reservoir contains 130 000 gallons.

a) Stating any necessary assumptions, sketch a graph of volume (*V* gallons) against time (*t* days). Find the equation of your graph.

b) Describe what information any numbers in your equation tell you.

c) After how many days will the reservoir be half empty?

---

a)

> Assume that the volume can be modelled by a linear equation in time *t*.

You only have values for volume and time.
Draw the axes, mark the points you know and join them with a straight line.

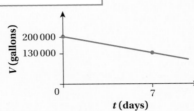

The graph of a linear equation is a straight line.

The phrase 'can be modelled by a linear equation' means 'can be modelled by a straight line graph'.

The gradient of the line is $\frac{130\,000 - 200\,000}{7 - 0} = -10\,000$.

The y-intercept is 200 000.

The equation of the line is $V = -10\,000t + 200\,000$ or $V = 200\,000 - 10\,000t$.

A graph that slopes down from left to right has negative gradient. ⊕ 1312

b) 200 000 is the initial volume.

The gradient of −10 000 shows that the reservoir is **losing** 10 000 gallons per day.

c) Half the initial volume is 100 000 gallons. Substitute this into the equation.

$$V = 200\,000 - 10\,000t$$
$$100\,000 = 200\,000 - 10\,000t$$

Solve to find *t*.

$$10\,000t = 100\,000$$
$$t = 10$$

Write a clear sentence to answer the question.

Assuming the drought continues, the reservoir will be half empty 10 days after the start of the drought.

Exercise 11A

1 The graph shows the daily sales of a soft drink for the period 2010 to 2015.

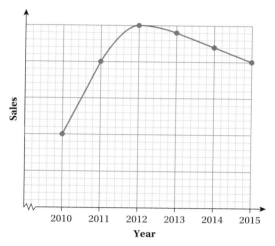

a) Criticise this graph.

Hint: Does it have all the information you need to be able to read values from it and decide what they mean?

b) When were the daily sales at a maximum?

c) In the year 2010, the daily sales were 6000 cans per day. What was the maximum number of daily sales?

2 The Highway Code gives the following thinking distances, braking distances and stopping distances for cars travelling at various speeds.

Speed	Thinking distance	Braking distance	Stopping distance
(V km/h)	(T m)	(B m)	(D m)
32	6	6	12
48	9	14	23
64	12	24	36
96	18	55	73
112	21	75	96

a) Can you spot and state a simple relationship between T, B and D?

b) Plot a graph of D against V.

c) Use your graph to estimate a safe distance to leave between you and the car in front when travelling at 80 km/h.

d) Estimate the safe distance in part c) as a number of car lengths.

3 a) Plot a graph of T against V for the data in question 2.

b) The thinking distance, T metres, is the distance the car travels before the driver reacts and starts to brake. How many seconds does the Highway Code assume it takes for a typical driver to react?

Hint: You will need to use the formula linking speed, distance and time.

4 The braking distance, B metres, is the distance the car travels before stopping once the driver has applied the brakes. Here is the graph of B against V.

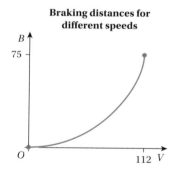

Braking distances for different speeds

The formula used to calculate B is
$$B = aV^2,$$
where a is a constant. Find the value of a.

Hint: Substitute the values you know, e.g. $V = 112$ and $B = 75$, into the formula.

5 The table gives the population of Britain for different dates in history.

Year (AD)	Population (millions)
200	2
400	1
1100	2
1300	5
1400	3
1700	5

a) Plot a graph for population against year.

Q b) What historical events coincide with the two periods when the population fell significantly?

Q c) The estimated population for 1100 AD is regarded as being reasonably reliable. Why is that?

d) If you were to extrapolate from the data shown on your graph, what would your estimate be for today's population? Why is this estimate so inaccurate?

6 A linear graph passes through the points (1, 5) and (4, 14).

a) Draw a pair of axes on squared or graph paper. Plot the points and draw a graph of the line.

b) Find the y-intercept.

c) Calculate the gradient.

d) Write down the equation of the line.

7 The graphs below are shown without any scales.

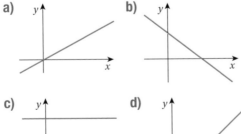

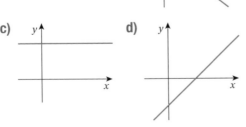

Match each graph to an equation from this list.

$y = -2x$ $y = 4$

$y = 2x - 9$ $y = 3x + 8$

$y = -x + 7$ $x = 2$

$y = 3x$ $y = -x - 5$

8

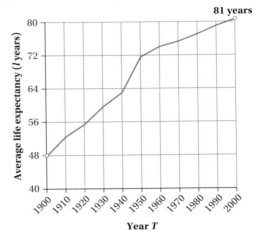

a) For females, from 1900 to 2000 the graph of average life expectancy against year was roughly linear. Write an equation to model the life expectancy (l years) in terms of years T since 1900.

b) Suggest some reasons for and against the idea that this trend in female life expectancy will continue throughout the 21st century.

9 Different weights are hung from a spring. The extension (increase in length) of the spring is measured for each load (weight). The graph shows the results.

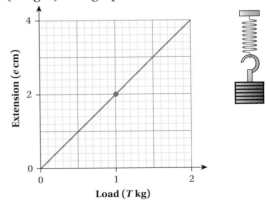

a) From the graph, find the extension of the spring for a load of 1.5 kg.
b) What might be a problem in extrapolating to much greater loads?
c) Find an equation connecting *e* and *T*.

10 The graph shows the results of one study into the probability of scoring a goal when shooting from a central position and faced only with a goalkeeper who is on the goal line.

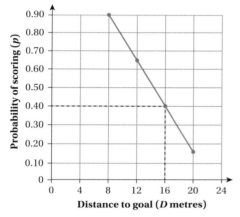

a) What is the gradient of the trend line?
b) Find the coordinates of the two points where the line crosses the axes.
c) What, if any, is the significance of these points?
d) Find the equation for *p* in terms of *D*.
e) What percentage of penalty kicks (11 metres from goal) can be expected to be successful?

11 *Parkrun* organises free, timed, 5 km runs all over the world. These are open to runners of all ages and abilities. Many of these runners try other longer distances, especially half and full marathons.

Parkrun has found that they can predict a runner's performance fairly accurately at these longer distances, assuming the runner does the necessary training.

The table shows three runners' times over the different distances.

5 km (minutes)	Half-marathon (hours:minutes)	Marathon time (hours:minutes)
15	1:08	2:27
18	1:23	2:58
30	2:18	4:58

a) Plot a graph of half-marathon times against 5 km times.
b) Jenny runs 5 km in 24 minutes. Use your graph to predict her half-marathon time.
c) Plot a graph for marathon times against 5 km times.
d) Ranjit's 5 km time is 21 minutes. Use your graph to predict his marathon time.

Without having to plot any points, you should now know that a graph with equation of the form $y = mx + c$ is a straight line. Two other types of graph you will need to recognise are quadratic and cubic graphs.

A quadratic equation has a term in x^2 and no higher powers of x.

A cubic equation has a term in x^3 and no higher powers of x.

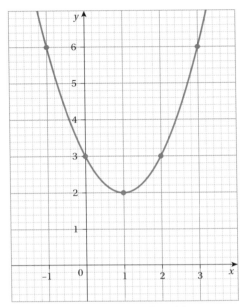

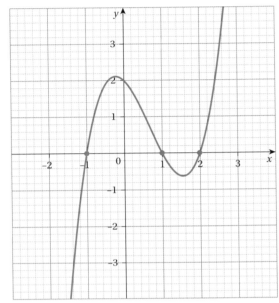

This is a quadratic graph, with equation $y = x^2 - 2x + 3$.

This is a cubic graph, with equation $y = x^3 - 2x^2 - x + 2$.

Any expression of the form $mx + c$ is called a **linear function** of x. Similarly, expressions such as $x^2 - 2x + 3$ and $x^3 - 2x^2 - x + 2$ are called quadratic and cubic functions of x, respectively.

Activity 5

Use a graph plotting package or graphical calculator to plot and investigate the graphs in this activity.

★ Investigate the shape of curves with quadratic equation $y = x^2$ and $y = x^2 + c$ (where c is a positive or negative number). Predict what the graphs of $y = x^2 + 5$ and $y = x^2 - 2$ will look like. Plot them to check your predictions.

★ For different values of a, b and c, investigate the shape of curves with quadratic equation $y = ax^2 + bx + c$. You should start by having b and c both zero and then by having just one of b and c zero. You should consider both positive and negative values of a. Can you now sketch any curve with equation $y = ax^2 + c$ without needing a graph plotter?

You can use a graph plotting package or graphical calculator to plot graphs from their equations. Make sure you know how to use one.

★ Investigate the shape of curves with cubic equation $y = x^3$ and $y = x^3 + c$ (where c is a positive or negative number). For different values of a, b, c and d, investigate the shape of curves with a cubic equation $y = ax^3 + bx^2 + cx + d$. Can you now sketch any curve with equation $y = ax^3 + d$ without needing a graph plotter?

Key point

A quadratic curve has an equation of the form $y = ax^2 + bx + c$. The y-intercept is c and the curve is a parabola.

If the coefficient of x^2 is positive, the curve is $\cup$ shaped.

If the coefficient of x^3 is negative, the curve is $\cap$ shaped.

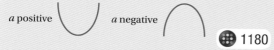

a positive a negative

⊞ 1180

A cubic curve has an equation of the form $y = ax^3 + bx^2 + cx + d$. The y-intercept is d.

The diagrams show the different shapes of the graph when the coefficient of x^3 is positive and when the coefficient of x^3 is negative.

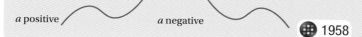

a positive a negative

⊞ 1958

> A parabola is a U shaped curve.

> The coefficient of a letter term is the number it is multiplied by.
> In the expression $3x^2 - 4x$, the coefficient of x^2 is 3 and the coefficient of x is -4.

Example 5 If a government set a tax rate of $T\%$, then the revenue £R billion raised through taxes can be modelled by a quadratic equation in T.

a) Draw the graph of $R = 100T - T^2$ for values of T between 1 and 100.

b) Explain whether or not this graph is a sensible representation of the revenue.

> $R = 100T - T^2$ has a term in T^2 and no higher power, so it is a quadratic equation, with variables T and R (instead of x and y).
> The graph axes will be labelled like this:

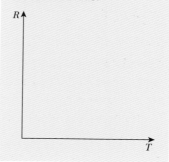

a) Look at the equation of the graph to predict its shape.

$R = 100T - T^2$ is a quadratic equation and the coefficient of the T^2 term is negative, so the graph will be $\cap$ shaped.

Make a table of values. Include $T = 0$ and $T = 100$, and some values in between. ⊞ 1168

T	0	20	40	50	60	80	100
R	0	1600	2400	2500	2400	1600	0

Plot the points and join them with a smooth curve.

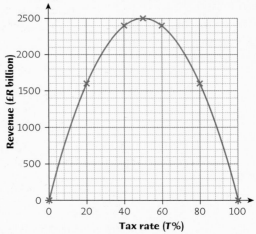

b) To see if the graph is a sensible representation, read the 'story' of the graph.

If a government sets a tax rate of 0% then tax revenue will be zero. If they set a tax rate of 100% they would expect to raise no revenue because there would be no incentive for people to work if all their income went in tax. So values of T less than 0 or larger than 100 have no meaning.

Otherwise the graph appears sensible. The revenue drops when $T > 50$ because as tax rates get higher people put more effort into tax avoidance measures.

Example 6 A manufacturer makes open-topped boxes from 40 cm by 30 cm sheets of cardboard. To make the net for the boxes, four same-sized squares of side x are cut from the corners as shown.

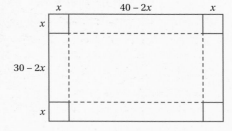

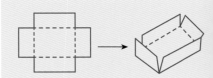

a) Find a formula for the volume, $V\,\text{cm}^3$.

b) Draw a graph of V against x.

c) Use your graph to estimate the value of x that gives the box with the maximum volume.

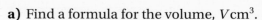

The volume of a cuboid is length × width × height.

$V = (40 - 2x)(30 - 2x)x$

To draw the graph, either make a table of values for V and x, or use a graphical calculator. ⊕ 1316

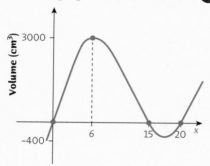

> If you expanded the brackets in $V = (40 - 2x)(30 - 2x)x$ you would get $V = 1200x - 140x^2 + 4x^3$, which is a cubic expression with positive coefficient of x^3.

The maximum volume occurs at $x \approx 6$.

For maximum volume, the squares cut out at the corners should have side 6 cm.

Example 7 For a basic rate taxpayer, draw a graph to illustrate the proportion of a person's income that is paid in tax.

> ▶ Assume that the personal allowance is £10 600 and that the basic rate of 20% is paid on taxable income of up to £31 785.

For an income of £10 600 no tax is paid and so the proportion is 0%.

For an income of £$(10\,600 + 31\,785)$ = £42 385 the tax paid is £31 785 × 0.2 = £6357

This is $\dfrac{6357}{42\,385} \times 100 = 15.0\%$.

You can check that the shape of the curve is as shown above by calculating the tax for different levels of income. It is best to set up a spreadsheet to do these repetitive calculations.

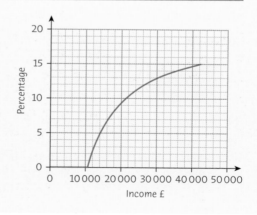

Activity 6

★ Set up a spreadsheet to calculate the proportion of a basic rate taxpayer's income that is paid in tax.

Exercise 11B

1 Decide if each graph is linear, quadratic, cubic or none of these.

a)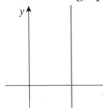

b)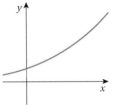

c)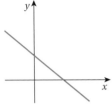

d)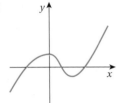

e)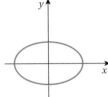

f)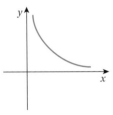

g)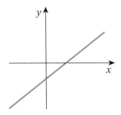

h)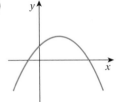

i)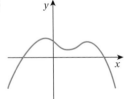

Hint: Compare them to the shapes of the graphs you investigated in Activity 5.

⊕ 1172

2 Sketch the curves.

a) $y = x^2$
b) $y = x^2 + 1$
c) $y = x^2 - 5$
d) $y = -5x^2$

3 a) Copy and complete this table of values for the graph of $y = x^3 - x^2 - 4x + 4$.

x	−3	−2	−1	0	1	2	3
y	−20						10

b) Hence draw the graph of $y = x^3 - x^2 - 4x + 4$.

Hint: Plot the coordinates from your table and join them with a smooth curve.

c) Use your graph to estimate the values of x for which $y = 3$.

Hint: Draw a line across the graph through the point $y = 3$. Where does it cross the graph line?

4 The path of a shot-put can be modelled by the equation

$$y = -0.08x^2 + x + 1.7$$

where x metres is the horizontal distance travelled and y metres is its height above the ground.

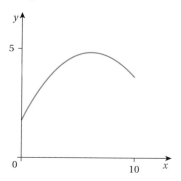

a) Make a table of values for $0 \leqslant x \leqslant 15$.

b) Plot an accurate graph of y against x.

c) Explain why the intercept is not at the origin $(0, 0)$.

d) Use your graph to find the distance of the throw.

e) What is the maximum height reached by the shot?

5 The number of items sold usually depends upon the selling price. For one item, market research predicted these sales.

Selling price (£P)	Number sold (N per year)
20	2000
30	1500
40	1000
50	500

a) Plot a graph of N against P. Draw the vertical axis up to £3000.

b) What type of relationship is shown by your graph?

c) Find a formula for N in terms of P.

6 Each of the items in question **5** costs £15 to produce and market.

a) Explain why the annual profit is $£N(P - 15)$.

b) Assume that the annual profit A is given by

$$A = £(3000 - 50P)(P - 15)$$

Plot a graph of annual profit A against selling price P.

c) Find the selling price that maximizes the annual profit.

For obvious reasons the clearance underneath a sea-going vessel is crucial for successful sailing.

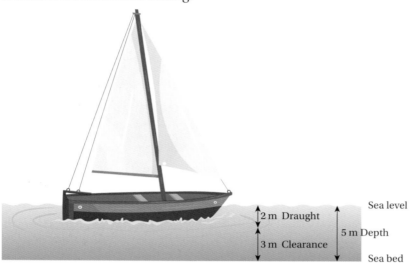

Sailors used to measure depth by lowering a rope with a weight on it over the side of the ship until the weight hit the seabed. Nowadays ultrasonic depth sounders can accurately map the seabed.

Nowadays, it is easy to measure depth and so to calculate

Depth − Draught = Clearance

The Costa Concordia cruise ship ran aground on the Italian coastline with more than 4 000 passengers and crew on 13 January 2012. Thirty two people died.

So, why do sailors still run aground, endangering both their vessels and crews? The simple reason is that they underestimate the effect of tides. Around the UK, the depth of the sea can vary by as much as 10 metres between high and low tides.

Tide tables show the times of high and low tides, and the depth at a given position.

Here is a tide table for the Solent on 19th April 2015. Your vessel may be in deeper or shallower water but sea depth will still be affected by the same tides.

	Depth (m)	Time (hr:min)
High tide	4.9	23:45
Low tide	0.4	05:26
High tide	4.8	12:05
Low tide	0.4	17.47

Tidal graphs given an even clearer picture. This one is for the same day as in the table.

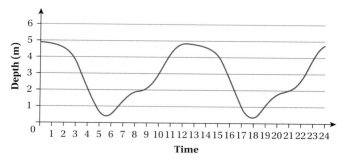

Activity 7

The diagram at the start of this section shows a yacht moored in the Solent at lunchtime on 19th April 2015. What would you advise?

Example 8
Sea-going vessels must time their arrivals and departures in ports so that they can enter and leave harbours with sufficient clearance. One vessel needs a depth of over 3 m. Use the tidal graph to find when it would be safe for it to enter or leave the harbour.

Draw a line across at Depth 3 m. Read off the times from the horizontal axis.

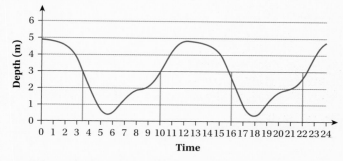

It will be safe to enter and leave the harbour when the depth is greater than 3 m: before 03:30, between 10:00 and 16:00 and after 22:00.

Key point
To find the approximate solution to two equations, find the x-coordinate where the two graph lines cross. ⊕ 1169, 1319

Example 9 A company's market share, M%, is modelled by the equation
$M = aT^2 + b$, where T is the number of months after December 2014.

Here is a graph of M against T

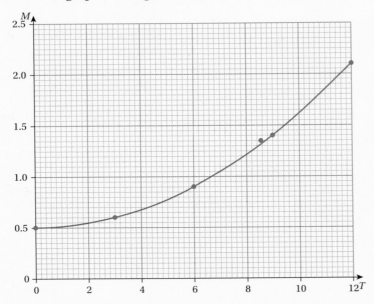

a) Use the graph to find when the market share reached 2%.

b) Find the values of a and b.

c) What is the prediction for the market share in January 2017?

Comment on your estimate.

---

a) Draw a line at $M = 2$ and read off the T value where it crosses the curve.

$T = 11.5$.

11.5 months after December 2014 is December 2015.

b) b is the y-intercept, read the value from the graph.

$b = 0.5$

Choose a point on the graph and read its coordinates.

$(3, 0.6)$ When $T = 3$, $M = 0.6$

Substitute these values into the equation $M = aT^2 + 0.5$ and solve.

$9a + 0.5 = 0.6$

$\qquad 9a = 0.1$

$\qquad a = 0.0\dot{1}$

> $0.0111111…$ can be written $0.0\dot{1}$. The dot over the 1 shows the 1 recurs.

c) In January 2017, $T = 25$

Substitute into the equation $M = 0.0\dot{1}T^2 + 0.5$

$M = 0.0\dot{1} \times 25^2 + 0.5 = 7.4$

The prediction may not be accurate. Extrapolating this far outside the range of the data given may not be justified.

Exercise 11C

1 The graph shows part of the curve $y = x^3$ and the lines $y = 110$ and $y = 100 - 20x$.

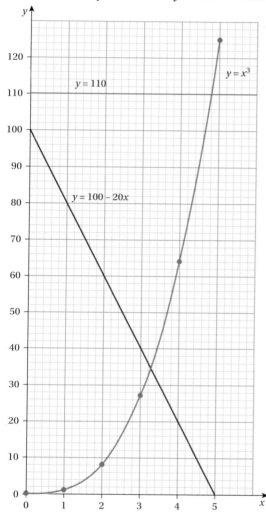

a) Use the graph to find approximate solutions to the equations
   i $x^3 = 110$
   ii $x^3 = 100 - 20x$
b) To solve the equation $x^3 + 10x - 50 = 0$, what line would you need to draw on the graph?

Hint: Rearrange the equation to $x^3 = \ldots$.

c) What is an approximate solution of $x^3 + 10x - 50 = 0$?

What is another name for the line $y = 0$?

2 a) Plot a graph of the equation $y = x^2$ for $-10 < x < 10$.
  b) Draw lines on your graph to solve the equations
     i $x^2 = 24 - 5x$
     ii $x^2 + 2x - 19 = 0$

3 a) Find three pairs of x and y values such that $x + y = 7$.
     Plot your points (x, y) and join them to draw the line $x + y = 7$. Extend your line to the edge of your coordinate grid.
  b) On the same grid, plot the line $2y = 3x + 5$.

You may wish to rewrite $2y = 3x + 5$ in the form $y = \ldots$.

c) Find values of x and y where $y + x = 7$ and $2y = 3x + 5$.

Two or more equations which must be satisfied at the same time are called **simultaneous equations**.

4 a) As the price of a commodity increases, the supply of the item also tends to increase. Explain why this is so in the case of oil.
  b) As the price of a commodity increases, the demand for the item tends to decrease. Explain why this is so in the case of oil.
  c) An item is priced at £P. The annual supply is S thousand, where
     $$S = -5 + P$$
     The annual demand is D thousand, where
     $$D = 100 - 0.5P$$
     By plotting suitable graphs, find the market price where demand exactly matches supply.

Consolidation exercise 11

1 Clifton Suspension Bridge

The main chains of the Clifton Suspension Bridge form two identical curves linking the tops of the towers. Stating any necessary assumptions, find the equation of one of these curves.

> Look back at Exercise 11A question **4**.

2
The graph shows the chess ratings of Magnus Carlsen, winner of the World Championship in both 2013 and 2014, and Viswanathan Anand, former World Champion and the first Indian grandmaster.

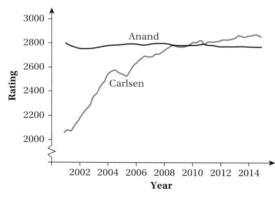

a) Describe how the chess ratings of these two players have changed over time.

b) Can you think of a reason which might explain the shape of Carlsen's graph?

c) Find equations that could model each player's rating (R) during this period in terms of the number of years (Y) before 2014.

d) Analyse the accuracy, or otherwise of your equation for Carlsen.

e) A player who is p rating points above another player is expected to score approximately $(50 + \frac{p}{8})\%$ in a match between them. What would this give for Carlsen's likely score in the 2014 World Championship match with Anand?

3
A manufacturer is considering producing a new low-cost sports bicycle. The set-up costs would be £80 000 and each bicycle would cost £100 to manufacture. Market research indicates that at a price of £150 the expected sales would be 1500 but at a price of £250 the expected sales would fall to 500.

a) Model the number of sales, S, as a linear function of the price £P.

b) Draw a graph of the manufacturer's costs against P.

c) Explain why the revenue from sales is given by $3000P - 10P^2$ and add a curve to represent this revenue on your graph.

d) Use your graph to find
   i   the lowest price for the bicycle at which the manufacturer would break even
   ii the price the manufacturer should choose to maximise profits.

4
The table gives the costs for two-year fixed interest mortgages from two companies.

Company	Annual interest	Arrangement fee
Afford Mortgages	2%	£1000
Safe Bank	2.25%	£500

a) To borrow £P from Afford Mortgages, the cost of the mortgage $C = £(1000 + 0.02P)$. Find an equation for the cost of the mortgage from Safe Bank.

b) On the same graph, draw lines representing the costs of the two mortgages.

c) Depending upon the size of the mortgage needed, which company would you recommend?

5 Samsung's market share of worldwide smartphones, between 2009 and 2012, featured in a legal action brought against the company by Apple. A good model for this market share, $M\%$, is a quadratic equation of the form

$$M = aT^2 + b,$$

where a and b are constants and T is the number of years after March 2009.

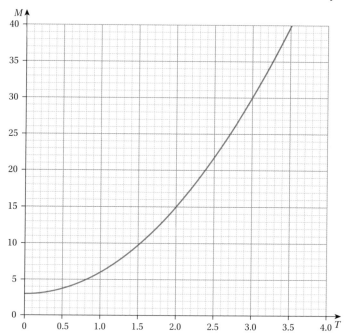

a) Use the graph to estimate when Samsung's share reached 10%.

b) Find the values of a and b.

c) What was Samsung's market share in mid-2015?

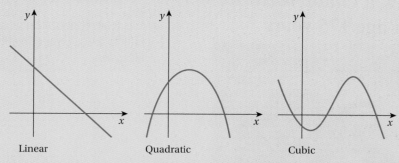

Investigation

Size matters

If you have ever held a small animal, such as a mouse or a bird, in your hand you will have noticed that its heart rate is much more rapid than your own.

🔍 Set up a spreadsheet for the weights and heart rates of various animals. For example:

	Animal	Weight (grams)	Heart rate (beats per minute)
1	Canary	20	1000
2	Human	70 000	72
3	Horse	450 000	38

Heart rate is just one of many general features of the natural world that seem to be related to size. For whatever features you choose, plotting graphs of pairs of features, such as heart rate against weight, may help you to spot relationships between them.

Cat survives fall from high-rise apartment

Ants are 50 times stronger than humans

Harriet, a Giant Tortoise, dies at age 175 years

You could investigate how size affects fall survival, strength, lifespan.

Many complicated physiological factors determine heart rate, terminal velocity in falls, strength and lifespan. However, some important general results can be based upon very simple models of animal sizes.

Consider two cubes, one with twice the side length of the other.

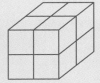

Length (L cm)	Surface area (A cm^2)	Volume (V cm^3)
1	6	1
2	24	8

When the length is doubled ($\times 2$), the surface area is multiplied by 4 ($\times 2^2$), and the volume is multiplied by 8 ($\times 2^3$).

The loss of heat from a warm-blooded animal is in proportion to its surface area.

The metabolic rate (the amount of 'fuel' it can burn to generate heat) is in proportion to its volume.

Using the cube model and units of centimetres:

- for the small cube the surface area is 6 times the volume
- for the cube twice the size, the surface area is 3 times the volume.

So it looks as if the smaller animal would need twice the metabolic rate of the animal twice its size in order to keep warm.

Put another way, one animal 'twice the size' of another could keep itself warm with a metabolic rate half of that of the smaller animal.

This suggests that size $\times$ metabolic rate might be constant.

For the cubes, a simple relationship is $V = L^3$, a cubic equation. Weight is proportional to volume, so this suggests that for two differently shaped animals, a good quantity for you to use for 'size' is $\sqrt[3]{\text{Weight}}$.

Metabolic rate is the speed of the chemical reactions in the body that release energy from food.

Similarly, you can use heart rate as a substitute for metabolic rate.

Use your spreadsheet to investigate if $\sqrt[3]{\text{Weight}} \times$ Heart rate is roughly constant.

Is it the same for some types of animal but not for others? Write a report of your findings or perhaps prepare a PowerPoint presentation.

Investigate other simple models connected with size, for example how size affects terminal velocity, strength and lifespan.

12 Rates of change

How does increasing the length of a pendulum change the time it takes to swing back and forth?

What is meant by instantaneous speed?

How large is the acceleration of a space rocket?

In this chapter you will study the connection between rates of change and the gradients of graphs.

When is a species predicted to be out of danger?

How can an airline minimise fuel usage?

When will a ball thrown straight up in the air reach its maximum height?

You should know how to:

- find the gradient of a straight line 1312
- find the equation of a linear graph (chapter 11) 1957
- recognise the shape of a quadratic graph (chapter 11). 1959

Understanding rates of change is the gateway to understanding many advanced applications of mathematics; for example, in economics, biology and physics. The study of rates of change starts with the idea of the gradient of a straight line.

Example 1 The graph shows an electrician's charges.

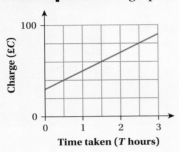

Explain what the *y*-intercept and gradient mean in this context.

The *y*-intercept shows how much the electrician charges for zero hours' work.

The gradient shows how much his charge increases every hour.

The *y*-intercept, £30, is the call-out charge.

The gradient represents the hourly rate, £20 per hour. 1312

The units of the gradient are pounds per hour.

Tradespeople who go out to workplaces or homes to install equipment or do repairs usually charge a callout fee, so they are paid for making the journey.

The hourly rate is the rate of change of *C* with respect to *T*.

Example 2 The graph shows the cooking time for a shoulder of lamb.

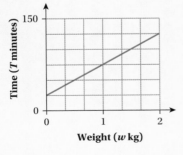

a) Explain the meaning of the gradient of the graph.

b) How would a cookery book describe this cooking time?

The units of the gradient are minutes per kilogram

a) The gradient is 50 minutes per kilogram. This is the time needed for each extra kilogram of weight.

b) Allow 50 minutes cooking time per kilogram plus an extra 25 minutes.

> The gradient is the rate of change of T with respect to w.

Example 3 For a particular department store, the graph shows how the number, n, of DAB radios sold per month depends upon the selling price, £S.

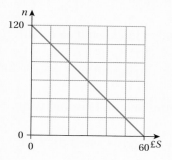

> A graph like this has negative gradient.
>
>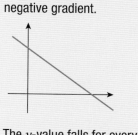
>
> The y-value falls for every increase in x.

a) Write down an equation for n in terms of S.

b) State the gradient of the line and explain what it means in this context.

---

a) Use $y = mx + c$. Read the y-intercept (c) from the graph and calculate the gradient. ⊞ 1312

$c = 120$

$\text{Gradient} = \frac{-120}{60} = -2$

$n = -2S + 120$ or $n = 120 - 2S$

b) The gradient is -2 and its units are 'sales per pound'. It means that sales fall by 2 for every £1 increase in price.

> The rate of change of n with respect to S is -2 sales per pound.

Key point

The gradient of a graph represents the rate of change of the quantity on the vertical axis with respect to the quantity on the horizontal axis.

Exercise 12A

1   Calculate the circumferences of a few circles of different radii, with all lengths in centimetres. ⊞ 1088

   a) Plot your results on a graph of circumference against radius.

   b) What is the gradient of your graph? What does this represent as a rate of change?

2   Kim's parents measured her height every birthday. The graph shows the results.

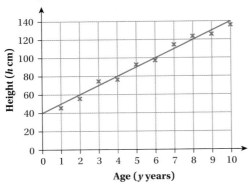

a) What is the gradient of the line that closely matches the trend of her height? Explain what this gradient represents.

b) The height of Kim's twin sister, Courtney, can be modelled by the equation $h = 9y + 48$.
   i   Compare Kim's and Courtney's heights.
   ii At one of their birthdays, Kim and Courtney were the same height. How old were they then?

c) Can you predict the difference between Kim's and Courtney's heights when they are both 30 years old?

3   This graph can be used for converting temperatures in °C into °F and vice versa.

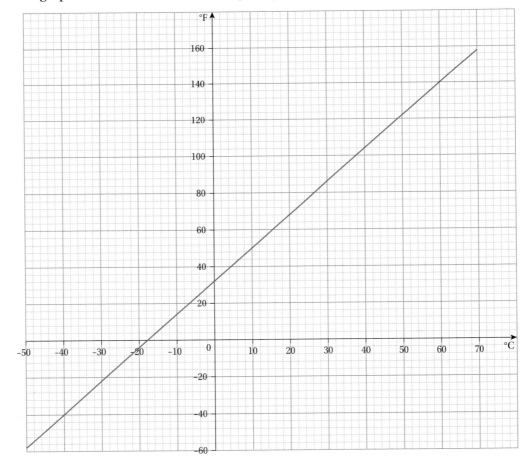

a) Use the graph to find the missing values in this table.

(i)	Coldest inhabited town (Siberia)		−50 °F
(ii)	Danger zone for food poisoning bacteria	5 °C–60 °C	
(iii)	Room temperature	20 °C	
(iv)	Air temperature in Bangkok		93 °F
(v)	Rock temperature in Mponeng gold mine		150 °F

b) What is the gradient of the graph? What does this gradient represent?

c) What would be the gradient of a graph of temperatures in °C plotted against temperatures in °F?

d) To convert a temperature in °C into a temperature on the Kelvin scale, you simply have to add 273.15. What would the gradient be for a graph of temperature in °F plotted against temperature in Kelvin?

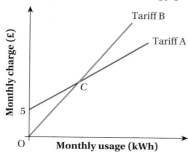

4 The graph shows an energy provider's two tariffs, A and B.

Energy use is measured in kilowatt-hours (kWh). 1 kWh represents using 1 kilowatt of energy for one hour.

a) What is measured by the gradient of the line for a tariff?

b) Use the intercepts and gradients to compare the two tariffs.

c) What is the significance of the point C? How would you decide which tariff a family should use?

d) By law, every energy tariff has to include a stated standing charge (the standing charge is the charge made even when no electricity is used). How can this energy provider satisfy this requirement?

> You could research the tariffs of different energy companies. Can all the tariffs be modelled with straight line graphs?

5 The graph of z against y is a linear graph of gradient 4.

The graph of y against x is a linear graph of gradient 3.

a) What is the gradient of the graph of z against x?

b) Explain how you found your answer, using the ideas of rate of change.

6 A satellite is in orbit x kilometres above the Earth's surface.

a) Stating any assumptions you make, sketch a graph to show how the length of the orbit, in kilometres, depends on x.

b) Write down the gradient of the graph.

Something that looks like a smooth sphere from a distance can look increasingly irregular as you zoom in.

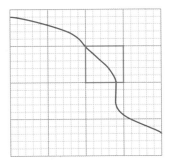

However, something very different happens when you zoom in on a quadratic or cubic curve.

Activity 1

Use a graphical calculator to check what happens when you zoom in on a quadratic or cubic curve. Check if the same is true for other functions such as $y = \sin x$ and $y = \cos x$.

Key point

When you repeatedly zoom in on many mathematical curves, they appear more and more like a straight line.

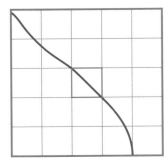
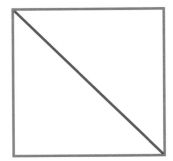

Here is an accurate graph of $y = x^2$.

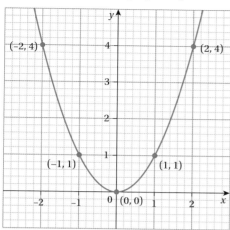

Activity 2

★ Estimate, by eye, the gradient of the curve at the five marked points.

Hold your ruler against the line. Move it until the angles appear equal, then draw the line.

One simple way to obtain a good estimate of the gradient of the curve at a point is to draw a line at that point, like this:

By eye, make the two marked angles roughly equal.

★ Copy or trace the graph of $y = x^2$.

★ Draw a tangent to the curve at the point $(1, 1)$.

★ Use your tangent to estimate the gradient of the curve at the point $(1, 1)$.

The line you have drawn is the **tangent** to the curve at the point $(1, 1)$. The gradient of this tangent line is the gradient of the curve at $(1, 1)$. ⊞ 1953

For example, for the $y = x^2$ curve at $(0, 0)$, the tangent is the x-axis, which has gradient 0.

You may have met tangents to circles before. A tangent to a circle touches the circle at one point only.

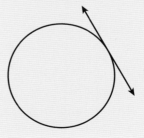

Activity 3

★ Draw a tangent to $y = x^2$ at $(2, 4)$. Use the tangent you have drawn to obtain an improved estimate for the gradient of $y = x^2$ at this point.

★ Why is there no need to draw the tangents at $(-1, 1)$ and $(-2, 4)$ to find the gradient at each point? What is an improved estimate for the gradient of $y = x^2$ at $(-1, 1)$ and $(-2, 4)$?

You can now apply the idea of zooming in to curves in general. You can find the gradient at any point by finding the gradient of the tangent at that point. Remember to look carefully at the scales on the axes when calculating gradients.

Example 4 Describe how the gradient of this curve changes as *x* increases.

Start at the left hand side of the graph and describe whether the gradient is positive, negative or zero, and whether it is increasing or decreasing.

From approximately zero, the gradient is positive and gradually increases, getting steeper. It then gradually decreases until it has value zero at the highest point on the graph. After this point the gradient is negative, and becomes increasingly steeper. ⊕ 1953

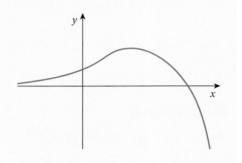

Example 5 The sketch graph shows the tax paid at different levels of income in the tax year 2015–16. (This is a simplification of the actual tax rules.)

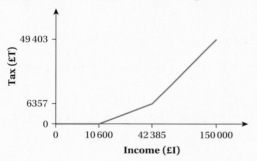

The personal allowance is the amount of money you can earn before you start to pay tax. Above the personal allowance, tax is charged at a percentage rate, called the basic rate. Higher earners pay a higher rate of tax above a certain income. The value of the personal allowance and the income where you start to pay the higher rate of tax change every year. The rates of tax may also change.

a) Find the gradients of each of the three straight line segments.

b) Describe the meaning of each of the three gradients found in part **a)**, in context.

c) Imagine zooming in at the point (42 385, 6357). What would you see happening?

A line segment is a straight line with definite beginning and end points.

a) Use the end points of the line segments to calculate their gradients.

$$0, \quad \frac{6357 - 0}{42\,385 - 10\,600} = 0.2, \quad \frac{49\,403 - 6357}{150\,000 - 42\,385} = 0.4$$

b) The line of gradient 0 represents paying no tax on any income up to the personal allowance of £10 600.

The line of gradient 0.2 represents the basic rate tax of 20%. For every extra pound earned in this income band you pay 20 pence tax.

The line of gradient 0.4 represents the higher rate tax of 40%. For every extra pound earned in this income band you pay 40 pence tax.

Tax is a percentage. 0.2 = 20%

c) No matter how far you zoom in, the two lines meeting at this point look exactly the same. There is *no* tangent at this point.

Example 6 Some clocks use a pendulum to measure time. The time period, T seconds, of a simple pendulum is the time taken for the pendulum to swing from one side to the other and back again. The graph shows how this time period depends upon the length, l metres, of the pendulum.

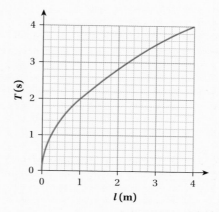

a) What, roughly, is the time period of a pendulum of length 2 metres?

b) Find the gradient of this curve at the point representing a pendulum of length 2 metres.

c) Explain the meaning of this gradient in context.

d) A pendulum, initially of length 2 metres, has its length increased by 2 cm. By how much is the time period changed?

Draw a line from 2 metres on the horizontal axis up to the graph line, and across to the vertical axis. Read off the value of T.

a) 2.8 seconds

b) Use two points on the tangent to calculate its gradient.

The tangent goes through $(0, 1.5)$ and $(2, 2.8)$

The gradient is $\frac{2.8 - 1.5}{2} = 0.65 \approx 0.7$ seconds per metre.

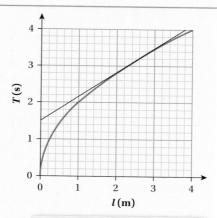

c) For *small* changes in the length, the time period of the pendulum increases by 0.7 seconds for every metre increase in length.

d) The increase of 2 cm is an increase of 0.02 m. As it is a small change in length, use the rate of change from part (c) to calculate the change in the time period.

$0.7 \times 0.02 = 0.014$ seconds

This only applies to small changes in length, because the gradient of the curve is changing.

Exercise 12B

1   An accurate graph of $y = x^3$ is shown below.

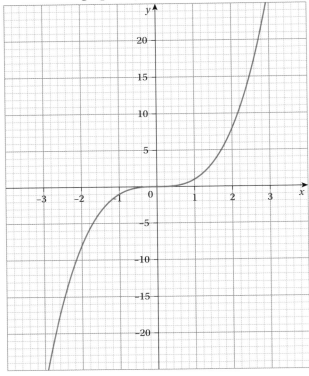

a) Describe how the gradient changes as x increases from -2.5 to $+2.5$

b) On a copy of the graph, draw a tangent at $x = 2$. Hence estimate the gradient of the curve at $x = 2$. Round your answer to the nearest integer.

c) Repeat part b) at $x = -1$.

d) Complete this table of gradients.

x	-2	-1	0	1	2
Gradient					

e) Can you spot a function of x which gives similar values to your gradients?

2   A bottle of milk is taken out of a fridge. The graph shows its temperature, $T\,°C$, t minutes after being taken out.

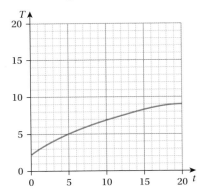

a) What is the temperature of the fridge?
b) What is the gradient of the curve at time $t = 0$?
c) Describe the meaning of your answer to part b).
d) By how much does the milk bottle warm up in 20 minutes?
e) Compare your answers to c) and d).

3   The height, h mm, of a seedling, t days after being planted, is modelled by
$$h = 3 + 2.1t - 0.03t^2$$

a) What was the height of the seedling when it was planted?

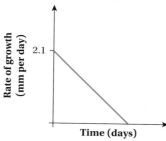
What is the value of t on the day it is planted?

b) By using a graphical calculator to plot the graph of h against t, or otherwise, find an accurate estimate of how fast the seedling is growing at time 10 days.
A graph of rate of growth against time is as shown.

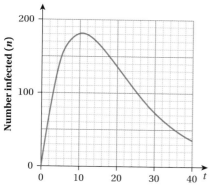

c) Copy this graph incorporating the point you found in part b).
d) Hence find when the rate of growth becomes zero.

4   The graph shows the number, n, of people in a village infected with a cold virus plotted against the number of days, t, since the start of the outbreak.

a) What is the initial rate of infection?
b) On a copy of the graph, draw a tangent at $t = 25$.
c) Find the gradient of your tangent and describe the meaning of your answer in context.
d) What would you expect to happen to the gradient for large values of t?

12.4 Optimisation

'It is the greatest good to the greatest number of people which is the measure of right and wrong.

Jeremy Bentham, 1748–1832

Much human effort goes into making something as great as possible (things like happiness, examination marks or profit) or as small as possible (things like pain or cost). Ideas of greatest and least are central to the physical laws of the universe. For example, physical systems tend to positions of maximum entropy and of minimum energy.

As an example of optimisation, consider this graph of aviation fuel use from Chapter 4.

Finding the greatest and least fuel use involves looking at the fuel use at just two types of point.

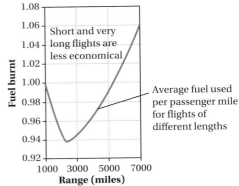

- The points at the extremes of the range. Here that is 1000 and 7000 miles. These points are called **boundary points**.
- The points in between where the graph turns round at a highest or lowest point. These points are called **turning points**.

For the fuel use graph it is easy to see that the least fuel per passenger mile is at the only turning point, for flights of approximately 2300 miles.

> A turning point in a graph can give a maximum ∩ or minimum ∪ value.

Activity 4

What is the connection between turning points and rates of change?

> Hint: Look at the rate of change either side of the turning point, and at the turning point itself.

Example 7

The size (n thousand) of one particular population of whales can be modelled by the equation $n = 2x^2 - 9x + 11$ where x is the number of decades since the first reliable count of the numbers of these whales.

a) Explain why this form of equation for n is likely to be a suitable one for the population.

b) Some values of the gradient of the graph of n against x are as shown.

x	1	2	3	4
Gradient	−5	−1	3	7

> Following years of declining numbers, humpback whales became endangered and were eventually protected by law. By 2015, numbers recovered and the National Oceanic and Atmospheric Administration (NOAA) proposed removing some populations of humpback whales from the protection of the Endangered Species Act.

Paper 2C

Use this table to predict when the size of the population was a minimum. Find this minimum size.

c) When is this model likely *not* to be a suitable one?

a) Think about the shape of a graph for this equation. Does the story of the graph match the story of the whales?

This is a quadratic equation which has a U-shaped curve. This matches the information that the number of whales decreased to a minimum and since then has increased.

b) Look for a pattern in the gradient values. They are going up in equal-sized steps.

The gradients go up by 4 every time x increases by 1. This shows that the gradient is a linear function of x.

$$\text{Gradient} = 4x - 9$$

At a turning point, the gradient is always zero.

At the turning point

$$4x - 9 = 0$$

so $\quad x = 2.25$

Find the value of n when $x = 2.25$

Substituting into $n = 2x^2 - 9x + 11$

$$n = 2(2.25)^2 - 9(2.25) + 11$$
$$= 0.875$$

Population of whales $= n$ thousand
$$= 0.875 \times 1000$$

The minimum size of population was (approximately) 875.

c) Think about very large and very small values of x.

The model will not be suitable for large values of x as whale numbers will not continue to rise forever.

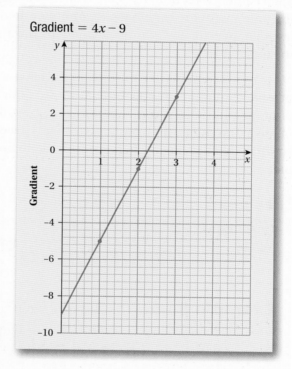

Gradient = 4x − 9

Key point

To find the greatest and smallest possible values of a function consider the boundary points and the turning points. The gradient at the turning points is zero.

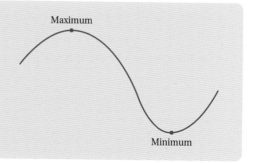

The camera on the left gives your speed as you pass the camera. Average-speed cameras, like the one shown on the right, are now commonly seen on many motorways.

Average-speed cameras are placed at least 200 metres apart. Each camera uses number plate recognition software to record the time that every car passes. A computer then calculates the average speed of each car by dividing the distance travelled by the time taken.

Most speed cameras like the one on the left are activated by passing cars and use conventional 35 mm film. They take two photographs half a second apart as a car covers a marked section of road. The computer calculates average speed over an interval so small that the speed can be considered to be the car's speed at that instant.

> Understanding the distinction between speed (or instantaneous speed) and average speed is important both in physics and for avoiding speeding tickets.

Key point

The average speed of an object over an interval of time is given by $\dfrac{\text{Distance travelled in that interval}}{\text{Time taken}}$ ⊕ 1121

Example 8 Here is the distance–time graph for a motorist. The first mile of her journey is in a 20 mph zone and the next mile is in a 30 mph zone.

a) What is the motorist's average speed

   i   in the first four minutes

   ii in the last minute

   iii overall?

b) Estimate her speed at time 4 minutes.

c) Does she break the speed limit at any time?

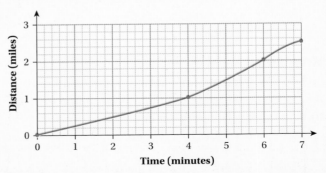

a) Find the distance travelled and time taken. The speed in mph is the number of miles travelled in 1 hour.

i She travels 1 mile in 4 minutes

This is equivalent to 15 miles in 60 minutes = 15 mph

ii In the last minute she travels $\frac{1}{2}$ mile

$\frac{1}{2}$ mile in 1 minute

30 miles in 60 minutes = 30 mph

iii $2\frac{1}{2}$ miles in 7 minutes

$2.5 \times \frac{60}{7}$ miles in 60 minutes ≈ 21 mph

⊞ 1322

b) In a distance–time graph, the speed is the gradient. Draw a tangent to the graph at $t = 4$ and calculate its gradient.

At 4 minutes, speed is approximately 20 mph.

c) Compare the speed (gradient) on the graph to the speed limit. Draw the lines to show average speed of 20 mph for the first mile, and average speed of 30 mph for the next 1.5 miles.

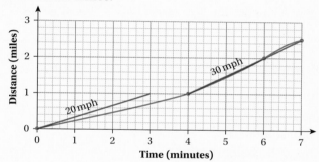

> Speed is the rate of change of distance with respect to time.

In places between 5 and 7 minutes the gradient of the graph is steeper than the gradient of the 30 mph line and so her speed is greater than the speed limit at those moments.

Key point

- The speed (or instantaneous speed) of an object at a particular time is the gradient of the tangent to its distance–time graph at that time. For units of metres and seconds, the units of speed are metres per second. You will often see 'metres per second' written in full and also written in shorthand form as m/s or ms^{-1}.

- When distance is measured vertically it is often given as 'height' or, less frequently, as 'depth'. The gradient of a height–time or a depth–time graph is still just speed but you will sometimes see it called 'vertical speed'.

- Velocity is speed in a given direction. For a distance–time graph of gradient $-5\,ms^{-1}$, mathematicians and physicists call $-5\,ms^{-1}$ the *velocity* but say that the *speed* is $5\,ms^{-1}$. In everyday use, speed is the same as velocity, ignoring any negative signs.

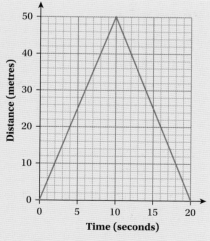

Exercise 12C

1   The speed–time graph given below shows how the speed of a roller coaster car varies with time, for the first 35 seconds of the ride.

   a)   What is the speed of the car after
      i   12 seconds   ii   24 seconds?
   b)   Describe the motion of the car during these 35 seconds.

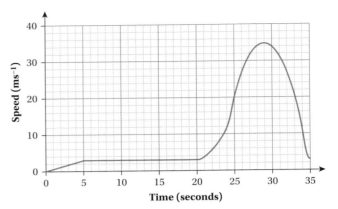

2   A ball is thrown vertically upwards. Its height–time graph is as shown.

   a)   What is the ball's initial speed?
   b)   Estimate the vertical speed of the ball 1 second after being thrown.

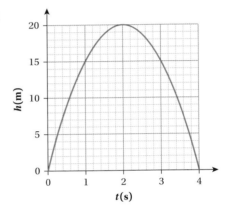

3   Here is a graph of velocity against time for a ball.

   a)   At what time is the ball instantaneously at rest?
   b)   Describe how the velocity of the ball changes.
   c)   Describe what could be happening to the ball over the time of this graph.
   d)   What is the special significance of the point where the ball is instantaneously at rest?

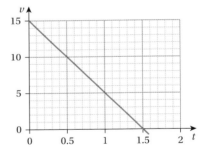

4   For lift-off, a space shuttle used two solid fuel rockets which burned for only two minutes.

   After t seconds, the height (h metres) of the shuttle could be modelled by the equation $h = 3.125t^2$.

a) To what height did the solid fuel rockets take the shuttle in the first two minutes?

b) Calculate the average speed in metres per second of the shuttle for the first two minutes of flight.

c) Plot a height–time graph for the shuttle in the first two minutes

d) Use your graph to estimate the speed of the shuttle at time two minutes.

5   The distance–time graph is for a car in a performance test.

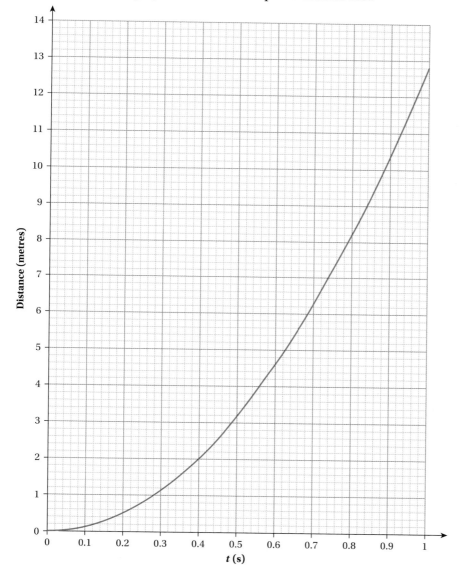

a) On a copy of the graph draw a tangent to the curve at $t = 0.5$

b) Hence estimate the speed of the car at $t = 0.5$

c) What is the initial speed of the car? Explain your answer.

d) The test was to measure the car's performance whilst accelerating from 0 to 60 mph. How is 60 mph represented on this graph?

12.6 Acceleration

One of the most widely quoted 'accelerations' is the acceleration due to gravity. This is approximately 10 metres per second per second, which can be written as $10\,\text{ms}^{-2}$. The units for this acceleration show that the downward velocity of any free-falling object increases by $10\,\text{ms}^{-1}$ every second.

Key point
- Acceleration is the rate of change of velocity.
- Average acceleration over a time interval is given by
$$\frac{\text{Change in velocity}}{\text{Time taken}}$$
- The acceleration at a particular instant is given by the gradient of a velocity–time graph.

Activity 6

A car is travelling at $10\,\text{ms}^{-1}$. In 5 seconds it accelerates to $30\,\text{ms}^{-1}$. What is the acceleration?

Activity 5

Q Falling objects have a 'terminal velocity'. This is a maximum velocity of fall which their downwards velocity approaches but does not exceed. Why do objects not continue to accelerate at $10\,\text{ms}^{-2}$?

Use the formula
$$\frac{\text{Change in velocity}}{\text{Time taken}}$$
Give the answer in ms^{-2}.

Example 9 A stone is dropped down a well and after 3 seconds you hear a splash.

In a velocity–time graph, the area under the graph is the distance travelled.

a) What is the velocity of the stone when it hits the water?
b) How deep is the well?

a) Before it starts to fall, the stone's initial velocity is 0.

The acceleration of a free falling stone is $10\,\text{ms}^{-2}$.

Substitute into acceleration $= \dfrac{\text{Change in velocity}}{\text{Time taken}}$

$10 = \frac{v - 0}{3}$ where v is velocity of stone when it hits the water.

$10 = \frac{v}{3}$

$v = 30\,\text{ms}^{-1}$

b) Sketch the velocity–time graph. At $t = 0$, $v = 0$, and at $t = 3$, $v = 30$.

The distance the stone falls is the area under the velocity–time graph.

distance $= \frac{1}{2} \times 3 \times 30 = 45$ metres

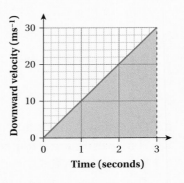

area of a triangle $= \frac{1}{2} \times$ base $\times$ height

Example 10 The take-off velocity of a commercial aeroplane is typically about 180 mph. To achieve this velocity, the aeroplane accelerates along a horizontal runway. The graph shows the velocity of an aeroplane t seconds after it enters the runway.

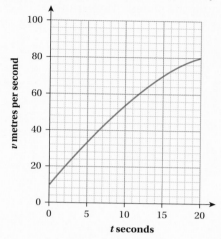

a) State the velocity of this aeroplane when $t = 0$. What does this velocity represent?
b) State the velocity when $t = 15$.
c) 1 mile is approximately 1609 metres. Show that the plane will take off when $t = 20$.
d) Estimate the acceleration when $t = 15$.
e) When is the acceleration a maximum? Estimate this maximum acceleration.

---

a) Read the value of v when $t = 0$.

When $t = 0$, velocity = 10 ms^{-1}. This is the aeroplane's velocity when it enters the runway.

b) When $t = 15$, velocity = 70 ms^{-1}.

c) At $t = 20$, speed = 80 ms^{-1}. Convert this to mph.

1 mile is approximately 1609 metres and there are 3600 seconds in an hour.

$80 \text{ ms}^{-1} = \frac{80 \times 3600}{1609} \approx 179 \text{ mph.}$ ⬤ 1970

d) Draw a tangent at $t = 15$ and find its gradient.

At $t = 15$, acceleration = 2.5 ms^{-2}

e) Find the steepest part of the line. Calculate its gradient.

The greatest acceleration is at $t = 0$. The gradient of a tangent at that point is approximately 5 ms^{-2}

Activity 7

A coin is dropped into another well, and hits the water after 5 seconds.

How deep is this well?

Acceleration is the rate of change of velocity with respect to time.

Consolidation exercise 12

1   A car joins a motorway at a speed of $25\,\text{ms}^{-1}$ and accelerates at $0.5\,\text{ms}^{-2}$ for 10 seconds.

   a)   What is the car's speed after 10 seconds?
   b)   Draw a speed–time graph for the car.
   c)   Find the distance it travels during the 10 seconds.
   You can assume that the distance travelled is the area under the speed–time graph.
   d)   Find its average speed over this period.

2   One year, a used-car dealer's stock of cars was as shown.

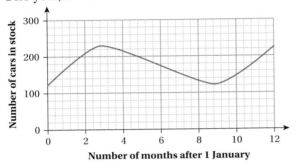

   a)   When was the number of cars in stock a minimum?
   b)   What was the net change in the dealer's stock during this year?

> The dealer will be constantly buying and selling cars. The **net change** measures just the overall effect. For example, if the dealer buys 5 cars and sells 8 cars one week, then the net change will be a loss of 3 cars.

   c)   What was the gradient of this graph at the point representing 1st January? Describe what this gradient means in context.
   d)   Repeat part c) for 1st July.
   e)   'Between 1st April and 1st October, the dealer sold 110 cars.' Criticise this deduction about the data shown on the graph.

3   An explorer in a distant settlement throws a stone vertically upwards. After t seconds, its velocity, $v\,\text{ms}^{-1}$, is given by $v = 20 - 3.7t$.

   a)   When does the stone reach its maximum height?
   b)   Draw a graph of v against t.
   c)   What is the gradient of this graph and what is measured by this gradient?
   d)   Where might this settlement be?

4   One day, at noon, a police squad raided an apartment. In the kitchen there was a mug of tea with a temperature of 80 °C. A forensic scientist obtained a cooling curve for a mug of tea in this apartment.

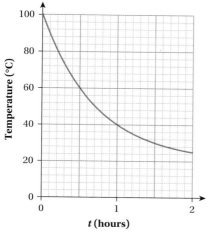

a) Use tangent lines to obtain estimates of the gradients of the curve at times 0 and 1 hours.
b) What do these gradients mean in context?
c) Describe how the mug of tea is cooling.
d) Estimate the time that the tea was poured. State any assumptions you make.

5   The vertical height, h metres, of a tennis ball, t seconds after being hit, is given by $h = 2.4 + 4t - 5t^2$.

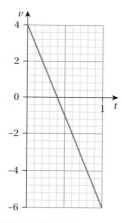

a) At what height was the ball hit?
b) By using a graphical calculator, or otherwise, find the time when the ball hits the ground.

The graph shows vertical speed, $v\,\text{ms}^{-1}$, against time, t seconds, for the first second of the ball's flight.

c) What is the acceleration of the ball?
d) At what time is $v = 0$? What, in context, is the significance of this point?

Investigation

Sidestepping Zeno's paradox

You will find many articles on the internet about Zeno, who was a Greek philosopher in the 5th century BC.

Zeno described a number of apparently paradoxical problems about motion. Two of the most well-known of these problems which you might like to research are *Achilles and the Tortoise* and *Zeno's arrow paradox*.

Zeno's arrow paradox is closely related to the ideas you have studied in this chapter.

An archer fires an arrow at an angle, so that its path is like this:

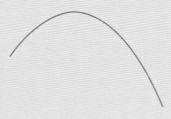

The rate of change at a point such as A is the gradient of the tangent at A. This is often called the **instantaneous rate of change** to distinguish it from the rate of change over an interval, such as from A to B.

However, Zeno pointed out the apparent absurdity of talking about a change at any one instant. Considering just that instant, there is no change at all! It is therefore perhaps best to think of the rate of change at A to be the rate of change over an extremely small (but not zero) interval about A.

This idea gives us a numerical way to find gradients at points without the need to draw tangents.

Example 11 Estimate the gradient of $y = x^2$ at the point $(3, 9)$.

Choose an x-value a little smaller than 3 and one a little larger. Work out the corresponding y-values.

At $x = 2.99$, $y = 2.99^2 = 8.9401$

At $x = 3.01$, $y = 3.01^2 = 9.0601$

Gradient $= \dfrac{9.0601 - 8.9401}{3.01 - 2.99} = 6$

Key point

To estimate the gradient at a point A on a curve, choose two points extremely close to A and find the gradient of the line through the two points.

Hint: Gradient $= \dfrac{\text{change in } y}{\text{change in } x}$

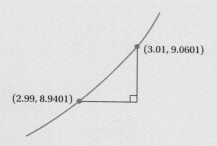

(3.01, 9.0601)

(2.99, 8.9401)

Repeat this calculation for other values of x and draw up a table of your results.

Value of x	Gradient of $y = x^2$
-1	-2
0	0
3	6
5	?

Notice that the gradient of $y = x^2$ at any point appears to be equal to twice the x-value.

a) Draw up a table of values for the gradient of another quadratic curve. Try to spot a formula for the gradient at any point on your curve.

b) Generalise your result as much as possible. Try to find a rule or description for how you can find the gradient function for a quadratic curve from its equation, without drawing its graph.

The techniques for finding the gradient at any point of some curves were initially developed by Leibniz (1646–1716) and Newton (1642–1747). This area of mathematics is known as calculus and is especially important in many applications of mathematics, especially physics.

13 Exponential functions

Why is modelling decay important for the doses of drugs to patients?

When do some populations grow exponentially?

In this chapter you will study how exponential functions are used to model growth and decay in a variety of contexts.

Why can debts spiral out of control?

What are pyramid selling scams?

You should know how to:

- use the index laws 🔘 1033, 1951, 1301
- calculate compound interest using a multiplying factor 🔘 1238
- find the gradient of a tangent 🔘 1953
- solve simple equations. 🔘 1154

Suppose that a single E. coli bacterium is provided with ideal conditions of temperature, pH and nutrients. Then, while these conditions last, the number of bacteria will increase, roughly as shown.

Time	Number
0	1
20 minutes	2
40 minutes	4
1 hour	8
...	...

Many kinds of E. coli bacteria are harmless to humans, but some kinds can cause serious infections.

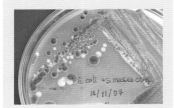

At this point these numbers appear unremarkable.

Activity 1

How many bacteria will be present after 5 hours and after 10 hours?

Find a formula for the number of bacteria after x hours.

Under ideal conditions, the number of E. coli bacteria doubles every 20 minutes.

This graph of number against time shows the typical shape of exponential growth. Growth may be slow to start with, but eventually it becomes extremely rapid. ⊕ 1070

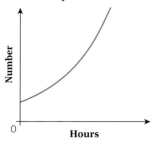

Activity 2

🔍 Check and comment on the following statement.

'In a single day, one cell of E. coli could produce a super-colony equal in weight to the entire planet Earth.'

Michael Crichton, The Andromeda Strain

Compound interest is an example of exponential growth.

Einstein is quoted as saying:

Compound interest is the eighth wonder of the world. He who understands it earns it, he who doesn't pays it.

You learned about compound interest in Chapter 2.

Einstein's point is well made by the following example.

Example 1

You owe £1000 on your Cardpro Credit Card. Assuming none of this debt is paid off, how much would you owe after 5 years and after 10 years? Find a formula for how much you would owe after x years.

CCJs are County Court Judgements for debt. Some companies specialise in providing credit to people who have had financial difficulties. The interest rates they offer are often extremely high.

The best way to tackle compound interest questions is to use a multiplying factor. ⊞ 1238
The interest added each year is 39.9% so the multiplying factor is 1.399.
Multiply the amount you owe by 1.399 once for each year.

After 5 years: £1000 $\times$ 1.399 $\times$ 1.399 $\times$ 1.399 $\times$ 1.399 $\times$ 1.399
                   = £1000 $\times 1.399^5$ = £5359.06

After 10 years: £1000 $\times 1.399^{10}$ = £28719.52

After x years: £1000 $\times (1.399)^x$

You can use index notation to write your calculation more simply. ⊞ 1033

Activity 3

If you are fortunate enough to receive annual interest of 100% then, over a year, each £1 of savings will become £2.

If you receive 50% every 6 months then each £1 will become £1 $\times 1.5^2$ = £2.25.

Now work out what happens if you receive interest even more frequently:

★ 25% every 4 months

★ 10% ten times a year

★ 1% one hundred times a year

★ and so on …

As the interest becomes more and more frequent, what number does the value of your savings become closer and closer to? Can you find a special number on your calculator that seems to equal the value when you are receiving interest extremely frequently?

You multiply by 1.5^2 because the multiplier is 1.5 and there are 2 six-month periods in a year.

The number that you may have just discovered was first discovered by the Swiss mathematician Jacob Bernoulli when studying compound interest in the same way that you have done. You will use this number a lot, later in this chapter

In section 13.1 you saw two examples of exponential functions: 2^{3x} and 1.399^x.

> **Key point**
>
> This chapter is about exponential functions. These have the form
>
>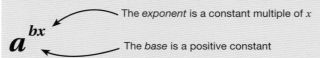
>
> The *exponent* is a constant multiple of x
>
> a^{bx}
>
> The *base* is a positive constant
>
> For example, 2^{3x} is an exponential function with base 2 and exponent $3x$.

3^{2x}, 5^{-x} and $\left(\frac{1}{2}\right)^{9x}$ are some other exponential functions.

The functions of Chapter 12 such as x^2 and x^3 are *not* of this type. For example,

The exponent is constant

x^3

The base is the variable

You can draw a graph of an exponential function by calculating the value of the function for different values of x and then plotting the points.

There are just two closely related types of graphs of exponential functions.

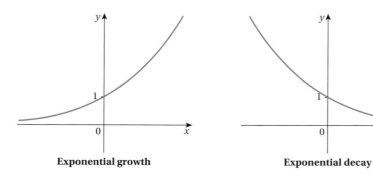

Exponential growth   **Exponential decay**

Activity 4

★ Use your graphical calculator or a computer to plot various exponential functions.

★ Summarise your findings.

★ Do any of the functions have graphs different from the two shown above?

★ How can you tell if the graph of an exponential function will show growth or decay?

Think about what happens when the exponent is a negative number or when the base is a fraction.

Paper 2C

Exercise 13A

1   State which of the following functions are exponential functions. For each exponential function, write down the base and the exponent.

   a) 2^{-x}
   b) x^3
   c) $\left(\frac{1}{2}\right)^{2x}$
   d) x^{-1}
   e) 10^x
   f) 10^7
   g) x^x

2   Use your calculator to find these values.

   a) 2^{37}
   b) $3^{2.5}$
   c) $8.9^{-1.7}$

3   Solve the equations.

   a) $2^x = 16$
   b) $5^x = 125$
   c) $7^x = 1$
   d) $2^x = \frac{1}{4}$

4   Suggest an exponential function that would have each graph.

   a)
   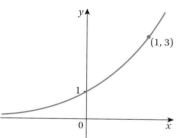

   b)
   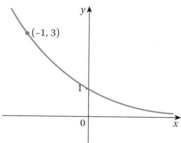

5   £1000 is invested at 4% interest per annum. How much will be in the account after

   a) 1 year
   b) 5 years
   c) x years?

6   The amplitude (or size) of horizontal ground movements at a distance of 100 km from the epicentre of an earthquake are roughly

   $3.6 \times 10^{M-11}$ metres,

   where M is the magnitude of the earthquake.

   A seismograph magnifies this amplitude by a factor of 28 000 to render the displacement more noticeable on the printout.

   a) What would be the magnified amplitude for an earthquake of magnitude 3?
       (Such an earthquake would be noticeable but unlikely to cause damage.)
   b) Repeat part a) for earthquakes of magnitude 4 and magnitude 5.
       What do you notice about your answers?

7   A drug was administered to a hospital patient at noon.
   The amount of drug remaining in the patient after t hours
   is given by $400 \times (0.8)^t$ mg.

   a) How much of the drug was administered at noon?
   b) Find the amount remaining at 3 pm.
   c) By trial and improvement, or otherwise, find the time
       when 100 mg of the drug is remaining.

A seismograph records the size of waves caused by an earthquake and produces an image of the information.

Activity 3 of Section 13.1 produced a sequence of numbers tending to a number 2.71828... This number is called e and can be found on your calculator. Like π, it is an irrational number that cannot be expressed as a fraction.

The number e is used as the standard base for exponential functions. The main reason for this will be seen in Activity 5 and in Section 13.4.

It is easy to calculate with powers of e since all scientific calculators have an e^x button.

> In mathematics the phrase 'a sequence tends to 2.71828...' means that 'a sequence gets closer and closer to 2.71828...'

> The number e was discovered by Jacob Bernoulli in 1683 but it was nearly 50 years before many of its properties were studied properly. The actual symbol e appears to have been first used in a letter Leonhard Euler wrote to Christian Goldbach in 1731.

> One of Euler's results, called Euler's identity, is widely regarded as one of the most beautiful equations in mathematics. This equation is
> $$e^{-\pi i} = -1$$
> and connects the apparently unrelated constants e, π, −1 and i (i is the symbol for the square root of −1).

Example 2 The probability of a colony of the ant species *Formica opaciventris* surviving t years is estimated to be $e^{-0.4t}$.

a) Find the probability of the colony surviving

   i 4 years

   ii 8 years

b) After how long is the probability of the colony surviving equal to $\frac{1}{2}$?

---

a) i Substitute $t = 4$ into the expression for the probability.

$$e^{-0.4 \times 4} = e^{-1.6} \approx 0.2$$

ii Substitute $t = 8$ into the expression for the probability.

$$e^{-0.4 \times 8} = e^{-3.2} \approx 0.04$$

b) You must solve $e^{-0.4t} = \frac{1}{2}$.

Substitute different values for t and use trial and improvement. ⊕ 1057

$t = 1$: $e^{-0.4 \times 1} = e^{-0.4} \approx 0.670$

$t = 2$: $e^{-0.4 \times 2} = e^{-0.8} \approx 0.449$

$t = 1$ gives you a probability greater than $\frac{1}{2}$ and $t = 2$ gives you a probability less than $\frac{1}{2}$ so you now know the value for t is between 1 and 2. Try $t = 1.5$

$t = 1.5$: $e^{-0.4 \times 1.5} = e^{-0.6} \approx 0.549$

$t = 1.7$: $e^{-0.4 \times 1.7} = e^{-0.68} \approx 0.507$

$t = 1.8$: $e^{-0.4 \times 1.8} = e^{-0.72} \approx 0.487$

So, by trial and improvement, $t \approx 1.7$ years.

Activity 5

By drawing tangents on a copy of the graph of $y = e^x$, or otherwise, find the gradients at $x = -1, 0, 1$ and 2.
Complete a table of results.

x	y	Gradient
-1		
0	1	
1		
2		

What do you notice?

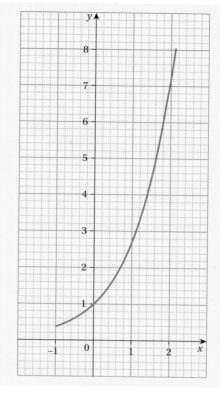

Exercise 13B

1 Find the values of

a) e^3 　　　 b) e^{-2} 　　　 c) $3e^2 + 4$

2 Using trial and improvement, or otherwise, find the value of x such that

a) $e^{5x} = 1$ 　　　 b) $e^{-x} = 1.2$ 　　　 c) $e^x = 5$

3 The temperature, $T°C$, of a cup of tea, t minutes after it was made, is modelled by the function $T = 76e^{-0.1t} + 22$.

a) What is the temperature of the cup of tea when it is first made?

b) What is the temperature of the cup of tea after 10 minutes?

c) What is the temperature of the cup of tea after a long time?

d) Sketch a graph of the function $T = 76e^{-0.1t} + 22$.

e) Suppose the temperature, $T°C$, of a cup of tea is modelled by the function $T = Ae^{-0.1t} + B$ where A and B are constants.

Interpret what is represented by each of
i 　 A 　　　 ii 　 B.

> Cooling curves are an important tool in a number of scientific areas. They are important in studies of the physical and chemical properties of objects. They are also used to help in the determination of the ages of stars, planets and the universe itself. On Earth, they have an important role in forensic science, determining the times of some crimes.

4 In a medical treatment, 500 milligrams of a drug are administered to a patient. At time t hours later, the amount X milligrams of the drug remaining in the patient is given by $X = 500e^{-0.25t}$.

a) Find X after 1 hour.

b) Find the value of t, correct to one decimal place, when $X = 200$.

The function e^x is known as the exponential function. As you discovered in Activity 5:

> **Key point**
> The gradient of the exponential function at *any* point is the same as the value of the function at that point.

Two graphs of $y = e^x$ are shown below. Each has a gradient line drawn.

You know that the gradient can be worked out using $\dfrac{\text{change in } y\text{-values}}{\text{change in } x\text{-values}}$. ⊞ 1312

In the first graph the gradient is drawn at the point $(2, e^2)$ and it also passes through $(1, 0)$.

The gradient of the line at point $(2, e^2) = \dfrac{e^2 - 0}{2 - 1} = e^2$. This is equal to the y-value.

Similarly the value of the gradient at point $(3, e^3) = \dfrac{e^3 - 0}{3 - 2} = e^3$.

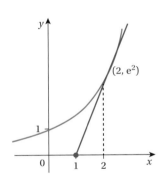
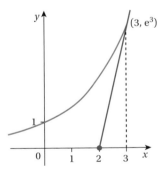

It is this property which explains why so many natural phenomena show exponential growth or decay.

In a population, the number of births will be roughly proportional to the size of the population. The rate of growth of the population is proportional to the population itself and so population can be modelled by a function based upon e^t.

Similarly, consider a large number of radioactive atoms. The number that will decay in some fixed time will be proportional to the total number. So, again, the number of atoms can be modelled by a function based upon e^t.

> **Key point**
> Consider two quantities N and x.
>
> If the rate of change of N with respect to x is proportional to N then
> $$N = N_0 e^{kx},$$
> for the constants N_0 and k.
>
> - N_0 is the value of N when $x = 0$.
> - k is a constant.

Example 3 A population of insects triples every 14 days.

a) Explain why the population can be modelled by an exponential function.

b) Show that the population P at time t days is approximately given by $P_0e^{0.078t}$, where P_0 is the initial population.

c) After how many days does the population double in size?

---

a) The population triples every 14 days however big or small the population. This means that the rate of change of the population is proportional to the size of the population. The rate of change of an exponential function is proportional to its value too, so it is a suitable function to use to model the population.

b) Using an exponential function, P can be modelled as $P = P_0e^{kt}$.

The expression you have been given tells you that $k = 0.078$

You also know that the population triples every 14 days.

Substitute $t = 14$ to test the function.

$P = P_0e^{0.078 \times 14} = P_0e^{1.092} \approx 2.98P_0$.

> Any situation where the rate of change of a population is proportional to the population size can be modelled by an exponential function.

This shows that after 14 days the population is approximately triple the initial population so the function is a good model.

c) If the population doubles after T days,

$2P_0 = P_0e^{0.078T}$

Dividing both sides by P_0 gives $2 = e^{0.078T}$

Use trial and improvement to find the value of T that solves the equation.

$T = 7$: $\quad e^{0.078T} = e^{0.078 \times 7} = 1.73$

$T = 8$: $\quad e^{0.078T} = e^{0.078 \times 8} = 1.87$

$T = 9$: $\quad e^{0.078T} = e^{0.078 \times 9} = 2.02$

By trial and improvement, $T \approx 9$.

Activity 6 Q

An exponential growth model for a population predicts the same percentage increase in each unit of time. The number of internet users *cannot* be modelled well by exponential growth because the growth rate from 1993 to 1996 was roughly 75% per year whereas it is now roughly 10% per year.

Research the number of internet users and find different functions to model the number of internet users in different periods. You may find that different exponential functions are needed for different periods.

In Example 3 you needed to solve the equation $e^{0.078T} = 2$ and you found an answer of approximately 9 using trial and improvement.

It can sometimes take a long time to find an answer using trial and improvement, especially if you need an accurate answer. Fortunately, there is a much faster method of solving this sort of equation with great accuracy using the ln (natural logarithm) button on your calculator.

Key point

The functions $\ln x$ and e^x are inverses of each other. For example,

$e^{1.61} \approx 5.00$ and $1.61 \approx \ln 5.00$

$e^0 = 1$ and $0 = \ln 1$

$e^a = b$ is equivalent to $a = \ln b$

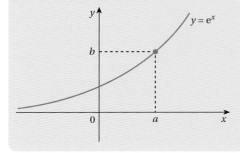

An inverse function is a function that reverses the action of another function.

Using the exponential function on a gives b.

Using the natural logarithm function on b reverses the process and gives a again.

Example 4 Solve $e^{0.078T} = 2$

Using the inverse, $0.078T = \ln 2$

Find T by dividing both sides by 0.078

$T = \dfrac{\ln 2}{0.078} \approx 8.89$

Example 5 Solve $3 + 5e^{-2x} = 4$ to find the value of x.

Subtract 3 from both sides	$5e^{-2x} = 1$
Divide both sides by 5	$e^{-2x} = 0.2$
Use the inverse	$-2x = \ln 0.2 \approx -1.609$
Divide both sides by -2	$x \approx 0.80$

Find what e^{-2x} is equal to before you use the inverse function, ln. ⊞ 1154

Paper 2C

Key point

The most difficult equation where you will use ln is likely to be of the form $A + Be^{kx} = C$, where A, B, C and k are known constants.

As in Example 5, before you use ln you should find what e^{kx} is equal to:

Subtract A from both sides	$Be^{kx} = C - A$
Divide both sides by B	$e^{kx} = \dfrac{C - A}{B}$

Example 6 Scientists have claimed that a cup of tea is best drunk 'six minutes after it has been made'.

Use the cooling curve equation $T = 76e^{-0.1t} + 22$ to work out how long it takes for a cup of tea to reach 60 °C, a sensible temperature at which to start drinking.

T is the temperature of the tea and t is the time in minutes after it has been made.

Substitute $T = 60$ into the equation

$$76e^{-0.1t} + 22 = 60$$

Solve the equation to find t

Subtract 22 from both sides.	$76e^{-0.1t} = 38$
Divide both sides by 76.	$e^{-0.1t} = 0.5$
Use the inverse.	$t = \dfrac{\ln 0.5}{-0.1} \approx 6.9$

So it takes nearly 7 minutes for a cup of tea to reach 60 °C.

The ln function is often needed when you are asked to find an exponential function to match a graph.

Consider, for example, this graph.

Suppose $y = A + Be^{kx}$, where you need to find values for the constants A, B and k.

The ln function is needed to find the value of k but first it is best to find A and B by much simpler means.

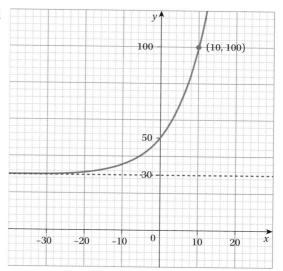

Key point

- When kx becomes a very large negative number then e^{kx} tends to zero.
- When $x = 0$, $e^{kx} = 1$.

Find A and B for the graph by applying the key points.

Make kx a large negative number.

e^{kx} tends to zero so the equation becomes $y = A$.

Use the graph to find the value when y is large and negative.

$A = 30$

Put some large negative values for kx into e^{kx} using your calculator and see what happens.

$y = 30$ can be seen from the graph.

Make $x = 0$.

$e^{kx} = 1$ so the equation becomes $y = A + B$.

Use the graph to find the y-value when $x = 0$.

$A + B = 50$

Substitute in $A = 30$ to find B. $B = 20$ so $y = 30 + 20e^{kx}$.

Finally, find k. Use the fact that $(10, 100)$ is a point on the curve.

Substitute in values for x, y, A and B. $\quad 30 + 20e^{10k} = 100$

Subtract 30 and divide by 20. $\qquad\qquad e^{10k} = 3.5$

Use the inverse function. $\qquad\qquad 10k = \ln 3.5 = 1.25$

Divide by 10. $\qquad\qquad\qquad\qquad k = 0.125$

So the equation is $y = 30 + 20e^{0.125x}$

(0, 50) is a point on the graph.

Example 7 A population, P thousands, is modelled by the function $P = 50 + Ae^{kt}$, where t is the number of years after 2000. The population in 2000 was 70 000 and in 2010 was 90 000. What does this model predict for the population in 2025?

Put $t = 0$. $\quad 50 + A = 70$

$\qquad\qquad\qquad A = 20$

> The population is 70 000, so $P = 70$.

Put $t = 10$. $\quad 50 + 20e^{10k} = 90$

Solve the equation to find k $\quad 20e^{10k} = 40$

$\qquad\qquad\qquad\qquad e^{10k} = 2$

$\qquad\qquad\qquad 10k = \ln 2 = 0.693$

$\qquad\qquad\qquad\qquad k = 0.0693$

> In 2010, $t = 10$ and $P = 90$.

So $P = 50 + 20e^{0.0693t}$

In 2025, $t = 25$.

So $P = 50 + 20e^{0.0693 \times 25} \approx 163$

The predicted population is 163 000.

> The equation of a particular exponential curve can be expressed in various forms.
> For example, consider:
> $y = 10 + 10e^{-kx}$, with k a positive constant
> $y = 10(1 + e^{-kx})$, with k a positive constant
> $y = 10 + 10e^{lx}$, with l a negative constant
> You should be able to spot that these are all the same curve, providing $l = -k$.

Activity 7

Find the equation of this curve in the form

$y = A(1 - e^{-kx})$,

where A and k are positive constants.

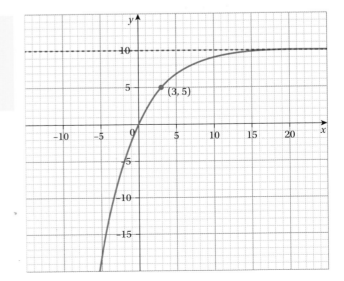

300 **Exponential functions**

Exercise 13C

You should compare your answers to questions **2** and **3** with those you obtained by trial and improvement in Section 13.3

1   Solve the equations.

   a) $5 - 2e^{-2x} = 4.8$ b) $3 + 7e^{0.05x} = 4$

2   The probability of a colony of the ant species *Formica opaciventris* surviving *t* years is estimated to be e$^{-0.4t.}$

   After how long is the probability of the colony surviving equal to $\frac{1}{2}$?

3   In a medical treatment, 500 milligrams of a drug are administered to a patient. At time *t* hours later, the amount *X* milligrams of the drug remaining in the patient is given by $X = 500e^{-0.25t}$.

   Find the value of *t*, correct to two decimal places, when $X = 200$.

4   In 1935 cane toads were introduced into Queensland, Australia, in an effort to control sugar cane beetles. The cane toads spread rapidly and are now themselves considered pests.

   A model for the area of Australia occupied by cane toads is given by $1600e^{0.078t}$ where *t* is the number of years after 1900.

   a) When does this model suggest that half a million km^2 were occupied?

   b) Can this model be used for small or large values of *t*? Explain your answer.

5   Using exponential functions, find equations for these two curves.

   a)
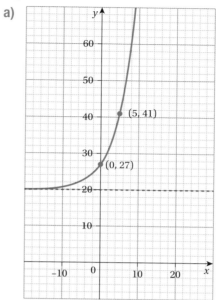

   b)
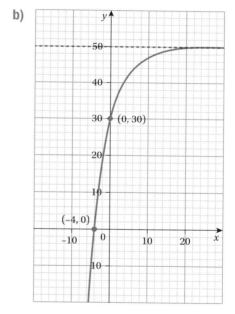

In the last section you found out that solving $e^x = b$ is easy using the ln function on your calculator.

If $e^x = b$ then using the ln function you know $x = \ln b$

Solving $a^x = b$ is almost as easy.

Key point

Using the ln function on $a^x = b$ gives $x = \dfrac{\ln b}{\ln a}$

Example 8 According to the 'rule of 70', an amount of money invested at 12% interest per annum will approximately double in value in $\dfrac{70}{12} \approx 6$ years. Use the ln function to check this estimate.

After t years, an amount £A will become £A $\times (1.12)^t$.

So $A \times 1.12^t = 2A$

Divide both sides by A. $1.12^t = 2$

Use ln. $t = \dfrac{\ln 2}{\ln 1.12} \approx 6.1$

It takes approximately 6.1 years for the amount of money to double so the initial estimate of 6 years is a good one.

The interest rate is 12% so the multiplying factor for compound interest is 1.12

 1060

Activity 8

Check that you fully understand the relationship between

    $a^x = b$ and $x = \dfrac{\ln b}{\ln a}$

by writing your own problems like the one below and checking the answers.

For example:

Think of two numbers	3 and 7
Use your calculator to find 3^7	$3^7 = 2187$
Use this result to write an equation	$3^x = 2187$
Then solve using ln	$x = \dfrac{\ln 2187}{\ln 3} = 7$

You should end up with the number you originally used as the exponent.

Key point

If you apply the $\dfrac{\ln b}{\ln a}$ rule to the equation $e^x = b$, you obtain

    $x = \dfrac{\ln b}{\ln e}$

This is the same as the standard equation you used earlier, $x = \ln b$, since $\ln e = 1$.

Example 9 Find the value of x such that $3^{7x} = 4000$.

In this example $\dfrac{\ln 4000}{\ln 3}$ gives $7x$ rather than x.

$$7x = \frac{\ln 4000}{\ln 3} \approx 7.55$$

Therefore, to find x, you must divide both sides by 7.

$$x \approx 1.08$$

Check this by substituting $x = 1.08$ in the initial equation

$$3^{7 \times 1.08} = 3^{7.56} \approx 4046$$

Example 10

> ## The number of E. coli bacteria can multiply one million fold in only 6 hours!

Is this sensational headline correct?

How long would it take, in ideal conditions, for the number of E. coli bacteria to multiply one million fold?

Assume that doubling occurs every 20 minutes. If the number of 20-minute periods that have elapsed is n, then the number of bacteria could multiply by a factor of 2^n.

Write an equation making the multiplying factor $2n$ equal to one million.

$$2^n = 1\,000\,000$$

Solve the equation by using the relationship between $a^x = b$ and $x = \dfrac{\ln b}{\ln a}$.

$$n = \frac{\ln 1\,000\,000}{\ln 2} \approx 19.93$$

It takes 19.93 periods of 20 minutes to reach one million bacteria.

Convert this to the number of hours.

It takes $\dfrac{19.93 \times 20}{60} \approx 6.6$ hours

The headline is (slightly) exaggerated.

Exercise 13D

According to one fable, the inventor of the game of chess asked for his reward to be

1 grain of rice for the first square on the board

2 grains of rice for the second square on the board

4 grains of rice for the third square on the board

8 grains of rice for the fourth square on the board and so on, up to and including the 64th square.

1   Make a Fermi estimate of the number of grains of rice in a sack of rice.

2   On which square of the chessboard would the reward be an entire sack of rice?

3   On which square of the chessboard would the reward be a million sacks of rice?

Consolidation exercise 13

1 Evaluate.

   a) 3^5

   b) $1.7^{2.9}$

   c) $10e^{-4}$

2 Solve the equations.

   a) $4^x = 8$

   b) $3^y = 5$

   c) $1.3^t = 5.7$

   d) $e^{5x} = 17$

3 Solve the equations.

   a) $5 + 7e^x = 9$

   b) $3 - 4e^{2t} = 1$

4 In 1801, the population of Scotland was 1.6 million. This had risen to 4.5 million by 1901. A model suggested for this population, P million, is

$$P = 1.6(1.11)^n,$$

where n is the number of decades since 1801.

   a) Show that the model is accurate for 1801 and 1901.

   b) Is it sensible to use this model to estimate the population in either 1851 or 1951? Explain your answer.

   c) The 1851 census gave a population of 2.9 million. How accurate is the model?

   d) What does the model give for the population in 2016? Use the internet to check the accuracy in this case.

> Population statistics and projections are crucial for governments when they are planning the future needs of a country.
>
> Rapid growth in the world's population poses difficult challenges for food and water supplies, healthcare and social cohesion.
>
> The 2015 world population was 7.3 billion and was growing at 1.14% per year. This rate of growth is slowing in most parts of the world except for sub-saharan Africa.

5 At h metres above sea level the air pressure, P pascals, is given by

$$P = P_0\,e^{-0.000124h}$$

   a) At the top of Everest (8848 metres above sea level) what is the air pressure as a fraction of that at sea level?

   b) At what height is air pressure half of that at sea level?

   c) What, apart from height, also affects air pressure?

6 The equation $\theta = 20 + 80e^{-kt}$ can be used to model the temperature of a large container of boiling water, t hours after heating was stopped.

   a) Show calculations to confirm that the water was initially boiling.

   b) Half an hour later, the water had a temperature of 60°C. Show that $k = 1.39$.

   c) How many minutes does it take for the temperature of the water to fall from boiling point to 30 °C?

   d) What is the significance of 20 in the formula for θ?

7 Carbon dating is based upon the fact that the mass, in grams, of carbon-14 in an organism, t years after its death, is given by $m = m_0 e^{-0.000121t}$.

   a) What does m_0 represent?

   b) How many years after the organism's death does it take for the amount of carbon-14 to be halved?

   c) Once you know the amount of carbon-14 in a fossil you have found, what additional information would you need to be able to date the fossil? How do you think this could be obtained?

d) After 50 000 years, how much would remain of 1 gram of carbon-14?

Q **e)** What happened in the mid-20th century that affected carbon-14 levels in living animals?

8 This bar chart shows the number of streaming subscribers (N millions) for an internet entertainment provider.

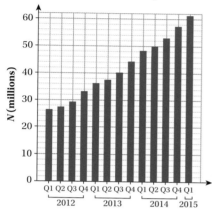

The tops of the bars appear to be following an exponential growth shape.

Suppose that a model of the form $N = N_0\,e^{kx}$ is to be fitted to these data, where x is the number of quarters since the first quarter of 2012.

The number of subscribers in the first quarter of 2012 ($x = 0$) is 26.5 million. The number in the first quarter of 2015 is 61.4 million.

a) Find N_0.

b) Explain why $e^{12k} = 2.32$

c) Hence find k to 2 decimal places.

d) What does the model suggest for the number of subscribers in the third quarter of 2013? Comment on your answer.

Review

After working through this chapter you should:

- be able to evaluate expressions of the form a^x
- know that graphs of exponential functions of x have shapes as shown

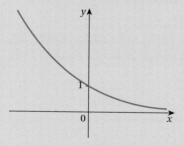
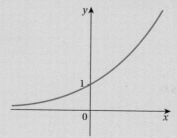

- be able to use exponential functions to model growth and decay in various contexts
- understand that e is the standard base for exponential functions
- know that the gradient at any point on the graph of $y = e^x$ is equal to the y-value of that point
- know that $\ln x$ and e^x are inverses of each other
- be able to use the ln function to solve equations of the form $Ca^x = D$ and $Ce^{kx} = D$.

Investigation

Understanding gifting circles

Women empowering women

Receive gifts of £24 000 in just a few weeks' time.

An amazing opportunity!

To get £24 000 to spend how you wish requires an outlay of only £3000.

- **Become financially independent**
- **Leave your money troubles behind**
- **Use your windfall to help other women**

> *This has been a golden opportunity to further my work in Zambia. Thank you to all the women who were courageous and generous enough to join my gifting circle.*
>
> *Dr Sandra A.*

> *It really works! But it's about so much more than money. I have benefitted tremendously – my new sense of sisterhood has completely altered my attitude to life.*
>
> *Zara D.*

Typically, a gifting circle starts with a group of 7 people. These are the 'bride' and 'bridesmaids' in the centre of the circle as illustrated here.

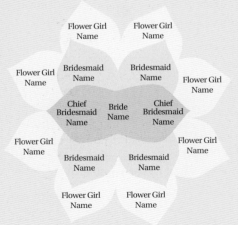

Different gifting circle schemes will have different names for the various positions.

'Gifting circles' are largely targeted at women. However, there are equivalent schemes targeted specifically at men.

Think about the language and contexts which might be used in these cases.

The original 7 people must recruit another 8 to fill the 'flower girl' places. To be allowed to join the scheme and, eventually, benefit from it themselves, the prospective flower girls must each give a gift of £3000 to the bride. The bride now leaves the circle and is free to use her £24000 of gifted money as she wishes. The circle then splits into two with everyone moving one step closer to the centre:

- each chief bridesmaid becomes a bride
- each bridesmaid becomes a chief bridesmaid
- each flower girl becomes a bridesmaid.

The reality

In practice, gifting circles operate as described for a short period of time. They then collapse, leaving most participants significantly poorer.

1 Investigate how this collapse is connected to exponential growth.

2 Research the internet for information about schemes of this sort. Related forms of scheme include pyramid selling, Ponzi schemes and chain letters.

3 What is the main feature which distinguishes pyramid selling from a genuine multi-level marketing operation?

4 Your friend is thinking of taking part in this scheme. Critically analyse the leaflet and use what you have learned in this chapter to explain why the scheme is a scam and why she shouldn't take part.

Gifting circles and related schemes continue to reappear in different countries and different guises. The terminology in which the schemes are described is designed to mask their true nature. Even the 'circle' description distracts from the actual pyramid structure.

The reality of needing to recruit more and more new members means that people strongly encourage their friends to join the circle. This may keep the scheme going long enough to enable them to more than recoup their outlay – but only at the expense of their friends who will face even greater difficulties in recruiting new members.

Pyramid schemes involve paying (some) participants for enrolling other people rather than undertaking any real business activity or service.

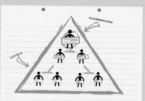

Investors in a **Ponzi scheme** receive returns on their investment from money paid in by new investors rather than from profit earned as part of a genuine business scheme.

THESE SCHEMES ARE SCAMS AND ARE STRICTLY ILLEGAL!

The first prosecutions in the UK saw six women convicted under the Unfair Trading Act 2008.

Committee members behind the scheme, 'Key to a fortune', pocketed up to £92000 each whilst nearly 90% of recruits lost out. Some joined several circles and lost up to £15000.

Practice questions

Paper 1

Preliminary Material

Going the extra mile

The pleasures first of passing your driving test and then owning your first car come with a variety of financial costs. These costs can be split into 'standing charges' and 'running costs'.

Standing charges

The cost of the car

This may be a one-off initial payment or some form of loan agreement. Even if the car is paid for outright, there is an opportunity cost in not having the cash available for earning interest or for some other purpose.

A standard way of considering the cost of a car is to use an annual figure for 'depreciation' and 'cost of capital'.

Other annual standing charges

Costs such as road tax, breakdown cover and (especially significant for young drivers) the cost of insurance must be added to the cost of the car.

Total annual standing charge

This total charge will, of course, vary for different drivers and different cars. The table shows approximate annual charges for a 'typical' driver according to the purchase price of the car when new.

Purchase price (£)	12 000	15 000	25 000	35 000
Standing charge (£ per year)	2 000	3 000	4 000	10 000*

*The jump in the total standing charge for the most expensive cars is partly caused by the higher insurance costs for these cars.

Running costs

These are often calculated as costs per mile. The main (and probably most obvious) cost is that of fuel. There are also smaller costs per mile for tyres, servicing and typical repairs.

Taxation 2015–2016

National Insurance 2015–2016

Percentage National Insurance Due	Minimum Weekly Income	Maximum Weekly Income	Minimum Monthly Income	Maximum Monthly Income
Non-contracted out				
Nil		Below £112		below £486
0%	£112	£155	£486	£672
12%	£155.01	£815	£672.01	£3532
2%	above £815		above £3532	
Contracted out				
10.6%	£155.01	£770	£672.01	£3337

Note: A person is '**contracted out**' if they are a member of a contracted out occupational pension scheme or personal/stakeholder pension. They pay a slightly lower rate of National Insurance as the state does not have to pay them as much pension.

A person with a weekly income of, say, £320 a week will pay 12% on the amount above £155.

A person with a weekly income of, say, £940 a week will pay 12% on the amount between £155.01 and £815 plus 2% of the amount above £815.

Income tax 2015–2016

Most people have a personal allowance. This is an annual amount of tax-free income. The personal allowance for 2015–16 was £10 600. The rates of income tax you pay depend on how much taxable income you have above your personal allowance.

Income tax rates and taxable bands 2015–2016

Rate	Taxable Income
Basic: 20%	£0 – £31 785
Higher: 40%	£31 786 – £150 000
Additional: 45%	Over £150 000

To calculate your income tax if your annual income is £100 000 or less

Find your taxable income by subtracting your personal tax allowance from your annual income.

You pay income tax at 20% on the first £31 785 of your taxable income.
You pay income tax at 40% on your taxable income over £31 785.

Student loans

Student loans are paid back at a rate of 9% of gross earnings in excess of £21 000 per year.
Interest is added to the loan from the time when earnings start.

Net pay

Your net pay or take-home pay is the money you earn after any deductions for income tax, National Insurance and student loan repayments.

Paper 1 questions

1 The numbers of cars licensed in Great Britain in 2014 were as shown in the table.

Numbers in thousands

Petrol	Diesel	Other	Total
19 054	11 209	251	30 513

(a) Explain why the individual numbers do not add up to the total.

[1 mark]

(b) A student intends to analyse the annual running costs of a car and decides to choose two cars of each type for the study.

(i) Give **two** reasons why this would not be a good sample to take.

[2 marks]

(ii) Give a full description of a better sampling method that could be used.

[4 marks]

2 (a) Which of the following are discrete data?
Choose **one** answer.

Colours of shoes   Shoe sizes   Lengths of feet

[1 mark]

(b) In order to write a report on shoe purchases, briefly describe how you might find relevant

(i) primary data

[2 marks]

(ii) secondary data.

[2 marks]

3 Salma invests £1500 in a savings account. The compound interest is fixed at 2% each year.

After 3 years, Salma withdraws £500.

After each following year she withdraws a further £500 (if possible) until the account is empty.

Create a table to show the amounts in Salma's account at the end of each year.

What is the total amount of money that she will have withdrawn?

[7 marks]

4 Estimate how many shots it would take to hit a golf ball the length of Great Britain.

Show details of all assumptions and calculations.

[6 marks]

5 Use **Taxation 2015–2016** from the Preliminary Material.

The spreadsheet shown below is used for the PAYE tax on 30th June 2015 for an employee of a company who has an annual salary of £24 000.

The Tax code shows that the employee has a tax-free allowance of £10 600.
The Tax to date shows the total tax paid in April, May and June 2015.

	A	B	C	D
1	Monthly salary	£2000	PAYE Tax	
2	Tax period	3	Tax to date	
3	Tax code	1060L		

(a) What do the letters PAYE stand for?   [1 mark]

(b) Calculate the value in cell D1.   [5 marks]

(c) Write down a formula which gives the value in cell D2.   [1 mark]

6 A newspaper asked its readers how much time they spent playing video games each week.

The responses of those who played video games are shown below.

The least time was 20 minutes and the greatest time was 900 minutes.

Length of time, t minutes	Number of readers
$0 \leqslant t < 20$	0
$20 \leqslant t < 60$	15
$60 \leqslant t < 120$	18
$120 \leqslant t < 300$	25
$300 \leqslant t < 600$	11
$600 \leqslant t < 1000$	11
$1000 \leqslant t$	0

The results of a similar survey of the readers of a magazine are shown as a box and whisker diagram below.

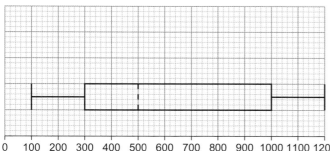

Number of minutes

(a) Compare the survey results for the newspaper readers with those for the magazine readers. You may wish to use a grid.   [8 marks]

(b) Explain possible reasons why the data for the two sets of readers might differ.   [2 marks]

7   Karl borrows £2000 at an APR of 15%.

He pays back the loan in two equal instalments, one at the end of the first year and a final instalment at the end of the second year.

(a) In the formula

$$C = \sum_{k=1}^{2} \left(\frac{A}{(1 + i)^{t_k}} \right),$$

state the values of C, t_1, t_2 and i.

[4 marks]

(b) Find the value of each instalment.

[4 marks]

8   **(a)** It is usually sensible to shop around for the best deal on purchases.

Give **two** good reasons why you might choose not to do this.

[2 marks]

(b) You may use **Going the extra mile** in the Preliminary material.

How far is it worth driving to fill up a car with petrol that is 1p per litre cheaper than at a nearby garage?

Show details of all assumptions and calculations.

[8 marks]

Paper 2

Preliminary Material

Make schools pay!

Schools should be fined up to £500 for every student who leaves without achieving at least a grade C in GCSE English or maths.

Policy Exchange, Aug 2015

According to Natasha Porter of the think tank *Policy Exchange,* this money could be put towards the costs of resit classes in further education colleges. This support is needed because:

Colleges are undertaking a large burden that is, in part, caused by the failure of some elements of the school system to have adequately ensured that 16-year-olds have achieved an appropriate baseline of qualifications.

These ideas have not found favour with Head Teachers and College Leaders who say that fining schools could be self-defeating and make the situation worse.

Schools are already facing real-term cuts in their budgets and unprecedented difficulties in recruiting staff, particularly maths teachers. A resit levy would potentially worsen this situation.

Brian Lightman, General Secretary of the Association of School and College Leaders

Some of the data that was used by *Policy Exchange* in their report *Crossing the Line* are given in the following tables concerning resits in 2013.

Resit figures 2013

	Schools	FE Colleges	Sixth Form Colleges
English GCSE	20 544	100 239	8 738
Maths GCSE	27 579	110 811	11 193

It is important to note that 'although schools educated a higher number of post-16 students overall than any other setting, FE colleges had much higher numbers of students who decided to retake maths or English GCSE'.

Pass rates for resits 2013

Percentages of resitting candidates achieving a grade C or better.

Schools	38%
Colleges	21%

A representative for one of the examination boards pointed out that when discussing the percentages of students achieving particular GCSE grades it must be remembered that the Government's regulator, Ofqual, ensures that the proportions of grades awarded from year to year only changes if the ability of the cohort taking the exam changes. This 'ability' is determined by the cohort's key stage 2 results. In effect, the quality of work produced by a cohort in the GCSE exam itself has little or no effect on the number of pupils receiving good grades.

Paper 2 Compulsory questions

1 The results of a survey of salaries are illustrated by the graph.

Salaries (£000s)

(a) Criticise this graph.

[2 marks]

(b) Making any necessary assumptions, which should be stated, draw a more suitable diagram to represent these data.

[4 marks]

2 Andy has borrowed £1000 which he must repay in two equal instalments, £600 at the end of the first year and a final payment of another £600 at the end of the second year. Andy makes the statement:

I will have to pay £200 interest over the two years of the loan. This is £100 per year and means that I am paying 10% interest per year.

(a) Which part of Andy's statement is correct?

[1 mark]

(b) Andy is not a mathematician. Using simple calculations, explain to him why the interest rate that he is paying is more than 10%.

[3 marks]

3 Use **Make schools pay!** on the Preliminary Material.

(a) The report *Crossing the Line* was discussed on a radio programme on 25th August 2015. Referring to the data used in the report, it was stated that

Further Education colleges take five times more English students and six times more maths students who don't make the grade than schools.

Comment fully on the numbers used in this statement.

[4 marks]

(b) Explain briefly why *'fining schools **could** be self-defeating'*.

[2 marks]

(c) Explain the observation that

The work that is done in secondary schools in the five years leading up to GCSEs may greatly affect the result of an individual student or the result of all the students in that school but does not affect the overall GCSE results of the nation's students.

[4 marks]

Optional Paper 2A

4 Typical carbon dioxide contents of air, in parts per million by volume (pmvv), are considered to be 350–1000 pmvv indoors and 250–350 pmvv outside.

An environmental health scientist takes measurements of the carbon dioxide content of air at street level in a town. The results for five samples of air were as follows.

395   384   389   393   391

These data can be assumed to be from a normal distribution with mean μ pmvv and standard deviation 4 pmvv.

Construct a 95% confidence interval for μ and hence comment on the street level measurements.

[7 marks]

5 An examination consists of two papers, I and II. The marks on these two papers are added together to give a candidate's overall percentage.

The table and graph show the results for 10 candidates. The graph also shows the line of best fit.

Candidate	Paper I	Paper II
A	7	9
B	10	19
C	16	22
D	15	29
E	20	33
F	25	35
G	22	38
H	28	38
I	25	50
J	38	52

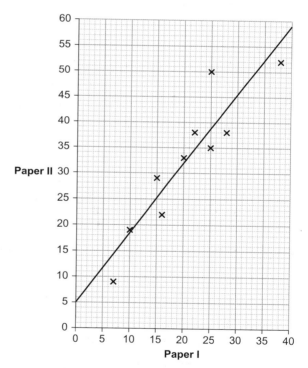

(a) Find and interpret the gradient of the line of best fit.

[3 marks]

(b) Describe the correlation between the two sets of marks. What does this indicate about the skills being tested by each paper?

[2 marks]

(c) One candidate has a medical certificate to show that his performance in Paper I had been affected by ill health. Which candidate is this likely to be? Justify your answer and suggest how the overall percentage mark for this candidate could be adjusted.

[3 marks]

6   Draw a histogram of the temperatures of a sample of healthy adults that would
    be consistent with the idea that body temperatures of healthy adults are normally
    distributed with mean 37 °C and standard deviation 0.4 °C. Justify your choice of
    histogram by commenting on **two** of its features.

    [6 marks]

7   The International Monetary Fund ranks countries according to their Gross Domestic
    Product (GDP). In 2014, the top six countries were, in order, USA, China, Japan,
    Germany, UK and France.

    The table shows approximate GDP per person and health expenditure per person for
    these countries. The amounts are in US dollars ($).

Country	GDP per person ($)	Health expenditure per person ($)
USA	55 000	9000
China	13 000	400
Japan	37 000	4000
Germany	46 000	5000
UK	40 000	3600
France	40 000	4900

(a) Plot a scatter graph of health expenditure per person (y) against GDP per
    person (x).

    [2 marks]

(b) Calculate the equation of the regression line of y on x and draw this line
    on your scatter graph.

    [5 marks]

(c) Italy has a GDP per person of $36 000 and India has a GDP per person
    of $6000. Use your regression line to estimate the expenditure on health
    in each of these two countries. Explain the likely accuracy of each answer.

    [4 marks]

8   The weights (in kg) of 1 kg bags of sugar packed by a particular filling
    machine have distribution $N(1.01, 0.01)$.

(a) What is the probability that a bag chosen at random is underweight?

    [4 marks]

(b) A random sample of 25 bags is selected. What is the probability that
    the average weight of these bags will be less than 1 kg?

    [4 marks]

Optional Paper 2B

4 Some of the data from a health survey of 1000 US citizens in 2013 are summarised in the Venn diagram.

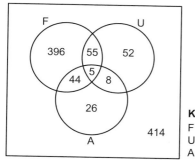

Key
F – Female
U – Under 15
A – Asthmatic

(a) Use these data to provide evidence for or against each of the following statements.

    (i) This was a stratified survey.

                                   [2 marks]

    (ii) Asthma affects approximately 1 in 12 people.

                                   [2 marks]

(b) Assuming that this sample is representative of the population as a whole, write a brief summary of how gender and age affect the proportions of people with asthma.

                                   [6 marks]

5 Marcus is preparing a vegetable curry for his flatmates. The table lists the separate activities, their durations and precedences.

Activity	Duration (minutes)	Immediate predecessor(s)
A: Lay the table	5	-
B: Chop vegetables	8	-
C: Mix spices	4	-
D: Fry spices and vegetables	3	B, C
E: Cook rice	15	-
F: Add water and cook curry	12	D
G: Serve chutneys	2	A
H: Microwave poppadoms	3	-
I: Serve meal	3	E, F, G, H

(a) Suppose that the meal has to be prepared in the shortest possible time.

    Construct an activity network and find the earliest start time and latest finish time for each activity.

                                   [7 marks]

(b) On the other hand, suppose that Marcus receives no help and that only activities E and F can be carried out at the same time as other activities.

    What is the latest that Marcus must start preparing the meal if he wishes it to be served by 7pm? Explain fully, in context, how this can be achieved.

                                   [4 marks]

6 Each ticket in a lottery costs £1. The chances of a ticket winning a prize are given in the table.

Lottery – £1 per ticket		
	Odds of Winning	**Prize**
Match 6	1 in 2 000 000	£100 000
Match 5	1 in 400 000	£10 000
Match 4	1 in 10 000	£250
Match 3	1 in 5000	£50
Match 2	1 in 250	£20
Match 1	1 in 250	£10

(a) What is a person's expected loss for each ticket they buy?

[4 marks]

(b) This lottery is advertised with the statement

> For only £1 you have a chance of 1 in 120 of winning a prize of up to £100 000!!!

Explain why this statement is factually correct but, nevertheless, misleading. You must show your working.

[4 marks]

7 (a) Briefly describe the 'cost-benefit principle'.

[2 marks]

(b) A company has an annual revenue of £500 000.

The company outsources approximately 50 hours of work each month at a cost of £60 per hour.

The owners of the company are considering an expansion which would enable all the work to be done in-house. The costs involved in this expansion are:

- salary and associated costs of £5000 per month

- leasing an extra 20 m² of office space at an annual cost of £100 per m².

The only other foreseeable benefit of the expansion is that it would enable the company to bid for one special contract which would increase the company's revenue by approximately 10% per annum.

What would the odds of winning the special contract need to be to justify the expansion? Fully justify your answer with numerical calculations.

[9 marks]

Optional Paper 2C

4 The graph shows how the installation cost, £S, of a solar panel system depends upon its capacity for generating electricity, E watts.

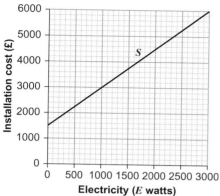

(a) Explain why you would not expect a graph of installation cost against electricity to go through the origin. **[2 marks]**

(b) The cost (£S) and capacity (E watts) are related by an equation of the form

$$S = mE + c.$$

Find the values of the constants m and c. **[3 marks]**

(c) The installation cost, £T, of a wind turbine system of capacity E watts is given by the formula

$$T = 0.75E + 2000.$$

Copy the graph of installation cost against electricity and on the same axes draw the graph of $T = 0.75E + 2000$. Hence comment on which system is cheaper to install and justify your answer. **[5 marks]**

5 Following a period of rapid growth, the populations of European countries are now roughly stable.

This same pattern appears to be being followed in most other countries in the world. United Nations (UN) predictions are that these countries may experience strong population growth until 2050 but that, thereafter, growth will slow significantly and will be stable or even falling by 2100.

African countries are an exception. On this continent, the UN is forecasting rapid population growth throughout the 21st century.

The table compares the populations of India and Africa, in millions, in 1950 and 2000.

	India	Africa
1950	360	230
2000	1020	811

(a) Assuming a linear model for the population of India, estimate its population by 2050. **[4 marks]**

(b) Assume an exponential model for the population of Africa. By solving the equation

$$811 = 230e^{50x}$$

estimate Africa's population by 2050. **[5 marks]**

6 Three newspaper cuttings which illustrate different aspects of the changing price of an ounce of gold are shown.

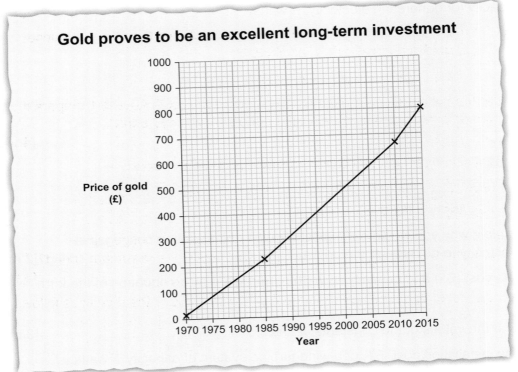

Gold proves to be an excellent long-term investment

A

Gold prices show long-term decline

B

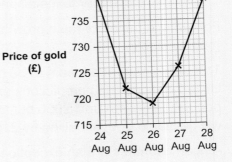

Gold prices bottom out

C

(a) State one way in which Graph A could be said to be misleading about changes in the price of gold. **[1 mark]**

(b) Make two criticisms of Graph B. **[2 marks]**

(c) (i) For the data shown in Graph C, explain why a quadratic graph could be
expected to give a good model of the prices over these few days.

[1 mark]

(ii) For these few days, find constants a and b such that the price per ounce,
£P, of gold is approximately modelled by

$$P = a + bt^2,$$

where t is the number of days after (and before) 26th August. Compare the
values given by your model with the data shown in Graph C.

[4 marks]

7 Usain Bolt's times for each 20 m in a 100 m race were as follows.

Distance (m)	0–20	20–40	40–60	60–80	80–100
Time (seconds)	2.93	1.76	1.66	1.61	1.67

(a) Describe how Bolt's speed changes during the race and support your
statements with numerical calculations. What was Bolt's maximum speed?

[4 marks]

A distance–time graph modelling the first two seconds of the race is as follows.

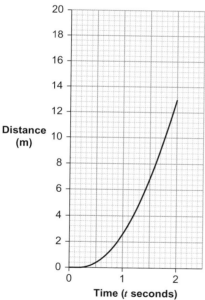

(b) What was Bolt's speed at each of the following times?

(i) $t = 0.1$

[1 mark]

(ii) $t = 1$

[3 marks]

(c) Showing details of your calculation, find Bolt's acceleration at time $t = 1$.
You may wish to use a grid.

[5 marks]

Answers

Chapter 1

Exercise 1A

Note that for some questions alternative answers may be acceptable.

1 a) ii, iv, x and xii
  b) i, v and ix
  c) iii, vi, vii, viii and xi
2 a) Carry out a survey of students.
  b) Look for information on the internet.
3 a) For example, questions about whether people are in favour of it, location, ticket cost, frequency of visit, etc.
  b) i Easy and cheap, but not random, excludes many people and likely to be biased.
     Only includes people who travel by bus on a Saturday evening and are happy to be interviewed. Groups travelling together may have similar views.
   ii Includes all local households, but return rate may be low. It excludes people who live more than 3 miles from the town centre whose views may be relevant.
   iii Very easy and inexpensive but excludes all those who do not listen to local radio or watch local TV. People may not be willing to phone or email to give their views.
4 The results may be very unreliable as people want to answer quickly to earn money and may not think about their response.
5 a) Group students from college according to gender, age, ethnicity and any other relevant characteristics (e.g. course attended, or part-time or full-time in some colleges). Randomly select students from each group in proportion to the number in the group. Interview each of the students or ask them to complete a questionnaire.
  b) Group people from electoral roll according to gender, age, area of town in which they live. Randomly select people from each group in proportion to the number in the group. Send a questionnaire to each person by post.
6 Use the electoral roll to find the proportions in sub-groups divided according to gender, age and any other relevant characteristics. Divide 500 into the same proportions, then go to a busy location (e.g. centre of city) and approach people who appear to be in each sub-group until the quotas have been reached.
7 a) Doctors' practices or towns/cities could be used as clusters (questionnaires sent out to all adults by post).
  b) No. People who are not on a doctor's list or do not live in the selected towns/cities have no chance of being part of the sample.
8 Increase the sample size, e.g. to 100, and make the sample representative.
  Use random numbers to select the required sample from each group, then ask them to complete a questionnaire. The return rate may be low unless he has some way of persuading the pupils to cooperate.
9 For example she needs to decide what age boundaries to use, whether to contact individuals or households. Patients may not give honest answers to questions asked by their dentist.
10 One possibility is to number individual square metres and use random numbers to select a sample of 20 of them.
11 a) Use random numbers and a list of all teenagers in the area.
  b) Divide teenagers into sub-groups according to gender and age, then use random numbers to select a sample in the same proportion.
  c) Use schools, colleges and large places of work as clusters. Select clusters at random, then interview every teenager or a random sample of teenagers from each.
  d) Divide required sample into quotas with proportions reflecting gender and age groups of the population. Go to a busy place (e.g. town centre on Saturday) to fulfil the quotas.
12 a) Total number of employees = 750.

	Male		Female	
	Full-time	**Part-time**	**Full-time**	**Part-time**
Factory	64	5	31	24
Warehouse	15	5	7	4
Office	8	2	11	15
Delivery	5	1	3	0

  b) Age, experience, type of work, etc.
  c) Reluctance to be truthful, poor return rate, etc.
13 a) i

	18	**19**	**20**	**Other**
Female	2777	1249	350	996
Male	2315	1190	354	769

   ii Courses, mode of attendance, type of HE, ethnicity, etc.
   iii Identifying, locating and contacting students, etc.
  b) Your report should summarise your findings.
14 There are many possible answers depending on your school or college and the questions you ask.

Exercise 1B

1 a) i £499 **ii** £500 **iii** £526.30 **iv** £715
  b) i £2.6 m **ii** £3.6 m **iii** £4.2 m **iv** £14.7 m
2 a) Stacy has taken the average of the two means without taking into account the number of workers in each group.
  b) £10.50
3 90 runs
4 a) mode = 23 mean = 20 median = 19
  b) mean (or median) with reason, mode too high.
5 a) i median = 1 minute IQR = 2 minutes
   ii median = 40.5 seeds IQR = 5 seeds
   iii median = £2.97 IQR = £1.29
  b) Each dataset has an extreme value which will affect the mean and range.
6 a) mean = 39.5 marks s.d. = 7.5 marks
  b) mean = 24 °C s.d. = 2.6 °C
  c) mean = 33.2 minutes s.d. = 6.4 minutes
  d) mean = 40.9 runs s.d. = 31.7 runs
7 a) marks, runs scored
  b) temperatures, journey times
8 a) modes 157 cm and 174 cm median = 166 cm
  mean = 164 cm
  range = 47 cm IQR = 17 cm
  standard deviation = 11.7 cm

b) Median and IQR as there are 2 modes and the other measures are all affected by the 134 cm extreme value. (Alternative answer: mean and standard deviation because they use all of the data.)

9 a) Time for which cars were left in a short stay car park

```
5 | 1 4 5 6 8 8 8 8 9 9 9 9 9 9
4 | 2 3 4 5 5 6 7 7 8
3 | 0 3 5 6 6
2 | 1 4 8
1 | 0 5 5 8
0 | 9
```

Key 5|9 means 59 minutes

b) i 59 min **ii** 50 min **iii** 45 min (accept 45.5)
iv 28 min (accept 26.5)

c) You should include relevant comments related to the context, for example, 'Nearly half of the times were over 50 minutes with a lot of these being just under 60 minutes. This suggests that the waiting time was limited to an hour.'

10 a) Route A   **i** 19 min   **ii** 26 min
               iii 27 min   **iv** 20 min
     Route B   **i** 31 min   **ii** 21 min
               iii 29 min   **iv** 9 min

b) i Include comments such as 'On average Route B takes longer, but the times are less variable than for Route A.'
ii Route B as only one morning took longer than 35 minutes, whilst there were 5 mornings when Route A took longer than 35 minutes.

11 Your report should include a stem-and-leaf diagram, an average and a measure of spread from those given below and relevant comments.

Age (in years) on marrying

```
      Women  |   | Men
     9 8 7 6 | 1 | 7 9
9 8 8 7 6 6 6 3 1 1 | 2 | 1 3 4 5 8 8 9
     8 5 4 2 0 | 3 | 0 2 2 3 4 5 5 5 8
             0 | 4 | 2 5
```

Key: 1|2|3 means age of woman 21, age of man 23
Men:
mode = 35 years   median = 31 years
mean = 30.25 years   range = 28 years
IQR = 10.5 years   standard deviation = 7.4 years
Women:
mode = 26 years   median = 26.5 years
mean = 26.7 years   range = 24 years
IQR = 10 years   standard deviation = 6.9 years
On average women married at younger and less variable ages than the men.

Exercise 1C

1 Answers should be based on the following results.

	Mode	Range	Median	Mean
Females	19 years	3 years	19 years	18.6 years (1 dp)
Males	17 years	4 years	18 years	18.7 years (1 dp)

2 a)

Runs scored by two batsmen

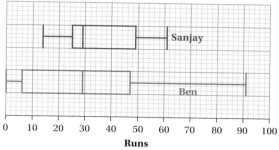

b) Ben's mean score is 32.9 with standard deviation 29.4, whilst Sanjay's mean score is 35.6 with standard deviation 15.8.

c) Both have same median. Sanjay is more consistent, but Ben occasionally manages a really good score. Sanjay is probably the wisest choice.

3 a) median = £21 000, interquartile range = £3000
b) UQ + 1.5 × IQR = £24 000 + £4500 = £28 500
£70 000 > £28 500, so £70 000 is an outlier.
c) £23 500 is the mean, but as it is higher than most of the wages, it is not a very good representative value – it has been inflated by the outlier.

4 a)

	Mean	SD	Median	IQR
Badley	22.1 (1 dp)	10.8 (1 dp)	19	17.5
Lootham	21.3 (1 dp)	10.1 (1 dp)	20.5	16

b)

Burglaries per month in two towns

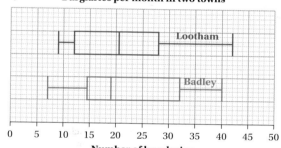

c) i The median supports Kylie's comment.
ii The mean supports Winston's point of view.

5 a) i 16 computers   **ii** 5 computers
b) i 17 computers   **ii** 2 computers
iii

Computers used in school library at lunchtime

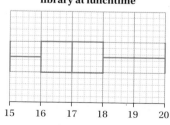

c) i 16.9 computers (1 dp)   **ii** 1.1 computers (1 dp)
6 a) i mode = 40 seeds, median = 40 seeds, mean = 39.4 seeds (1 dp)
ii We checked packets of seeds in our stocks and found that they contained on average 40 seeds.

b) LQ = 39, UQ = 40, IQR = 1
44 is 4 above 40 – this is more than 1.5×1, so 44 is an outlier.

c) Choose samples at random from throughout their stock. Use a bigger sample. (Other answers are acceptable.)

7 Include comparison of estimates similar to the following.

	Minimum	LQ	Median	UQ	Maximum
Turnover	£8 m	£13 m	£15.3 m	£22.7 m	£35 m
Wage Bill	£6.2 m	£12 m	£17 m	£24.4 m	£37.4 m

For example:
On average the clubs are paying more in their wage bill than their turnover.
Some clubs are paying more than their turnover.
The wage bill is more varied than turnover.
The largest turnover is more than 4 times the lowest.
The largest wage bill is more than 6 times the lowest.
The clubs' turnovers and wage bills are both more spread in the upper half of the distribution than the lower half.

8 a) i For the mode Gemma has just given the highest frequency.
For the range she has used 18 as the highest value.

ii mode = 16 years   range = 5 years

b) i Mark has found the mean and the standard deviation of the frequency values.

ii mean = 14.3 years (1 dp)
standard deviation = 1.6 years (1 dp)

c) i Sunita has ignored the frequencies and used the values halfway, a quarter of the way and three-quarters of the way along the age list.

ii median = 14 years   IQR = 3 years

iii **Ages of students on exchange trip to Germany**

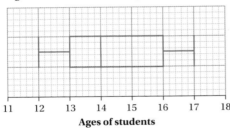

Ages of students

Comments including more spread in the third quarter of the distribution than the others.

9 a) **Errors made by clerical workers**

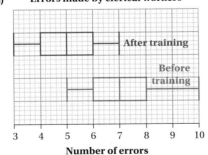

Number of errors

b) Your comments may include:
- reduction in median implying improvement
- reduction in overall spread meaning workers are now more consistent.

10 a) **Results of memory test on consecutive days**

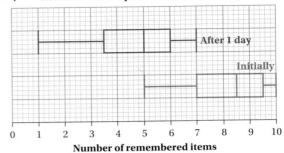

Number of remembered items

b) Initially   mean = 8.2 items
standard deviation = 1.6 items (1 dp)

After 1 day   mean = 4.8 items (1 dp)
standard deviation = 1.7 items (1 dp)

c) Your comments may include:
reduction in median and mean indicating that the volunteers remember fewer items
increase in spread indicating that results are a little more variable after 1 day
lower scores are more spread than the upper scores

11 a) i mode = 1 dog, median = 1 dog,
mean = 1.4 dogs (1 dp)

ii The calculation for the mean includes all of the dogs.

iii The mode and median are both possible values for the number of dogs whereas the mean is not.

b) i The frequencies (number of families who have each number of dogs).

ii The median and quartiles.

12 a) i Sonja is not correct because in this case the median and the upper quartile are both 2, so they are both shown by the same vertical line.

ii The maximum value should be at 6, not 5.

b) i At present the bar chart does not have a title or label on the vertical axis, so it does not give as much information as the box and whisker plot.

ii If the labels were added, the bar chart would probably be more useful to the manager as it shows how many customers used the shop.

13 a) i **Results of 50 metre swimming races**

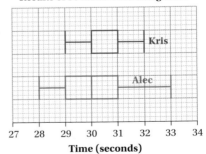

Time (seconds)

ii

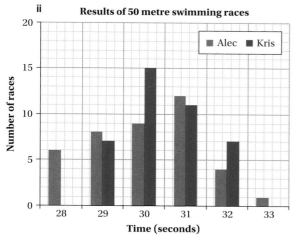

Results of 50 metre swimming races

iii The bar chart. Possible reasons:
It shows all of the times.
It shows that Alec had the best time on 6 of the races.
The box and whisker plot also shows that Alec had the best time, but it does not show on how many races.
Most people are likely to understand the bar chart whilst box and whisker plots are less well understood.
The times for Kris have the same lower quartile and median, so the box and whisker plot does not show these clearly.

b)

	Mode	Range	Median	IQR	Mean	SD
Alec	31 s	5 s	30 s	2 s	30.1 s	1.3 s
Kris	30 s	3 s	30 s	1 s	30.5 s	0.99 s

The mode gives Kris the shorter average time whereas the mean gives him a longer average time and the median is the same for both. All the measures of spread indicate that Kris has more consistent times.
So Kris could use the mode and range.

14 Your report should include some of the values given below (or the equivalent in centimetres) which support your comments.

Shop	Dress size 12 (inches)		Dress size 14 (inches)	
	Bust	Hips	Bust	Hips
Mode	35.4 & 36.2 & 37.0	40.2	38.2	40.0
Range	2.4	2.6	3.2	2.9
Mean	36.4	39.025	38.3	41.275
SD	0.85	1.09	1.00	1.19
Median	36.2	39	38.2	41.35
IQR	1.35	2.2	1.2	2.3
BS	34.65	36.61	36.2	38.18

15 Your answer should state which estimates you used and a comparison of your results with those from the whole class.

Exercise 1D

NB Allow approximate answers from graphs.
1 a) $200 < m \leq 250$ kg
b)

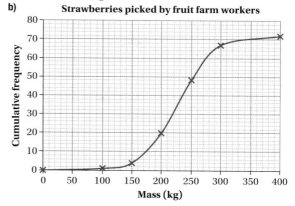

Strawberries picked by fruit farm workers

c) i 229 kg   **ii** 194 kg   **iii** 262 kg
   iv 68 kg   **v** 217 kg   **vi** 269 kg
d) i 229 kg (3 sf) **ii** 56.0 kg (3 sf)
2 a) 30 000–40 000 people
b) i

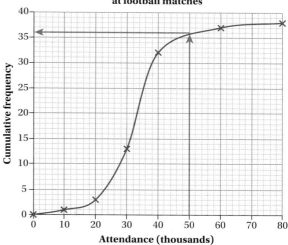

Cumulative frequency graph of attendances at football matches

ii Median ≈ 33 000 people   IQR ≈ 10 000 people
iii Yes. The graph suggests that 36 matches had *under* 50 000 attending. % with attendance *over* 50 000 = $\frac{2}{38} \times 100 = 5.3\%$ (1 dp)
c) i 33 400 (3 sf)   **ii** 11 600 (3 sf)
d) Last year the attendance was on average higher, but more variable. The average attendance has decreased this year.
3 a) Before: median ≈ 19 minutes   IQR ≈ 16 minutes
After: median ≈ 8.5 minutes   IQR ≈ 11 minutes
b) Your comments should include:
The time taken for the apprentices to diagnose the fault is on average quicker after the training (by about 10 minutes). The performance is also less variable.
4 a) mean = 14.35 kg (2 dp)
standard deviation = 1.46 kg (2 dp)

b) i

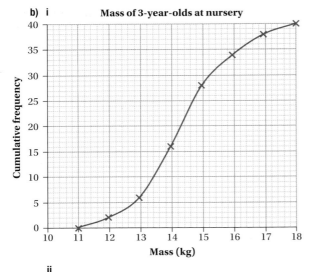

Mass of 3-year-olds at nursery

y-axis: Cumulative frequency
x-axis: Mass (kg)

ii

Percentile	5	10	25	50	75	90	95
Mass from book (kg)	11.8	12.3	13.1	14.1	15.3	16.5	17.3
Mass (kg) to 1dp	12.0	12.6	13.4	14.3	15.2	16.4	17.0

The percentile weights are similar to those given in the book, but those up to and including the 50th percentile (the median) are a little higher and those above the median are lower.

5 a) Online modal class = 30–59 minutes
 TV modal class = 60–89 minutes

b) i Online mean = 56.0 min (3 sf)
 standard deviation = 26.5 min (3 sf)
 TV mean = 76.5 min (3 sf)
 standard deviation = 28.6 min (3 sf)

ii On average the students spent about 20 minutes longer watching TV.
 The times spent watching TV were more variable than those spent online.

c) i

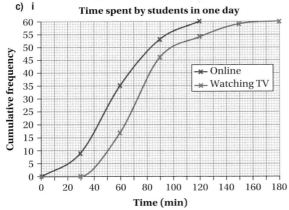

Time spent by students in one day

y-axis: Cumulative frequency
x-axis: Time (min)

Legend: Online, Watching TV

Online median ≈ 54 minutes IQR ≈ 36 minutes
TV median ≈ 72 minutes IQR ≈ 33 minutes

ii The difference between the medians (at 18 minutes) is similar to the difference between the means. However, the IQR is higher for the times online whereas the standard deviation is higher for the times spent watching TV.

6 a) Comments making use of the following values read from graph or box and whisker plot:
Minimum 5 years, Maximum 89 years, Median ≈ 34 years, LQ ≈ 21 years, UQ ≈ 48 years, IQR ≈ 27 years
For example: Distribution is more widely spread in the upper half.
The ages vary between 5 years and 89 years with an average of 34 years.
50% of the users are between 21 and 48 years old with the ages of those in the upper half of the distribution more spread than those in the lower half.

b) i The CF graph allows information such as percentiles to be found and used.

ii The box and whisker plot shows the minimum, median, quartiles and maximum age more clearly.

c) The diagrams do not give the total number of people who took part, the way the data was grouped, the frequencies or any information about how the survey was carried out and how the question was worded.

7 a) i

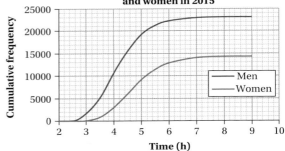

Cumulative frequency graph showing London Marathon times for men and women in 2015

y-axis: Cumulative frequency
x-axis: Time (h)

Legend: Men, Women

ii Comments making use of:
Men: median ≈ 4.1 hours LQ ≈ 3.5 hours
  UQ ≈ 4.7 hours IQR ≈ 1.2 hours
Women: median ≈ 4.7 hours LQ ≈ 4.1 hours
  UQ ≈ 5.3 hours IQR ≈ 1.2 hours
Men were on average about 0.6 hours (roughly 35 mins) quicker. Their times varied in a similar way to the women's.

b) i Men: median ≈ 4.2 hours LQ ≈ 3.6 hours
  UQ ≈ 4.9 hours IQR ≈ 1.3 hours
Women: median ≈ 4.8 hours LQ ≈ 4.2 hours
  UQ ≈ 5.5 hours IQR ≈ 1.3 hours
The fastest runner was a man who took about $2\frac{1}{4}$ hours (or 2.3 hours), about 20 minutes (or 0.3 hours) quicker than the fastest woman. Men were on average about 0.6 hours (roughly 35 mins) quicker than the women. The men's times varied in a similar way to the women's with the times for the faster half of the runners closer together than those of the slower half. At least one woman took over 9 hours.

ii Both the men and women were on average a little quicker in 2015 and their times were a little less variable.

8 a) i The mean rate of pay is £11.62 per hour for men and £11.11 per hour for women. The difference between these averages is 51 pence and this explains Tracy's statement.
Josh has used the median pay for men. At £8.01 per hour this is 45 pence less than the £8.46 per hour, the median for women.

ii Some men at the higher end of the distribution earn much more than the women – this inflates the mean for the men.

b) Men: LQ = £6.57 per hour   UQ = £12.50 per hour
IQR = £5.93 per hour
Women: LQ = £6.87 per hour   UQ = £12.28 per hour
IQR = £5.41 per hour
The male rates of pay are more variable than the female rates of pay.

c) Your comments should make use of:
Men: Mean = 17.4 hours   Median = 17.8 hours
LQ = 9.6 hours   UQ = 24.1 hours
IQR = 14.5 hours
Women: Mean = 18.4 hours   Median = 19.3 hours
LQ = 12.5 hours   UQ = 24.6 hours
IQR = 12.1 hours
On average women work more hours part-time. The hours men work part-time are more variable.
Many more women work part-time than men.

d) No minimum or maximum values are given.

9 a) Assuming the last class is 70–79, that is using the same width as adjacent classes
(A different assumption such as 70–89 makes very little difference to the values.)
Males 1981   mean = 25.7 years (3 sf)
standard deviation = 6.02 years (3 sf)
Males 2011   mean = 32.8 years (3 sf)
standard deviation = 7.34 years (3 sf)
Females 1981   mean = 23.3 years (3 sf)
standard deviation = 5.52 years (3 sf)
Females 2011   mean = 30.7 years (3 sf)
standard deviation = 6.96 years (3 sf)

b) Males 1981   median ≈ 24 years (nearest year)
IQR ≈ 5 years (nearest year)
Males 2011   median ≈ 31 years (nearest year)
IQR ≈ 10 years (nearest year)
Females 1981   median ≈ 22 years (nearest year)
IQR ≈ 5 years (nearest year)
Females 2011   median ≈ 29 years (nearest year)
IQR ≈ 8 years (nearest year)

c) On average both genders were married for the first time at a much later age in 2011 (in both cases about 7 years older). The ages were more variable in 2011 than 1981. In both years the females were on average 2 years younger than the males.

d) A variety of sampling methods is possible including a range of voluntary methods using questionnaires and cluster sampling using records from registry offices. A major problem with the latter may be data protection laws.

10 A variety of statistical methods can be used.

11 a) Your comments may refer to the following results.
Hours of Sunshine
Eastbourne   modal class = 220 < t ≤ 240 hours
mean = 238 hours (3 sf)
standard deviation = 42.2 hours (3 sf)
median ≈ 230 hours   IQR ≈ 50 hours
Whitby   modal class = 120 < t ≤ 160 hours
mean = 182 hours (3 sf)
standard deviation = 35.7 hours (3 sf)
median ≈ 180 hours   IQR ≈ 53 hours

Cumulative frequency graph showing hours of sunshine per month in August at Eastbourne and Whitby

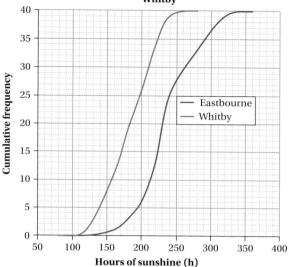

Box and whisker plots of hours of sunshine in Eastbourne and Whitby in August

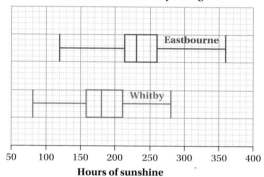

Rainfall
Eastbourne   modal class = 80 < x ≤ 100 mm
mean = 55.3 mm (3 sf)
standard deviation = 35.3 mm (3 sf)
median ≈ 44 mm
IQR ≈ 59 mm
Whitby   modal class = 30 < x ≤ 40 mm
mean = 60.1 mm (3 sf)
standard deviation = 34.6 mm (3 sf)
median ≈ 50 mm
IQR ≈ 49 mm

Cumulative frequency graph showing rainfall in Eastbourne and Whitby in August

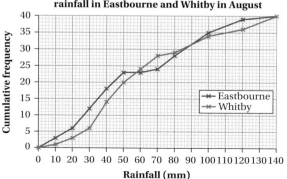

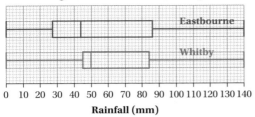

Box and whisker plots of rainfall per month in August at Eastbourne and Whitby

Exercise 1E

1 a) Assume that the last class width is the same as the previous class width (or other sensible assumption).
Missing values:

	3
	3.88
	3.30
50	1.72
100	0.23
100	0.08

b)

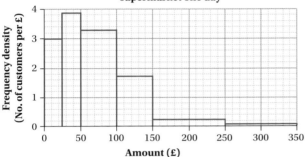

Amounts spent by customers at supermarket one day

2 a) Assume that the last class width is the same as the previous class width (or other sensible assumption).

Age (years)	No. of people (thousands)	Lower class boundary	Upper class boundary	Class width	Frequency Density (thousands per year)
0–4	4014	0	5	5	$4014 \div 5 = 803$
5–15	11 179	5	16	11	$11\,179 \div 11 = 1016$
16–44	21 453	16	45	29	$21\,453 \div 29 = 740$
45–64	16 328	45	65	20	$16\,328 \div 20 = 816$
65–74	6031	65	75	10	$6031 \div 10 = 603$
75–89	4574	75	90	15	$4574 \div 15 = 305$
90+	527	90	105	15	$527 \div 15 = 35$

b)

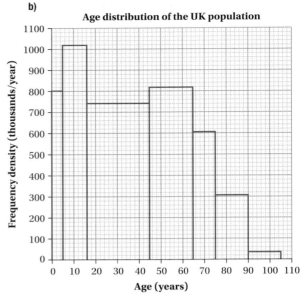

Age distribution of the UK population

3 a) i mean IQ = 107.7 (1 dp)
standard deviation = 8.5 (1 dp)

ii This year's students have on average a higher IQ than last year's students.
The IQs of this year's students are more variable than last year's students.

b) i

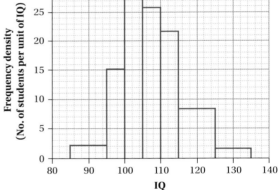

IQ of students joining Gainsby College this year

ii 46% (nearest %).

4 a) i The histogram uses frequency rather than frequency density on the vertical axis. This means that the area will not represent the frequency as it should do on a histogram. Also the values above 360 minutes are not shown.

ii Assumptions: Early planes have all been taken as 0 min delays.
No planes were later than 540 minutes
Varying group sizes and frequencies make it difficult to show each group accurately.

Histogram showing delays at Gatwick in one month

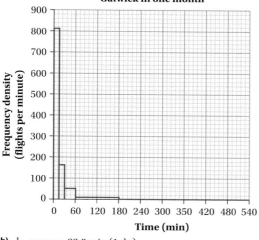

b) i mean = 23.8 min (1 dp)
standard deviation = 43.4 min (1 dp)
ii The values are estimates because the data is grouped so the individual values are not known.

5

External door widths on public buildings

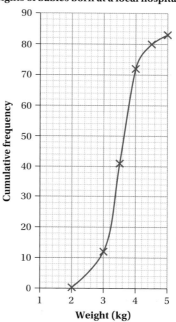

a) 14
b) 38% (nearest %).

6

Time t (hours)	No. patients	
$0 < t \leqslant 1$	1×78	78
$1 < t \leqslant 2$	1×95	95
$2 < t \leqslant 4$	2×28	56
$4 < t \leqslant 6$	2×15	30
$6 < t \leqslant 8$	2×12	24
$8 < t \leqslant 12$	4×2	8
$12 < t \leqslant 24$	12×0.5	6

% less than 4 hours = 77% (nearest %).
Mean visit time = 2.88047 hours = 173 minutes
The hospital is falling far short of the 95% target and its mean visit time is much worse than other hospitals.

1 a) **Errors on Helen's cumulative frequency graph:**
Points are plotted at the mid-interval values rather than upper boundaries of each group.
The vertical axis is labelled as frequency rather than cumulative frequency.
The graph does not have a title. The weight units are not given.
Errors on Liam's histogram:
There is no label on either axis.
Liam has actually used frequency on the vertical axis rather than frequency density, so frequency is not represented by the area of each column as it should be.
The title does not give enough information.

b) **Weights of babies born at a local hospital**

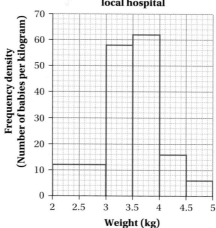

2 Your report should include graphs and comparisons of statistical measures.

Consolidation exercise 1

1 a) i Sample chosen in such a way that each member of the population has an equal chance of being included.
ii Sample chosen to reflect the number in each subgroup of the population where the subgroups are defined by characteristics which may include gender, age or other relevant features.
b) Any random method is acceptable.
c) i **Male:** 4 from Year 1, 4 from Year 2 and 5 from Year 3.
Female: 3 from Year 1, 2 from Year 2 and 2 from Year 3.
ii Draw names out of a hat for each year.
Assign each person a number and use a random number generator to select from each year.

2 a) i **Height of plants (in centimetres)**

Outdoors		Indoors
9	0	
9 8 7 7 6 5 3 1 1 0	1	0 2 4 5 7 8 8
7 6 5 3 2 1 0	2	2 2 3 7 7 7 8 9 9
2 1	3	0 2 4 5

Key 1|3|0 means
Outdoors 31 cm
Indoors 30 cm

ii Outdoor plants modes = 11 cm and 17 cm,
Indoor plants mode = 27 cm
iii Outdoor plants range = 23 cm,
Indoor plants range = 25 cm

b) i **Height of plants grown outdoors and indoors**

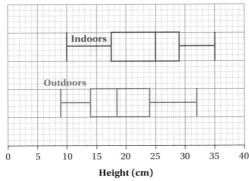

ii Your description may include:
On average the plants grow taller when they are indoors.
The heights of indoor plants are more variable than those of outdoor plants.
The distribution of the heights indoors is more spread in the lower half of the distribution whilst the heights of the outdoor plants are more spread in the upper half.

3 a) i **Marks achieved by college students in an A level French test**

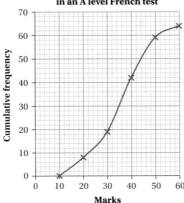

ii 40th percentile ≈ 33 marks. 40% of the students achieved marks up to 33 marks.
iii 53 students

b) **Marks achieved by college students in an A level French test**

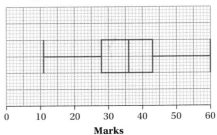

4 a) No
Julie's data is primary as she collected it herself. Also the data is discrete since money can only take particular values.

b) **Amounts spent by students on clothes in one month**

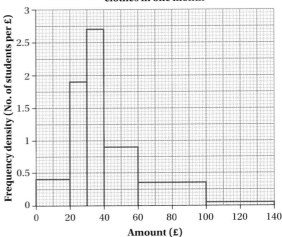

c) mean = £43 (nearest £)
standard deviation = £24 (nearest £)

5 a) Tanya's graph needs a key and the consumption units.
It would be better if the vertical axis of Oliver's chart started at zero, as his scale exaggerates the scale of the increase/decrease in consumption of the different fuels, the title 'Year' could be added to the horizontal axis and 'UK' added to the title.

b) Consumption of petrol has decreased whilst that of diesel increased over this period.

6 a) Main points:
This sample involving only 6 teachers is too small to reflect the opinions of all teachers.
Also it will not be representative because of the different numbers of schools in each group and the different numbers of teachers in each school and college.

b) Suggestions should involve a much larger sample and stratification.
Problems of difficulty and cost in interviewing a large number of teachers could be overcome by telephone interviews or the use of questionnaires (although the return may be lower).

7 a) modal class = £(0–100) thousand
b) mean ≈ £1.8 million (1 dp)
standard deviation ≈ £5.5 million (1 dp)

c) median $\approx$ £80 000

interquartile range $\approx$ £670 000

Problems arise because of the very large frequency in the first group. It contains the minimum, the lower quartile and the median.

d) Your comments may include:

Well over half of the films took less than £100 thousand at the box office.

The mean amount taken at the box office, £1.8 million, was inflated by the small number of films that took over £20 million. (In fact over 80% of the films took less than the mean.)

8 Your answer may include:

95% of the waiting times met the target, but 5% did not.

Modal class 16–17 weeks

Mean $\approx$ 13.9 weeks, standard deviation $\approx$ 4.2 weeks (1 dp)

Median $\approx$ 15.8 weeks, IQR $\approx$ 6.1 weeks (1 dp)

Cumulative frequency graph, box and whisker plot and/or histogram

9 There are a variety of appropriate questions.

10 Your report should include a summary of your findings supported by statistical measures and diagrams.

11 Your comparisons could include charts and statistical measures of central tendancy and spread.

Chapter 2

Exercise 2A

1 Taxable salary   £10 988

Tax   £10 988 $\times$ 0.2 = £2197.60

2 Taxable salary   £40 988

Basic rate   £31 785 $\times$ 0.2 = £6357

Higher rate   £9203 $\times$ 0.4 = £3681.20

Total tax   £10 038.20

Exercise 2B

1 a) £(4230 $\times$ 12 $-$ 9000) = £41 760

b) £31 785 $\times$ 0.2 = £6357

£9975 $\times$ 0.4 = £3990

£10 347

2 a) A   A person starts paying NI at 12%.

B   They now start paying income tax at 20% as well as NI at 12%.

C   Income tax increases to 40% but NI drops to 2%.

b) A   0%

B   2.9%

C   24.7%

D   34.7%

3 First split the weekly salary into the bands.

0%   £155

12%   £815 $-$ £155 = £660

2%   £950 $-$ £815 = £135

Calculate the percentage amounts and add them to find the total NI.

12% of £660 = £660 $\times$ 0.12 = £79.20

2% of £135 = £135 $\times$ 0.02 = £2.70

Total = £79.20 + £2.70 = £81.90

Ann paid £81.90 each week.

4 £(435 $-$ 155) $\times$ 0.106 = £29.68

5 a) £21 588 $\div$ 12 = £1799

£(1799 $-$ 672) $\times$ 0.12 = £135.24

b) £(21 588 $-$ 10 600) $\times$ 0.2 = £2197.60

£2197.60 $\div$ 12 = £183.13

c) £(21 588 $-$ 17 335) $\times$ 0.09 = £382.77

£382.77 $\div$ 12 = £31.90

d) £(1799 $-$ 135.24 $-$ 183.13 $-$ 31.90 $-$ 50) = £1398.73

e) Rate 1799.00; Amount 1799.00; PAYE Tax 183.13; PAYE NI 135.24; Student loan 31.90; Total deductions 400.27; Total gross pay 1799.00; Total gross pay TD 10794.00; Tax paid TD 1098.78; NI TD 811.44; Pension TD 300.00; Student loan TD 191.40; Net Pay; 1398.73

6 a) £8000 $\times$ 0.02 = £160   **b)** £8000 $\times$ 0.4 = £3200

c) £8000 $\times$ 0.09 = £720 extra is repaid on the loan.

He takes home an extra £3920, i.e. 49% of the extra pay.

Exercise 2C

1 a) $m = 3$, $A_1 = A_2 = A_3 = 437.98$, $1 + i = 1.15$, $t_1 = 1$, $t_2 = 2$, $t_3 = 3$

b) £437.98 $\left(\frac{1}{1.15} + \frac{1}{1.15^2} + \frac{1}{1.15^3} \right) \approx$ £1000

2 £1651.88 $\left(\frac{1}{1.3} + \frac{1}{1.3^2} + \frac{1}{1.3^3} \right) \approx$ £3000, as required

3 a) $750 = \frac{1000}{(1 + i)^2}$

$(1+i)^2 = \frac{1000}{750} = 1.\dot{3}$

$1 + i \approx 1.155$

APR = 15.5%

b) £450 $\left(\frac{1}{1.131} + \frac{1}{1.131^2} \right) =$ £749.67, as required

4 a) $1500 = A \left(\frac{1}{1.25} + \frac{1}{1.25^2} \right)$

$1500 = 1.44A$

£1041.67

b) $1500 = \frac{2000}{(1 + i)^2}$

$(1 + i)^2 = 1.33$

APR = 15.5%

5 a) Substituting values for i, A and t into the formula gives

$C = \frac{1000}{1.215} + \frac{1000}{1.215^2} = 1500.44878$

This is very close to the value of £1500 so APR is approximately 21.5%.

b) £981.82

Exercise 2D

1 b) 20 years 1 month (approximately)

2 11 years 1 month (approximately)

3 £241 $\times$ 1000 = £241 000

£133 $\times$ 1500 = £199 500

Paying back higher amounts makes a large difference to the total payments.

4 a) Interest is 0.6% or 0.006 times the debt.

New debt is old debt + 0.006 of old debt, ie 1.006 of old debt.

b)

n	A_n (£)
0	120 000
1	119 800
2	119 598.80
3	119 396.39
4	119 192.77
5	118 987.93
6	118 781.86

c) £(120 000 $-$ 118 781.86)

£1218.14

Exercise 2E

1 a) £46.54, £3892.52

b) $r = (1 + 0.0121)^2 - 1 = 0.0243$

2.43%

2 a) $\frac{2}{4} = 0.5\%$

b) A   4.02   B   808.02   C   808.02

D   4.04   E   812.06

c) $r = (1 + 0.005)^4 - 1 = 0.02\,015$

2.02%

3 a) £4000 $\times (1.0198)^3 \approx$ £4242.34

b) The nominal rate is $i = 2 \times 1.98\% = 3.96\%$ (or 0.0396).

The AER is given by $r = \left(1 + \dfrac{i}{n}\right)^n - 1$

Substitute $i = 0.0396$ and $n = 2$ into the formula.

$r = \left(1 + \dfrac{0.0396}{2}\right)^2 - 1$

$r = (1 + 0.0198)^2 - 1$

$r \approx 0.04 \quad$ or $\quad 4\%$

Exercise 2F

1 £90

2 $\pounds\dfrac{1625}{0.65} =$ £2500

3 a)

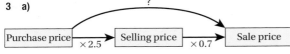

A profit of 150% means:

selling price = purchase price + 1.5 × purchase price
$\qquad\qquad = 2.5 \times$ purchase price

Reducing the price by 30% means the sale price is 70% of the selling price.

So sale price = 0.7 × selling price

To convert the purchase price to the sale price multiply by 2.5 and then by 0.7.

Purchase price × 2.5 × 0.7 = 1.75 × purchase price

1.75 = 175% so the dealer makes a profit of 75%.

b) Reverse the diagram.

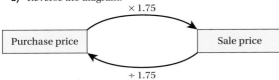

To work out the purchase price when you know the sale price you have to divide by 1.75.

The dealer paid $\pounds\dfrac{2100}{1.75} =$ £1200

4 No. $\quad\pounds\dfrac{1010.50}{117.5} \times 1.2 =$ £1032

Exercise 2G

1 £124.35

2 500 × 1.25 × 1.2 = $750

3 a) €512, £410

b) £1 $= \dfrac{512}{410} \approx$ €1.249

4 $\dfrac{191.91}{1.5278} =$ ¥125.61

5 a) £220 × 1.0275 = £226.05

b) €1 = 83.4 pence

Consolidation exercise 2

1 £35 760

2 $\dfrac{1000}{1.04} = 961.5\,\text{cm}^3$

3 a) £2227.80

b) £896.80

c) Borrow £5000

4 a) £500 × 1.02⁴ = £541.22

b) $0.02 = \left(1 + \dfrac{i}{12}\right)^{12} - 1$

$\sqrt[12]{1.02} = 1 + \dfrac{i}{12}$

$1.00165 = 1 + \dfrac{i}{12}$

$i = 0.0198$

1.98%

5 Let R be the multiplying factor for reducing the price.

$2 \times R = 1.2 \Rightarrow R = 0.6$

A reduction of 40%.

6 a) A 400 × 3.7382 = 1495.28 lira

B 395 × 3.8517 = 1521.42 lira. B is the better deal.

b) £169.68

7 a) i 500 × 1.0017¹² ≈ £510.30

ii $\dfrac{510.30}{500} = 1.0206$, 2.06%

b) $P = S \times 1.019^{2n}$

8 $R = \sqrt[8]{\dfrac{5000}{4000}} - 1 = 0.028\,29$

2.83%

9 Your answers will very depending on your estimates in part **a)**.

a) Any reasonable amount, for example:
8 pints of milk, 3 kg of potatoes, $\frac{1}{2}$ kg of sugar, 3 loaves.

b) 665.5, 738.5, 751, 771.5, 797, 782, 800, 853.5, 1065, 1177

c) 11.0%, 1.7%, 2.7%, 3.3%, −1.9%, 2.3%, 6.7%, 24.8%, 10.5%

10 a) 80 000 × 1.04 − 18 000 = 65 200

b) She has only taken off 1 month's repayment.

$A_{n+1} = 1.04A_n - 18\,000$

c) i B6 = B5*1.04 − 18 000

ii 4 17 152.33

  5 −161.57

She pays it off in the last month of year 5.

11 Canada $4.64

UK $4.37

USA $4.79

The USA was the most expensive country.

12 a) $800 = \dfrac{1000}{(1 + i)^2}$

$i = 0.291$

29.1%

b) $600 = A\left(\dfrac{1}{1.3} + \dfrac{1}{1.3^2}\right)$

$A =$ £440.87

c) With the second loan, Mary has use of the full amount for much less time.

13 a) 1.019¹² − 1 = 0.253,   25.3%

b) Because interest is being paid on interest.

14 a) 1512.75   12.86   1525.61

  1525.61   12.97   1538.58

b) 1.70%

c) 1.71%

Exercise 3A

1 a) 9×10^3 **b)** 2×10^{-6} **c)** 8.5×10^7 **d)** 1.5×10^{-5}

2 a) 2.4×10^4 **b)** 3.6×10^7 **c)** 8×10^2 **d)** 3×10^3

3 a) 6×10^{11} **b)** 2000 or 2×10^3

c) 1.44×10^{11} **d)** 50 or 5×10

e) 1.8×10^{-2} **f)** 2×10^{11}

4 Your answers will vary. A body weighing 50 kg has a volume of approximately 50 000 cm³.

5 a) 6.626×10^{-34} **b)** 7×10^{-34} or 10^{-33}

c) 10^{-99}

6 Assume that the distance is 1000 km and that you walk at 5 km/h. Then you would need to walk for 200 hours. Assuming you walk for 8 hours per day, it would take you 25 days. (The record is just over 9 days but most people take a couple of months.)

7 a) 9.5×10^{15} metres

b) $2.4 \times 9.5 \times 10^{15} = 2.3 \times 10^{16}$ metres

c) $\dfrac{2.3 \times 10^{16}}{6.1 \times 10^7} \approx 3.8 \times 10^8$ hours ≈ 40 000 years

8 **a)** If you weigh $50\,kg$, your volume of blood is roughly
$\frac{7}{100} \times 50 = 3.5$ litres.

b) Assume that you live 80 years.
$3.5 \times 60 \times 24 \times 365 \times 80 \approx 1.5 \times 10^8$ litres.

9 What speed does she walk at going to the well?
What speed does she walk at returning from the well?

10 **a)** Drinking, washing, cooking, (possibly drinking water for animals and irrigation).

b) The recommended amount of water an adult should drink in a temperate climate like that of the UK is 2–3 litres per day. In fact, Grace carried approximately 20 litres so this would cover the family's requirements for drinking water but not much more.

11 $5\,km/h$ is a brisk walking pace and so an outward journey of 3 hours would be reasonable. Returning with a weight of $20\,kg$ would be much harder work and take much longer. In fact, Grace took 10 hours for the return trip. Each night she set off at midnight and returned at approximately 10 a.m.

12 **a)** $\frac{7000 \text{ million}}{9} \approx 800$ million

b) $\frac{800 \text{ million}}{60 \text{ million}} \approx 13$
The number of people without access to safe water is roughly 13 times the population of the UK.

13 **a)** In order of usages: personal washing, toilet flushing, other washing, garden, drinking.

b) Your own estimate of your family's water use.

c) 500 litres is the figure given by Waterwise.

Exercise 3B

1 **a)** Approximately $7500\,km$
b) Approximately $7000\,km$
c) Approximately $7000\,km$

2 Your own estmate of your lifetime heartbeats.

3 **a)** Greenland looks shorter than it 'should' because of the curvature of the Earth
b) Too small
c) She could use a picture of the globe that has Greenland nearer the centre of the picture or simply make an allowance for the curvature effect.

4 The distance is roughly $5500\,km$.
The speed is $\frac{5500}{20 \times 24} \approx 11\,km$ per hour.

5 **a)** The Gulf Stream. The Gulf Stream's effect on travel times was not understood properly until it was studied by the scientist Benjamin Franklin. Captains who were unaware of this would find themselves sailing against the current for a large part of the journey.
b) They avoided travelling in the Gulf Stream although they did have to cross it several times.

6 At, say, $6\,km$ per hour, the Gulf Stream could cut a ship's speed from $11\,km$ per hour to $5\,km$ per hour. However, it is unlikely that a captain would be so unlucky as to travel directly against the current all the way. If we use two-thirds of this effect, then at a speed of $11 - 4 = 7\,km$ per hour the journey would be increased by 12 days. This appears to be a reasonable estimate since it was reported that British ships took up to two weeks longer.

Consolidation exercise 3

Many estimations are possible for the questions in this exercise, depending on your chosen set of assumptions. Ensure you state your assumptions clearly and show all your working. Compare the different approaches adopted by your class.

Chapter 4

Question A

You may or may not agree with the author's point of view but what you should have recognised is that it is just a point of view. It is a statement of two related assertions not supported by *any* evidence or reasoning.

Question B

a) The failure of anti-obesity campaigns aimed at the young.
Her own research on obese mothers.
She concludes that efforts and research should be devoted to initial causes of obesity in babies.

b) Her research seems to have been focused on mothers with gestational diabetes and may not have wider implications.
The potential obesity of an adult may be affected by what happens *both* in utero *and* in early childhood.

Question C

a) Yes. 2 out of 10 patients with the disease have a negative X-ray and 2 out of 10 is the same as 1 out of 5.

b) No. Only 2 out of 983 patients who test negative have the disease. This is $\frac{2}{983} \times 100 = 0.2\%$, not 20%.

Question D

a) Emotive: 'Blatantly obvious', 'no sane person'.
Vague: 'so many'.

b) The effects of Salbutamol.
The reasons for deaths of cyclists.
Lance Armstrong's ban.

c) The first two sentences imply that all performance enhancing drugs should be legalised. The final sentence implies the opposite.

d) That it is performance enhancing.
That the cyclists' deaths were caused by taking performance enhancing drugs.

Question E

a) Employers must encourage people to work beyond pensionable age.

b) It is damaging to the UK for people to retire abroad.
It gives some support but if this were in an employer's interests they would do this anyway.

c) The reasoning is based both upon the affluence of pensioners and their potential poverty.

Question F

a) Conclusion – it is wrong not to eat everything on your plate.
Reason – people are starving in Africa.

b) No. Even the idea of not putting so much food on your plate in the first place has only tenuous links with the reason given.

Question G

Your answers should include:

Evidence:	Over 5% of land has been built on.
Reason:	We must protect green belt.
Conclusion:	Do not support the old dairy proposal.
Self-contradiction:	What little remains appears to be 95% of the land.
Assumptions:	Is the old dairy on green belt land?
Vague term:	Over 5%.
Emotive terms:	Beautiful; unspoilt; no reasonable person; over-priced 'executive'; make the developers rich; so badly need.

Question H

The number of possibilities for naming 10 Zener cards one after the other is $5 \times \ldots \times 5$ (10 terms) $= 5^{10} \approx 10$ million.
One sequence made up at random therefore has a 1 in 10 million chance of being correct.

Question I

The lottery is bound to be won (eventually) by at least one person. Picking that person after the event because they have won the lottery proves nothing. [A similar effect occurs with DNA matching. If a suspect's DNA has a 1 in a million match with DNA at a crime scene, then there could be roughly 70 people in the UK similarly matched and so *unless there is other evidence*, the DNA evidence is far from being convincing.]

Question J & K

a) The bias is amongst professors, *not* male professors.
b) The phrase 'The gender of the faculty participants did not affect responses…' was simply ignored in the newspaper article.
c) You probably found the newspaper article easier to read. It has more commonly used words and a lively, chatty style.

Question L

It has no significance.

Question M

Although he had met a large number of British people, these were *not* representative of those living of the Isle of Wight.

Question N

a) There are two control groups, one taking the placebo and the other taking Acetazolamide. Control groups are under exactly the same conditions as the test group apart from not having the treatment being trialled. Any differences in the results of the test group and trial group are then occurring because of the new treatment (or by chance).
A drug with no active ingredient is called a placebo. This tests whether the psychological effects of simply being on the trial are significant.
b) It would be unethical to give a patient a treatment that may have side effects without getting their consent first.

Question O

a) The increase is roughly 20% but for such small numbers cannot be said to have soared.
b) There may be more 20 mph zones.

Question P

a) It will unnecessarily worry parents who, in any case, are unlikely to be able to detect a difference.
b) 'intervention can be useful', 'a small number', 'among those which had the highest…'.
c) Unnecessary extensive practical, emotional and financial commitment for the family.
d) The research paper has no value and the newspaper article has a negative value.

Question Q

a) Optimists are probably more likely to enter the lottery. You could survey the personalities of people buying lottery tickets.
b) 'scientific', 'convincingly demonstrated'.

Question R

a) No-one remembers the many occasions when odd animal behaviour is not followed by earthquakes (or the earthquakes *not* preceded by odd animal behaviour).

b) Friction can cause static electricity.
Underground rock movements cause static electricity.
Animals can detect this static electricity.
Animals behave oddly.
Friction may or may not cause static electricity so direct detection of static electricity is necessary. The effects of static electricity on animal behaviour could also be tested.

Question S

a) As you have already seen, correlation does not necessarily imply causation. In this case the fact that atmospheric CO_2 increases *after* the temperature rises completely undermines Al Gore's argument.
b) The fact that someone might state an incorrect reason for their conclusion might well give you cause to doubt their judgement but actually tell you nothing about whether their conclusion is true or false! In this case, there is plenty of scientific justification for saying that CO_2 is a greenhouse gas, i.e. one that can cause less heat to be radiated out from Earth. What the graphs show is that changes in atmospheric CO_2 did not *trigger* historic fluctuations in the Earth's temperature.

Consolidation exercise 4

1 Since the symbol for 2014 is twice the height and width of the other symbol it gives the impression of more than doubling. The accepted convention in such an example is to double the *area*.

2 a) Evidence: Graduates earn more than non-graduates.
   Conclusion: Students should pay tuition fees.
b) Correlation does not necessarily imply causation. A major effect is likely to be that the qualities which help to achieve a university place also achieve a higher salary.

3 a) They are biased towards being satisfied. People tend not to use the extreme boxes on questionnaires. The difference between very satisfied, good and excellent is very vague.
b) $\frac{7 + 5 + 6 + 3}{25} = \frac{21}{25} = 84\%$
c) Because of the bias caused by the category wording and also the fact that the sample is biased towards Aysha's own particular course.
d) $\frac{15}{16} \times 70 + \frac{6}{9} \times 180 \approx 186$
   $\frac{186}{250} \times 100 \approx 74\%$

4 a) $\frac{133}{600} \approx 22\%$, $\frac{642}{6400} \approx 10\%$
b) The figure of 20% has been based upon one figure from part a) and applied to the total budget of £50 billion.
   50 billion $\times$ 0.2 $= 10$ billion.
c) There is no reason to apply a figure based essentially upon one minor cost area (that of mobile phones) to the entire range of services.
d) The total saving is £775k on a budget of £7000k.
   $\frac{775}{7000} \approx 11\%$
   50 billion $\times$ 0.11 $= 5.5$ billion
e) No. For the reasons given in c).

Chapter 5

Exercise 5A

1

Weight (kg)	64	44	**52**	**80**	56
No. of s.d. above mean	0.5	−2	−1	2.5	**−0.5**

2 a) No, distribution stops at 24.
  b) No, not symmetrical.

3 **a)** $\mu=40$, $\sigma=1$ **b)** $\mu=76$, $\sigma=2$
c) $\mu=120$, $\sigma=20$ **d)** $\mu=1080$, $\sigma=20$

4 **a)** Women are generally shorter (by 8 cm on average). Their heights are also slightly less variable.

b) Both heights are 2 s.d. above the mean for their distributions. 5% of the distributions are further than 2 s.d. from the mean and, by symmetry, this means that 2.5% are above the mean.

c) $\frac{2.5}{100} \times 1000 = 25$

5 **a)** The positive value indicates that, on average, there have been positive returns on stock market investments.

b) This is much greater than the mean. So, although the overall trend is positive, the daily change is almost as likely to be negative as positive.

c) **i** This is 1 s.d. below the mean. $\frac{1}{6}$ of values are below this, roughly 61 days per year.

ii This is 2 s.d. above the mean. $2\frac{1}{2}$% of values are above this, roughly 9 days per year.

6 **a)** Too little stock means that a business may not be able to meet customer demand.

b) Too much stock ties up capital invested in purchasing or manufacturing that stock. It also requires paying for warehousing and, in some cases, the stock may become out of date or unfashionable.

c) This is more than 2 s.d. below the mean and so has a $2\frac{1}{2}$% chance.

7 **a)** 5% should be outside 2 s.d. from the mean, roughly 10 packets.

b) $\frac{2}{3}$ of packets should be within 1 s.d. of the mean, roughly 133 packets.

c) Half the number in **b)**, roughly 67 packets.

8 **a)** Highly unlikely

b) $\frac{1}{6}$ or 16.7%

c) $100\% - 16.7\% - 2.5\%$
Roughly 81%

Exercise 5B

1 **a)** **i** 0.68268 **b)** **i** $0.68268 \approx \frac{2}{3}$
ii 0.9545 **ii** $0.9545 \approx 95\%$
iii 0.9973 **iii** $0.9973 \approx 1$

2 **a)** 0.04457 **b)** 0.0968 **c)** 0.15245

Exercise 5C

1 **a)** 7 **b)** 6 **c)** $\sqrt{40} \approx 6.32$

2 **a)** 0.6 **b)** -0.0625 **c)** -1 **d)** 0.667
e) 0.378 **f)** -0.894

3 **a)** 0.87286 **b)** 0.02118 **c)** 0.15866 **d)** 0.99865
e) $0.99379 - 0.93319 = 0.0606$
f) $0.97725 - 0.15866 = 0.81859$
g) $0.99202 - 0.94062 = 0.0514$

4 **a)** These tails of the distribution are mirror images of each other in the y-axis.

b) Let each tail have area A.
LHS $= 1 - 2A$
RHS $= 2(1 - A) - 1 =$ LHS

5 **a)** 0.14 **b)** 2.33 **c)** -1.28
d) 1.96 **e)** 0.44

6 **a)** $\frac{c}{10} = 0.44$ so $c = 4.4$. The interval is $63.6 < R < 72.4$.

b) $\frac{c}{10} = 1.16$ so $c = 11.6$. The interval is $56.4 < R < 79.6$.

7 **a)** This is a shape which arises when two different populations are combined. The difference might be age, gender or race.

b) This could be for people on an adventure park ride that has a height restriction.

Consolidation exercise 5

1 **a)** $P\left(z > \frac{4.3 - 3.5}{0.5}\right) = 1 - 0.9452 = 0.0548$

b) $P\left(z > \frac{W_{max} - 3.5}{0.5}\right) = 0.873$

$\frac{W_{max} - 3.5}{0.5} = -1.14$

$W_{max} = 2.93$ kg

2 $P\left(z > \frac{100.5 - 98.3}{0.8}\right) = P(z > 2.75) = 1 - \Phi(2.75) = 1 - 0.99702$
$= 0.00298$

3 **a)** $P\left(z > \frac{152 - 149.3}{12.7}\right) = P(z > 0.213) = 1 - \Phi(0.213)$
$= 1 - 0.58317 = 0.41683$

b) $P\left(z > \frac{154 - 149.3}{12.7}\right) = P(z < 0.370) = \Phi(0.370) = 0.64431$

c) $P\left(z < \frac{145 - 149.3}{12.7}\right) = P(z < -0.339) = 1 - \Phi(0.339)$
$= 1 - 0.63307 = 0.36693$

d) Standardised measurement of 148 cm is
$z = \frac{148 - 149.3}{12.7} = \frac{-1.3}{12.7} = -0.102$
Standardised measurement of 151 cm is
$z = \frac{151 - 149.3}{12.7} = \frac{1.7}{12.7} = 0.134$
Probability between 148 cm and 151 cm is
$\Phi(0.134) - (1 - \Phi(0.102)) = 0.55172 - (1 - 0.53983)$
$= 0.09155$

4 **a)** $P(X < 90)$ is $P\left(z < \frac{90 - 65}{20}\right) = P(z < 1.25) = 0.89435$

b) $P(X > 60)$ is $P\left(z > \frac{60 - 65}{20}\right) = P(z > -0.25) = 0.59871$

5 **a)** $P(X < 60)$ is $P\left(z < \frac{60 - 48}{20}\right) = P(z < 0.6) = 0.72575$

b) $P(30 < X < 60) = P(X < 60) - P(X < 30)$
$P(X < 30)$ is $P\left(z < \frac{30 - 48}{20}\right) = P(z < -0.9) = 1 - \Phi(0.9)$
$= 1 - 0.81594$
$P(30 < X < 60) = P(X < 60) - P(X < 30)$
$= 0.72575 - (1 - 0.81594) = 0.51469$

c) $\Phi(1.28) = 0.9 \Rightarrow z = 1.28$
$z = 1.28 = \frac{k - 48}{20} \Rightarrow k = 48 + 20 \times 1.28 = 73.6$

6 **a)** $P(X < 205)$ is $P\left(z < \frac{205 - 184.5}{13.6}\right) = P(z < 1.507) = 0.93448$
$= 93.4\%$ (3 sf)

b) $P(145 < X < 190)$ is $P\left(\frac{145 - 184.5}{13.6} < z < \frac{190 - 184.5}{13.6}\right)$
$= P(-2.904 < z < 0.404)$
$= \Phi(0.404) - (1 - \Phi(2.904))$
$= 0.65542 - (1 - 0.99813) = 0.65355$

7 **a)** **i** $P(X > 75)$ is $P\left(z > \frac{75 - 68}{13}\right) = P(z > 0.538)$
$= 1 - \Phi(0.538) = 1 - 0.74540$
$= 0.2946$

ii $P(58 < X < 72)$ is $P\left(\frac{58 - 68}{13} < z < \frac{72 - 68}{13}\right)$
$= P(-0.769 < z < 0.308)$
$= \Phi(0.308) - (1 - \Phi(0.769))$
$= 0.62172 - (1 - 0.77935) = 0.40107$

b) $\Phi(1.28) = 0.9 \Rightarrow z = 1.28$
$z = 1.28 = \frac{p - 68}{13}$
$p = 68 + 13 \times 1.28 = 84.6$

8 $P(X < 25)$ is $P\left(z < \frac{25 - 31.6}{4.3}\right) = P(z < -1.53) = 1 - \Phi(1.53)$
$= 1 - 0.93699$
$= 0.06301$

9 $P(\text{reject}) = P(X < 52) + P(X > 57)$
$= P\left(z < \frac{52 - 55}{1.5}\right) + P\left(z > \frac{57 - 55}{1.5}\right)$
$= (1 - \Phi(2)) + (1 - \Phi(1.33))$
$= (1 - 0.97725) + (1 - 0.90824)$
$= 0.11451$
No. reject nails $= 0.11451 \times 11\,000 = 1260$
No. good nails $= 11\,000 - 1260 = 9740$

10 **a)** $P(X > 27) = P\left(z > \frac{27 - 24.9}{1.05}\right)$
$= P(z > 2) = 1 - \Phi(2)$
$= 1 - 0.97725 = 0.02275$

b) $P(22 < X < 25)$ is $P\left(\frac{22 - 22.8}{0.89} < z < \frac{25 - 22.8}{0.89}\right)$
$= P(-0.899 < z < 2.47)$
$= \Phi(2.47) - (1 - \Phi(0.899))$
$= 0.99324 - (1 - 0.81594) = 0.80918$
$= 80.9\%$ (3 sf)

11 a) $P(X < 200) = P\left(z < \frac{200 - 185}{10}\right)$
$= P(z < 1.5) = 0.93319$

b) $P(X > 175) = P\left(z > \frac{175 - 185}{10}\right)$
$= P(z > -1)$
$= P(z < 1) = 0.84134$

c) From the above, probability is **a)** $- (1 - $ **b)**$) = 0.77453$

12 a) i $P(X < 145) = P\left(z < \frac{145 - 140}{2.5}\right)$
$= P(z < 2) = 0.97725$

ii $P(138 < H < 142)$ is $P\left(\frac{138 - 140}{2.5} < z < \frac{142 - 140}{2.5}\right)$
$= P(-0.8 < z < 0.8)$
$= \Phi(0.8) - (1 - \Phi(0.8))$
$= 0.78814 - (1 - 0.78814) = 0.57628$

b) $\Phi(-1.04) = 0.85 \Rightarrow z = -1.04$
(height exceeded by 85%)
$z = -1.04 = \dfrac{h - 140}{2.5}$
$h = 140 - 2.5 \times 1.04 = 137.4\,\text{cm}$

Chapter 6

Exercise 6A

1 The mean is 175.6 g.
2 174.5 g, because it is based on a larger size of sample.
3 $\frac{5 \times 175.6 + 25 \times 174.5}{30} \approx 174.7\,\text{g}$
4 $\Phi\left(-\frac{5}{3}\right) \approx 0.05$

Exercise 6B

1 a) 0
b) $\overline{X} \sim N\left(60, \left(\frac{4}{\sqrt{15}}\right)^2\right)$
$\overline{z} = \frac{62 - 60}{4/\sqrt{15}} \approx 1.94$
$1 - \Phi(1.94) = 0.02619$
2 $\overline{X} \sim N\left(100, \frac{80}{5}\right) = (100, 4^2)$
a) $1 - \Phi\left(\frac{7}{4}\right) = 1 - \Phi(1.75)$
$= 0.04006$
b) $\Phi\left(\frac{9}{4}\right) - \Phi\left(\frac{1}{4}\right)$
$= \Phi(2.25) - \Phi(0.25)$
$= 0.38907$
3 a) $\overline{X} \sim N\left(90, \frac{90^2}{7}\right)$
$P(\overline{X} > 120) = 1 - \Phi\left(\frac{30}{90/\sqrt{7}}\right) = 1 - \Phi(0.88)$
$= 0.18943$
$< 20\%$
b) The distances driven each day in one particular week need to be independent of each other.
4 a) $z = \frac{5.08 - 5.1}{0.2} = -0.1$
$\Phi(-0.1) = 1 - \Phi(0.1) = 0.46017$
b) $\overline{z} = \frac{5.08 - 5.1}{0.2/\sqrt{10}} = -0.32$
$\Phi(-0.32) = 1 - \Phi(0.32) = 0.37448$
c) $\overline{z} = \frac{5.08 - 5.1}{0.2/10} = -1$
$\Phi(-1) = 1 - \Phi(1) = 0.15866$
5 a) $P(\overline{H} \leq 20) = 0.5$ and $P(20 < \overline{H} < 21) = 0.34134$.
b) $\Phi(1) = 0.84134$ and so 21 is 1 standard error above the mean that is, $20 + \frac{3}{\sqrt{n}}$.
c) $\frac{3}{\sqrt{n}} = 1$ and so $\sqrt{n} = 3$ that is $n = 9$.

Consolidation exercise 6

1 a) i 0
ii $P\left(z < \frac{425 - 421}{2.5}\right) = \Phi(1.6) = 0.94520$
iii $P\left(-\frac{3}{2.5} < z < \frac{3}{2.5}\right) \Phi(1.2) - \Phi(-1.2) = 0.76986$
b) $z = 2.05$
$\frac{x - 421}{2.5} = 2.05$
$x \approx 426$
2 $40.5 \pm 1.96 \frac{5}{\sqrt{100}}$
$[39.5, 41.5]$
3 $946 \pm 2.896 \times \frac{69}{3}$
$[879, 1013]$
4 a) $\overline{x} = \frac{497.5}{25} = 19.9$
$19.9 \pm 2.3263 \times \frac{0.4}{\sqrt{25}}$
$= [19.7, 20.1]$
b) 20 is within the CI. There is no reason to doubt the claim.
5 a) $168.1 \pm 1.96 \frac{8.3}{\sqrt{6}}$
$[161.5, 174.7]$
b) 175 is not in the CI. He is right to be suspicious.
6 a) $251.1 \pm 2.0537 \times \frac{1.94}{\sqrt{50}}$
$[250.5, 251.7]$
b) $CI > 250$
The claim appears reasonable.
7 $758 \pm 1.96 \times \frac{10}{\sqrt{12}}$
$[752, 764]$
750 is lower than the CI so he should be concerned.
8 a) $\overline{L} \sim N\left(\mu, \frac{\sigma^2}{n}\right)$
b) $\overline{l} = \frac{978}{10} = 97.8$
95% value gives $z = 1.96$
$97.8 \pm 1.96 \frac{4.69}{\sqrt{10}}$
$= 97.8 \pm 2.9068$
$= (94.9, 100.7)$

Chapter 7

Exercise 7A

1 a) Graph B – the higher the temperature, the more ice creams people are likely to buy.
b) Graph A – no correlation between height and IQ.
c) Graph C – runner is likely to improve after training (take a shorter time for the marathon).
d) Graph D – the heavier the mass, the longer the extension.
e) Graph C – the more powerful the engine, the quicker the car should accelerate (not an exact relationship as weight of the car will also affect this).
f) Graph B – the taller the tree, the greater its circumference is likely to be.
2 a) Positive correlation with justification (taller people tend to have larger feet and shorter people smaller feet).
b) Negative correlation with justification (those households that consume a lot of butter are likely to consume a small amount of margarine and vice versa).
c) Positive correlation with justification (the heavier the load, the longer the lorry will take to gain speed).
d) Positive correlation with justification (the higher the drop, the higher the ball will rebound).
e) Negative correlation with justification (the higher the average speed, the shorter the time it will take to travel between the towns).

3 a)

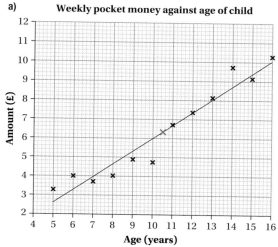

Weekly pocket money against age of child

b) Strong positive correlation.

c) Approximately 70 pence.

4 Graphs – Both need to include a title, labels and the mean point.

Lines – Both need to follow the trend of the points.

Ahmed's line should be turned clockwise so that there are an equal number of points above and below the line (it should not pass through the origin).

Kayleigh's graph should be turned anti-clockwise to give a steeper gradient that more closely follows the trend.

5 a) i The higher the price the fewer will be sold. (Negative correlation).

ii Scatter graph shows strong negative correlation.

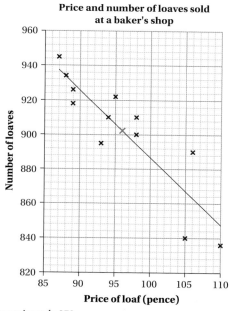

Price and number of loaves sold at a baker's shop

iii Approximately 870

b) i

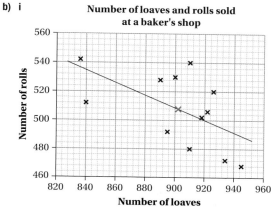

Number of loaves and rolls sold at a baker's shop

ii Borderline strong/weak negative correlation. The more loaves people buy, the fewer rolls they want/need.

iii 521 (approximately 520)

c) i

Price of loaves and rolls sold at a baker's shop

ii The price of wheat, wages etc.

6 a) For the area,

LQ = 37 033   UQ = 121 230   IQR = 84 197

UQ + 1.5IQR = 247 525.5 (sq km)

For the population,

LQ = 4181.83   UQ = 11 622.82   IQR = 7440.99

UQ + 1.5IQR = 22 784.305 (thousand)

Mexico's area, 1 964 375 (sq km), and population, 120 286.66 (thousand), are bigger than these values, so Mexico's area and population are both outliers.

b)

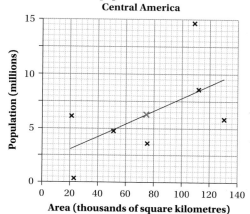

Area and populations of countries in Central America

c) Positive correlation (borderline strong/weak)

7 Scatter graphs show that there is strong positive correlation between each pair of variables. So larger farm animals tend to have larger brains and live longer.

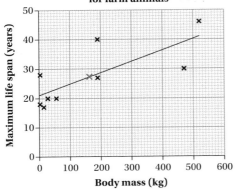

Maximum life span against body mass for farm animals

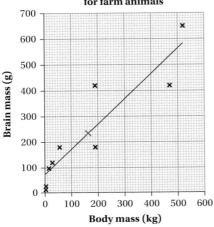

Brain mass against body mass for farm animals

Maximum life span against brain mass for farm animals

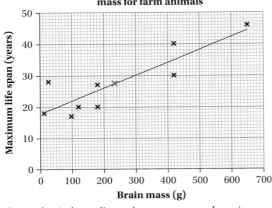

8 a) London is the outlier as the average annual earnings there are far higher than in the other regions. Also allow that the point representing London lies far from the general trend or use of the definition of an outlier given in Q6 as follows:

For average annual earnings,
LQ = 24.898, UQ = 26.9375, IQR = 2.0395
UQ + 1.5IQR = 30.00 (£thousands)
The average annual earnings in London at 35.069 (£thousands) is greater than this, so is an outlier.

Sonja should disregard the outlier (as London is a special case where earnings are much higher than in the regions and this will inflate the mean value), but make this clear when reporting the results of the investigation. Perhaps draw another scatter graph in which the point representing London is omitted.

b) i The scatter graph could exclude or include London (as the point representing London does follow the general trend even though the values are large).
Also the line of best fit could be included or omitted as it has not been requested.

ii Strong positive correlation. Regions with higher average earnings have higher house prices.

iii Population, demand for housing, job availability and prospects, attractiveness of the area etc.

9 Your investigation should include a scatter graph with a line of best fit and a description of the correlation.

Exercise 7B

Answers from calculators and spreadsheets may differ slightly from those given.

1 a)

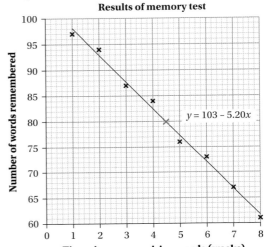

Results of memory test

$y = 103 - 5.20x$

b) i $y = 103 - 5.20x$ (3 sf)

2 a) $y = 11.5 + 0.547x$ (3 sf)

b)

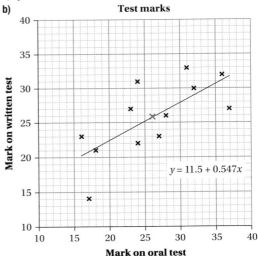

Test marks

$y = 11.5 + 0.547x$

c) 30 (nearest mark)

3 a) $y = -181 + 2.00x$ (3 sf)

b) -181. For example, this would be the price given for the most expensive season ticket when the cheapest was £0. A negative value below the cheapest price does not make sense.

c) For each increase of £1 for the cheapest ticket, the most expensive increases by £2.

4 a) The point $(320, 245)$ is incorrect.
Amy has reversed the coordinates of the point for Cardiff.

b) $y = -9.29 + 1.29x$ (3 sf)

c)

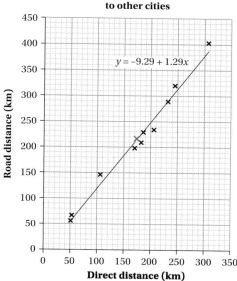

Direct and road distances from Sheffield to other cities

$y = -9.29 + 1.29x$

d) For each extra kilometre on the direct distance, the road distance increases by about 1.3 kilometres.

e) This would be extrapolation. The direct distance for Aberdeen is outside the range of the given data.

5 a)

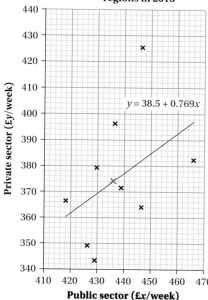

Median gross weekly earnings in UK regions in 2013

$y = 38.5 + 0.769x$

b) South East

c) i $y = x$

ii $y = x$ would pass through $(0, 0)$ and have a steeper gradient (for every extra £1 in public sector there would be an extra £1 rather than 77 pence in the private sector).

6 a) i $C = 1220 - 17.0n$ (3 sf)

ii Cost reduces by £17 for each year older.

b) i £540

ii

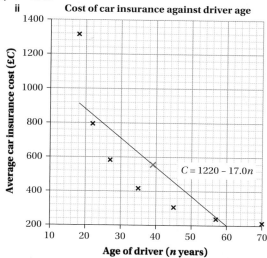

Cost of car insurance against driver age

$C = 1220 - 17.0n$

iii Linear regression is not appropriate in this case because the points appear to lie on a curve, rather than a straight line. The cost falls rapidly at first but then more slowly as age increases.
The 40-year old driver is likely to pay much less than £538 (a regression curve would suggest under £400).

7 a)

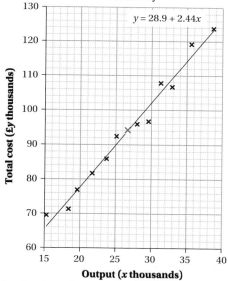

Total costs against monthly output of factory

$y = 28.9 + 2.44x$

b) i Cost for each extra unit of output $\approx$ £2.44

ii Fixed costs (costs not dependent on output) $\approx$ £28 900

c) £96 000 (3 sf)

d) This would be extrapolating beyond the range of the given data.

e) 25 000 (nearest thousand)

8 a) It is expected that students will find that the relationship between height and stride length is approximately $h = 2.5l$ where h represents height and l represents stride length.

b) 108 cm and 73 cm (nearest cm)

Exercise 7C

Answers from calculators and spreadsheets may differ slightly from those given.

1 a) i $R = 46.0 + 0.134P$ (3 sf)

ii $r = 0.981$ (3 sf)

b)

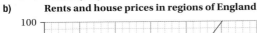

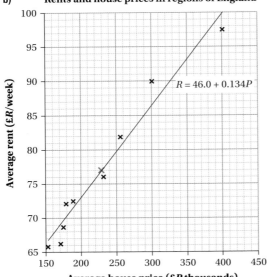

Rents and house prices in regions of England

$R = 46.0 + 0.134P$

2 a) i $y = 0.192 + 2.37x$ (3 sf)

ii $r = 0.851$ (3 sf)

b)

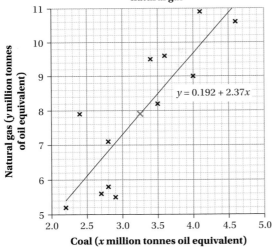

UK monthly consumption of coal and natural gas

$y = 0.192 + 2.37x$

3 a) i $P = 8860 - 720n$ (3 sf)

ii $r = -0.898$ (3 sf)

iii

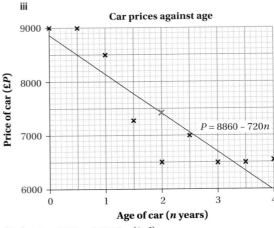

Car prices against age

$P = 8860 - 720n$

b) i $P = 8380 - 0.0745m$ (3 sf)

ii $r = -0.798$ (3 sf)

iii

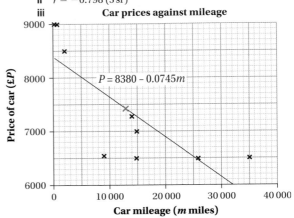

Car prices against mileage

$P = 8380 - 0.0745m$

c) Comments could include:
the price is more strongly related to the age than to the mileage
the price reduces by £720 each year and reduces by about $7\frac{1}{2}$ pence for each extra mile travelled
the prices implied for a new car are £8860 and £8380. These are both incorrect as the table gives the price of a new car to be £8995.

d) The graphs suggest that curves may give a better fit than straight lines.

4 The bust and hip measurements for size 12 have stronger correlation than the bust and hip measurements for size 14. pmcc for size 12 is 0.766 (3 sf), pmcc for size 14 is 0.408 (3 sf)

5 Using UK extraction as x and exports as y gives
$r = 0.890$, $y = 46.5 + 0.282x$ (3 sf)
Using UK extraction as x and imports as y gives
$r = -0.974$, $y = 159 - 0.416x$ (3 sf)
As UK extraction of oil and gas increases exports increase and imports reduce. In both cases the correlation is strong, particularly that between extraction and imports.

6 a) Mean values and line of best fit.

b) The years and trend over time.

c) On Amber's graph the closeness of the points to the regression line indicates strong positive correlation. On Harry's graph the similarity of the shapes of the graph indicates strong positive correlation.

d) $r \approx 0.97$ (2 sf)

7 a) i pmcc for buses and cars = -0.958 (3 sf)

ii pmcc for buses and rail = -0.392 (3 sf)

iii pmcc for cars and rail = 0.562 (3 sf)

b)

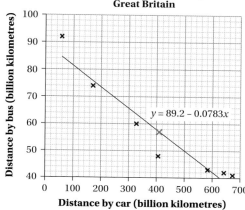

Distances travelled by passengers in Great Britain

$y = 89.2 - 0.0783x$

Distance by bus (billion kilometres) vs Distance by car (billion kilometres)

c) Your description should include:

strong, negative correlation for buses and cars
weaker negative correlation for buses and rail
fairly strong positive correlation for rail and cars
bus travel decreases as car and rail travel increase
rail travel increases with car travel – the table shows that both have increased over the period shown (may be due to increased population as well as increased mobility)
linear regression may not be appropriate as the points on the graph seem to lie near a curve

8 a) i $w = 24.5 + 1.63l$ (3 sf)

ii

Seabirds
Graph of wingspan against length

$w = 24.5 + 1.63l$

Wingspan w (cm) vs Length l (cm)

b) i $r = 0.814$ (3 sf)

ii Fairly strong positive correlation. Wingspan increases with length.

iii Wingspan and length are more closely related for garden birds than for sea birds.

c) Rashid is correct. The pmcc for mass and length is 0.943 (3 sf) and the pmcc for mass and length cubed is 0.959 (3 sf)

9 a) i True – all values of the pmcc are positive.

ii True – in each case the pmcc value for males is greater than that for females.

iii True – for both males and females the pmcc value for foot length on hand length is greater than that for foot length on height ($0.779 > 0.743$ and $0.661 > 0.561$).

iv True – the regression equation, $y = 31.46 + 0.251x$ has a gradient of 0.251 which means that an extra 1 cm in foot length predicts an increase of 0.251 cm in foot width and this is approximately $\frac{1}{4}$ cm.

v True – the regression equation, $y = 22.17 + 0.095x$ has a gradient of 0.095 which means that an extra 10 cm in height predicts an increase of $10 \times 0.095 = 0.95$ cm in hand length and this is approximately 1 cm.

vi False – the regression equation, $y = -24.53 + 0.165x$ has a positive gradient of 0.165

b) There are many possible body measurements to investigate.

10 a) i $r = 0.992$ (3 sf)

ii $G = -84.1 + 3.43E$ (3 sf)

b)

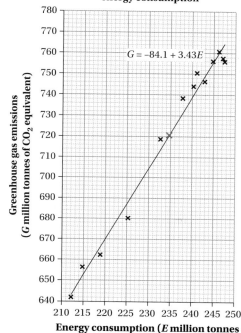

UK greenhouse gas emissions against energy consumption

$G = -84.1 + 3.43E$

Greenhouse gas emissions (G million tonnes of CO_2 equivalent) vs Energy consumption (E million tonnes of oil equivalent)

c) Greenhouse gas emissions are reducing and the target for 2027 will be less and outside the data range. Extrapolating does not give reliable predictions.

11 Data values are ranked in order to find this correlation coefficient.

Consolidation exercise 7

1 **a) b)**

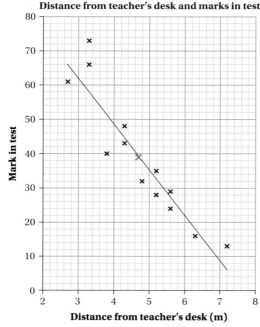

Distance from teacher's desk and marks in test

c) Very strong negative correlation. Those students who sit nearer have got better marks, but correlation does not necessarily imply causation. The teacher's theory may be correct, but there may be another reason for the better marks, the students' ability – the cleverer students may have decided to sit near the teacher's desk.

2 There is strong positive correlation of body mass with height. The taller the dog, the heavier it tends to be. The gradient of the regression line is 0.864. This suggests that the mass increases by about 0.9 kg for each extra centimetre of height.
There is fairly strong negative correlation of life span with body mass. This suggests that lighter dogs tend to live longer. The gradient of the regression line is -0.072, suggesting that an increase in mass of 1 kg corresponds to a reduction in lifespan of 0.072 years (about 26 days).

3 **a) i**

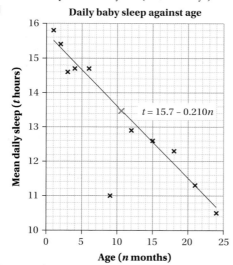

Daily baby sleep against age

$t = 15.7 - 0.210n$

ii Daisy (9 months, 11 hours)

b) i $t = 15.7 - 0.210n$ (3 sf)
ii 13.6 hours
iii 30 months is not within the range of the given data. Extrapolating does not give reliable predictions.
iv 15.7 hours.
This predicts the mean daily sleep for a newborn baby, but is also extrapolating beyond the given data. The points on the graph near $n = 0$ suggest that the actual mean daily sleep may be higher than this.
v The gradient of -0.21 indicates that a baby needs 0.21 hours (about 13 minutes) less sleep each month.

4 **a) i** A scatter graph shows that, in general, households with higher rates of unemployment have lower incomes. However London does not follow the general trend.
The data for London may be included for completeness or excluded as a special case (with the reason given).
Excluding London gives
$y = 52.0 - 2.37x$ (3 sf) where $x =$ unemployment rate and $y =$ average household income.
$r = -0.931$ (3 sf).
Including London gives
$y = 50.8 - 1.96x$ (3 sf), $r = -0.549$ (3 sf).
There is strong negative correlation between average household income and unemployment (particularly when London is excluded). The greater the level of unemployment, the lower the average household income.
ii Correlation does not necessarily imply causation. There does seem to be a relationship between unemployment and average household income, but the data for London shows that it is possible to have a high average household income without a low unemployment rate. So unemployment is not the only factor that has an effect on average household income in a region. Size and characteristics of the population and availability of work (supply of and demand for labour) will also be important. (Katie is wrong in saying unemployment *causes* low incomes.)

b) Regression of entry to university rate on annual household income:
including London $y = 19.3 + 0.285x$ $r = 0.688$ (3 sf)
or omitting London $y = 24.2 + 0.142x$ $r = 0.355$ (3 sf)
Regression of entry to university rate on unemployment rate
including London $y = 29.9 - 0.00844x$
                                   $r = -0.00571$ (3 sf)
or omitting London $y = 30.3 - 0.149x$
                                   $r = -0.146$ (3 sf)
Stacy is wrong – there is some correlation between entry to university rate and annual household income (fairly strong when London is included), whereas there is very little correlation between entry to university rate and unemployment rate (especially when London is included).

5 **a)** $y = -32.2 + 1.46x$ (3 sf), $r = 0.987$ (3 sf)

b)

Braking distances against speed for new car

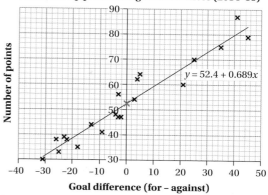

$y = -32.2 + 1.46x$

c) i 92 m (nearest metre)
ii 143 m (nearest metre)
The prediction for 85 mph is likely to be reasonably accurate as points on the graph before and after 85 mph also support this finding. The extrapolation used for 120 mph is not likely to be so accurate as 120 mph is outside the range of known data. Also the trend from points near the end of the graph has a steeper gradient.

d) The points appear to lie on a curve, rather than a straight line, so a model based on a curve would be better than linear regression in this case.

6 Regression of medal points on GDP:
$y = 41.9 + 0.0111x$ $\qquad r = 0.847$ (3 sf)
or (omitting China & USA)
$y = 19.4 + 0.0224x$ $\qquad r = 0.644$ (3 sf)
Regression of medal points on population:
$y = 68.9 + 0.113x$ $\qquad r = 0.607$ (3 sf)
or (omitting China)
$y = 25.2 + 0.660x$ $\qquad r = 0.866$ (3 sf)
or (omitting China & USA)
$y = 18.5 + 0.781x$ $\qquad r = 0.720$ (3 sf)
The correlation is strong for medal points on both GDP and population, so there is a relationship with both. The rest of the answer will depend on whether or not outliers were excluded when using the data.
Either:
Including outliers: the correlation coefficient is stronger between medal points and GDP than medal points and population, so Adam is wrong.
or
Excluding outliers: the correlation coefficient is stronger between medal points and population than medal points and GDP, so Adam is correct.

7 a) Using x to represent the goal difference and y the points,
$y = 52.4 + 0.689x$ with $r = 0.964$ (3 sf)
There is very strong positive correlation between the goal difference and points.

Premiership points and goal differences (2014–15)

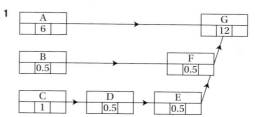

$y = 52.4 + 0.689x$

b) Some possibilities for investigations are:
Using x to represent the goals for and y the points:
$y = 2.13 + 1.03x$ with $r = 0.931$ (3 sf)
There is very strong positive correlation between the goals for and points.
Using x to represent the goals against and y the points:
$y = 117 - 1.33x$ with $r = -0.823$ (3 sf)
There is strong negative correlation between the goals against and points.
Using x to represent the goals for and y the goals against:
$y = 71.4 - 0.466x$ with $r = -0.682$ (3 sf)
There is fairly strong negative correlation between the goals for and goals against.
Using x to represent the games won and y the points:
$y = 12.1 + 2.81x$ with $r = 0.991$ (3 sf)
There is extremely strong positive correlation between the games won and points.
Using x to represent the games lost and y the points:
$y = 95.6 - 3.02x$ with $r = -0.952$ (3 sf)
There is very strong negative correlation between the games lost and points.
Using x to represent the games won and y the games lost:
$y = 25.9 - 0.807x$ with $r = -0.902$ (3 sf)
There is very strong negative correlation between the games won and games lost.

8 Your own research and report on consumption of chocolate and number of Nobel prizes won which should include a comment about correlation not implying causation.

Chapter 8

Exercise 8A

1

A				G	
6				12	

B				F	
0.5				0.5	

C		D		E	
1		0.5		0.5	

2

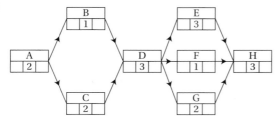

3 Your estimates may differ from those given here.

Activity	Duration (minutes)	Immediate predecessors
A: Prepare and test bath water	3	–
B: Get towel and talcum powder (if used) ready	1	–
C: Get clean clothes ready	1	–
D: Remove nappy	0.5	A, B, C
E: Dispose of nappy	0.5	D
F: Clean baby's bottom	1	D
G: Lower baby into water and gently wash baby	3	F
H: Play with baby in bath	5	G
I: Place baby on towel and dry thoroughly	3	H
J: Put new nappy and clothes on baby	2	I

4

Activity	Duration	Immediate predecessors
A	1	–
B	3	–
C	1	A
D	2	A, B
E	1	B
F	4	C, D
G	3	D, E
H	2	E
I	2	F, G, H

5

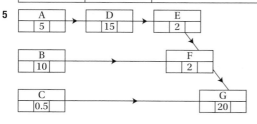

6

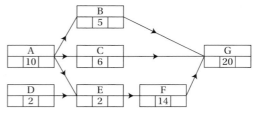

Exercise 8B

1 0 for C and 2 for F. C is critical because it has zero float.

2 **a)** An essential activity is one which has to be completed if the project is to be completed.

  b) A critical activity is one which has to be completed on time if the project is to be completed on time.

3 **a)** A, C and E.

  b) **i** 2 people

   ii Either person does A and either person does E. One person does C whilst the other person does both B and D.

  c) 12 weeks. This is simply the total of the durations of all the activities.

4 **a), b), c)**

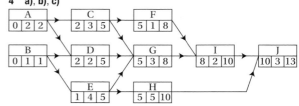

  d) 13 weeks; ACGIJ, BEGIJ and BEHJ

5 **a), b)**

Activity	Immediate predecessors	Early time	Late time
A	–	0	3
B	–	0	1
C	A, B	2	6
D	B	1	6
E	B	1	6
F	C	5	11
G	D	6	10
H	D, E	6	10
I	F, G	10	15
J	G, H	10	15
K	I, J	15	16

  c) BDHJK, 16 days

  d) E, float 2 days

6 a), b), c)

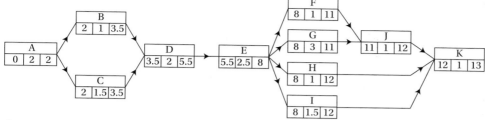

d) A, C, D, E, G, J, K

e) **i** $3.5 - 2 - 1 = 0.5$ days

ii H, $12 - 8 - 1 = 3$ days

f) 14.5 days (extra 1.5 days in I absorbed in its float)

Exercise 8C

1 a), b)

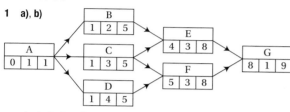

c) A, D, F and G

d)

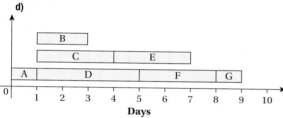

e)

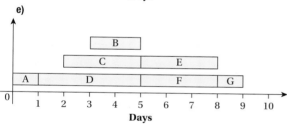

f) **i** It may be easier to deal with unexpected delays later.

ii Paying for materials and labour is delayed.

2 a), b), c)

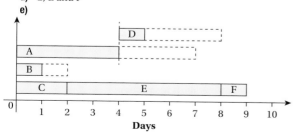

d) C, E and F

e)

f) No. The critical path has length 9 days irrespective of how many workers are employed.

3 a) A, B, C and D

b) B or B and G

c), d), e)

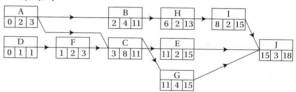

For the second possibility add an arrow from G to H.

4 a), b), c)

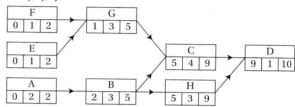

d) 18 weeks; DFCGJ

e)

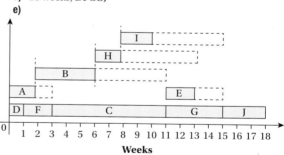

f) Reduced by 2 weeks (to 16 weeks)

Consolidation exercise 8

1 a), b), c)

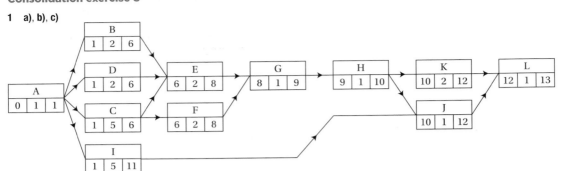

d) A, C, E, F, G, H, K, L

e)

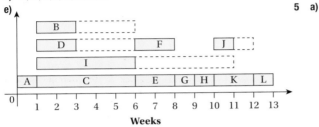

Weeks

2 a) Start (0, 0), A(0, 7), B(0, 5), C(4, 7), D(7, 13), E(4, 13), End(13, 13)

b) $t \geqslant 3$

3 a), b), c)

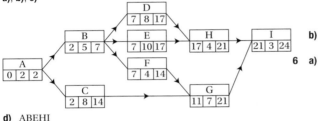

d) ABEHI

e) C(4), D(2), F(3), G(3)

f)

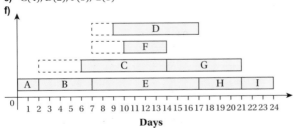

Days

4 a), b)

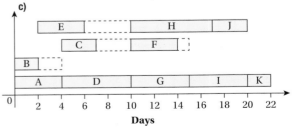

c) 22 days, BEGHIK

d) i H(15), I(16)

ii 2 days

5 a)

Activity	Earliest start	Latest finish
A	0	2
B	2	8
C	2	8
D	2	13
E	8	17
F	9	17
G	17	20
H	20	23
I	20	23
J	22	26
K	23	26
L	26	27

b) i ACEGIKL

ii $13 - 2 - 7 = 4$

6 a)

Activity	Earliest start	Latest finish
A	0	4
B	0	4
C	4	11
D	4	10
E	2	10
F	10	15
G	10	15
H	10	17
I	15	20
J	17	20
K	20	22

b) ADGIK and ADHJK; 22 days

c)

d) F starts 2 days later, I starts 1 day later. Minimum completion time is now 23 days.

7 a), b)

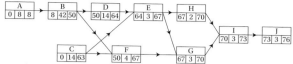

c) 76 days

d) **i** 13 days NB Activity F would become critical.
   ii Planning permission may not be granted.

8 a), b)

Activity	Early time	Late time
A	0	2
B	2	10
C	2	4
D	4	10
E	4	7
F	7	11
G	7	11
H	7	11
I	11	12

c) 12 weeks; ACEFI

d) B 6 weeks, D 3 weeks

e) Programmers were switched between D and F so as to take 8 weeks on D and 6 weeks on F. This only increased the completion time by 2 weeks because of the 3 weeks float on D and the deadline was met.

Chapter 9

Exercise 9A

1

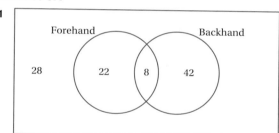

Tennis players

42 out of 70 i.e. 60%.

2 a) 0.5% of the people have the disease.

b) Out of 1000 tests, 997 give the correct result. So it is 99.7% accurate.

c) The test did not indicate the disease in 1 out of 5 sufferers. $\frac{1}{3}$ of the people indicated as having the disease have been falsely diagnosed and worried needlessly.

d) To achieve this accuracy requires no test; simply tell everyone that they do not have the disease. This example indicates how, for relatively rare diseases, tests have to be exceptionally accurate to have value.

Exercise 9B

1 a) 26 in 52 or 1 in 2

b) 4 in 52 or 1 in 13

c) 13 in 52 or 1 in 4

d) 2 in 6 or 1 in 3 (The dice must show 3 or 6)

2 a) Zero   **b)** 1 in 36   **c)** 6 in 36 or 1 in 6

3 Repeat the experiment many times. Instead of having a double-headed coin, you could simply use ink to mark both faces of one coin and one face of a second coin.

4 a) Unless the players are very evenly matched it is unlikely that the result of the match will be approximated very well by equally likely events.

b) The symmetry of this experiment means there are 52 equally likely events.

c) There is no symmetry here and so the events will not be equally likely.

5 a) **i** 12 in 30 or 2 in 5
   ii 5 in 30 or 1 in 6
   iii 2 in 30 or 1 in 15

b) 2 in 5

6 a) 50%   **b)** 54.4%   **c)** 31.8%

Exercise 9C

1 HH ⎤
   HT ⎦ 4 heads obtained
   TH
   TT – 4 tails obtained
   $\frac{1}{4}$

2 a) A fair amount would be £$\left(\frac{1}{2} \times 100\,000 + \frac{1}{2} \times 100\right)$ = £50 050.

b) A person might feel that the benefits to them of £30 000 are so great that it is not worth a 50:50 shot at £100 000.

3 a) 111, 112, 121, 122, 211, 212, 221, 222
   (Where 1 means Karpov wins and 2 means Kasparov wins.)

b) Only the last of these has player 2 (Kasparov) winning the match. His chances would therefore have been 1 in 8 or 12.5%.

Exercise 9D

1 a) 0.2

b) **i** 0.35   **ii** 0.65   **iii** 0.1   **iv** 0.9

c) 0.2

d) $\frac{0.3}{0.35} \approx 0.86$

e) 1

2 a) Equally likely events

b) $\frac{100}{205} \approx 0.49$

Exercise 9E

1 a) 0.5

b) **i** 0.8   **ii** 0.7   **iii** 1

2 a) $a = 0.3$, $b = 0.8$, $c = 0.6$

b) **i** 0.56   **ii** 0.68   **iii** 0

3 a) $\frac{8}{13}$ **b)** $\frac{8}{13} \times \frac{7}{12} + \frac{5}{13} \times \frac{8}{12} = \frac{8}{13}$

c) $\frac{14}{39}$ **d)** $\frac{20}{39}$

e) Both cards are black.

4 $\frac{6}{13} \times \frac{7}{12} \times \frac{6}{11} + \frac{7}{13} \times \frac{6}{12} \times \frac{6}{11} + \frac{7}{13} \times \frac{6}{12} \times \frac{6}{11} = \frac{63}{143}$

5 a) Assuming that results on the tests are independent of each other.

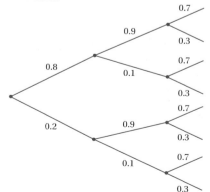

b) 0.504

c) $0.8 \times 0.9 \times 0.3 + 0.8 \times 0.1 \times 0.7 + 0.2 \times 0.9 \times 0.7$
$= 0.398$

d) $0.7 \times 0.8 \times 0.9 \times (0.3 + 0.1 + 0.2) = 0.3024$

6 a) i $0.1 + 0.6 \times 0.1 = 0.16$
ii $0.3 + 0.6 \times 0.1 = 0.36$

b) $(0.6 \times 0.8)^4 = 0.48^4 \approx 0.05$

7 a)

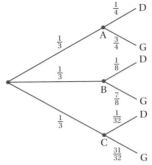

D = Defective
G = Good

b) $\frac{1}{3} \times \frac{1}{4} = \frac{1}{12}$ **c)** $\frac{1}{3} \times \frac{3}{4} = \frac{1}{4}$

d) **b)** and **c)** exhaust the possibilities for a bulb from A.

e) $\frac{1}{3} \times \left(\frac{1}{4} + \frac{1}{8} + \frac{1}{32}\right) = \frac{13}{96}$

8 a) $0.35 \times 0.1 = 0.035$

b) $0.65 \times 0.75 = 0.4875$

c) $0.4875 + 0.35 \times 0.9 \times 0.55 \approx 0.661$

Exercise 9F

1 a)

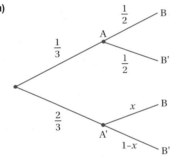

b) $\frac{1}{6} + \frac{2}{3}x$
$\frac{1}{6} + \frac{2}{3}x = \frac{1}{4} \Rightarrow x = \frac{1}{8}$

c) $1 - x = \frac{7}{8}$

2 a) $\frac{1}{13}, 1$ **b)** Yes, $P(A \cdot C) = p(A)$

c) No, $P(B \cdot C) \neq P(B)$

Exercise 9G

1 $\pounds(0.06 \times -1000 + 0.24 \times 2700 + 0.35 \times 1800 + 0.35 \times -100)$
$= \pounds1183$

2 She loses £100 with probability $\frac{19}{20}$ and wins £900 with probability $\frac{1}{20}$.
Her expected loss is $\pounds\left(100 \times \frac{19}{20} - 900 \times \frac{1}{20}\right) = \pounds50$.
This result is an obvious one because each ticket costs £1 but only wins 50p on average. The loss is therefore £0.50 per ticket. [This method of analysis is a good way of analysing the national lottery.]

3 a) A
A first: $\pounds\left(\frac{3}{5} \times 160 + \frac{2}{5} \times 280\right) = \pounds208$
B first : $\pounds\left(\frac{2}{5} \times 120 + \frac{3}{5} \times 280\right) = \pounds216$

b) £8

Consolidation exercise 9

1 a) Dependent since compatibility of beliefs is a factor of importance to couples.

b) Independent. (Although toothache could possibly be triggered by a common cause such as cold weather or a particular meal.)

c) Dependent. (Although they would be independent if, for example, I never walked to work.)

2 a) $\frac{5}{26}$ **b)** $\frac{9}{26}$

3 a)

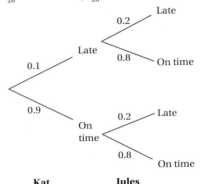

Kat   **Jules**

b) 0.02

c) $0.08 + 0.18 = 0.26$

d) $2 \times 0.02 + 1 \times 0.08 + 1 \times 0.18 = 0.3$

4 $\frac{1}{4}(1+2+3+4) = 2.5$ This assumes that the dice is unbiased.

5 a) $\frac{1}{4} \times \frac{1}{20} + \frac{1}{4} \times \frac{1}{40} + \frac{1}{2} \times \frac{1}{40} = \frac{1}{32}$

b) $\frac{6}{32} = 0.1875$

6 a) Yes, the odds can be made almost equal to 75%.

b) If 1 pot contains a single white ball and the other pot contains the remaining balls, the probability of choosing a white ball will be $\frac{1}{2} + \frac{1}{2} \times \frac{49}{99} = \frac{74}{99}$.

7 a) $\frac{1}{5} \times \frac{1}{20} + \frac{4}{5} \times \frac{1}{4} = \frac{21}{100}$ i.e. 21%.

b)

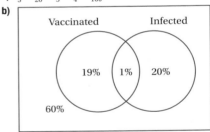

c) $\frac{1}{21}$

8 a)

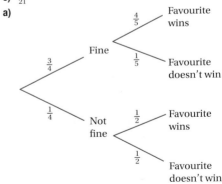

b) $\frac{3}{4} \times \frac{4}{5} + \frac{1}{4} \times \frac{1}{2} = \frac{29}{40}$

9 $\frac{1}{10} \times 1 + \frac{1}{5} \times 3 + \frac{2}{5} \times 5 + \frac{1}{5} \times 3 + \frac{1}{10} \times 1 = 3.4$

10 a) 38%

b) $P(\text{vaccinated} \cdot \text{infected}) = \dfrac{P(\text{vacinnated} \cap \text{infected})}{P(\text{infected})}$

$= \frac{0.03}{0.38} = \frac{3}{38}$

11 $£\left(\frac{3}{4} \times 2500 + \frac{1}{4} \times 1500\right) = £2250$

12 a) $\left(\frac{1}{6}\right)^3 = \frac{1}{216}$

b) $£\left(\frac{5}{6} \times 300\,000 + \frac{1}{6} \times \frac{5}{6} \times 60\,000 + \frac{1}{6} \times \frac{1}{6} \times \frac{5}{6} \times 125\,000 \right.$
$\left. + \frac{1}{6} \times \frac{1}{6} \times \frac{1}{6} \times 300\,000\right) \approx £37\,616$

13 Assume the two problems occur independently. Then the probability of one or both is 0.28. The expected penalty is £0.28 × 25 000 = £7000.

Chapter 10

Exercise 10A

1 a) The company should not recall the batches but should pay compensation as necessary. That is what the company did.

b) The extremely bad publicity the company received forced them, eventually, to recall the batches *and* pay compensation *and* spend money to rebuild their reputation. It was just as appropriate in this case, as in many others, to apply cost–benefit analysis. However, the analysis is of little value if major costs or benefits are not included.

2 a)

	£	
Extra annual revenue	20 000	(100 000 × 0.2)
Extra revenue per month	1667	(20 000 ÷ 12)
Reduced costs	2400	
Total benefit	**4067**	
Salary	4000	
Office space	150	(1800 ÷ 12)
Overheads	300	
Total extra costs	**4450**	

Costs exceed benefits by £383.
Assuming there are no other benefits (such as improving the quality of the company's work) the expansion is not justified.

b) Further revenue required £383 × 12 = £4596
$\frac{24\,596}{100\,000} \approx 24.6\%$

3 Unless something has changed to affect your judgement of the value of going, the logical decision is to buy another ticket.

4 If the *extra* costs are greater than the *extra* benefits of the project then it is the spending of any further money which would be a misuse of taxpayers' money. [The project did continue. The valuations given for this project were questionable and so the project may have been worth continuing – but not for the reason given by Senator Dalton.]

Exercise 10B

1 The expected extra cost of bad weather is £0.1 × 100 000 = £10 000. This is less than the control measure costs of £15 000. So the company should not change the way the work is set up. However, if the company could not afford the £100 000 penalty of the delays then its decision might well be different.

2

Control measure	Cost of the measure (£)	Expected penalty (£)	Total cost (£)
X only	4000	9000	13 000
Y only	8000	3000	11 000
Both X and Y	12 000	0	12 000

Take measure Y only because this has the least expected cost.

3 No. The expected loss in 1 year is £1.50 which is less than the extra premium.

Consolidation exercise 10

1 a) A Because, on average, this gives the greatest gain.
B They are attracted to the potential of becoming rich.
C They are concerned to minimise their possible loss.

b) Typically, gambling advertisements, for the lottery for example, stress the maximum possible gain.

c) Typically, insurance advertisements will stress the effects on you of the maximum possible loss.

2 Because, on average, he will win from these positions.

3 a) E is not a critical activity and has a float of 1 week.

b) Assuming delays occur independently.

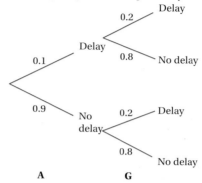

Probability of delay = 0.1 + 0.9 × 0.2 = 0.28

c) £9000 × 0.28 = £2520

d)

Control measures	Expected cost (£)
None	2520
A only	1000 + 0.2 × 9000 = 2800
G only	500 + 0.1 × 9000 = 1400
Both A and G	1500

Recommend control G only as it gives the lowest expected cost.

4 The original expected value is $£\frac{1}{4} \times (1 + 10 + 5000 + 20\,000)$ = £6252.75
This becomes $£\frac{1}{3} \times (1 + 10 + 20\,000) = £6670.33$. Although most participants would be disappointed by this outcome, the value actually increases by £417.58.

5 a) Freedom from worry about a possible large loss in the future.

b) Because once people have set up a direct debit they are less likely to cancel the insurance in later years.

c) Assuming you can afford to replace any food, even the smaller premium of £25 would require high values for the food and probability of failure. £250 and 0.1 would, for example, be the break-even point.

Chapter 11

Exercise 11A

1 a) There is no vertical scale.

b) 2012

c) $6000 \times \frac{5}{2} = 15\,000$ cans per day.

2 a) $D = T + B$

b)

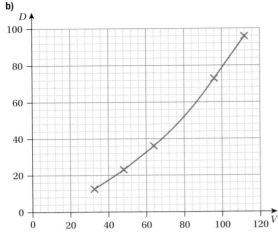

c) Approximately 54 m

d) 13.5 (The average car length is 4 metres)

3 a)

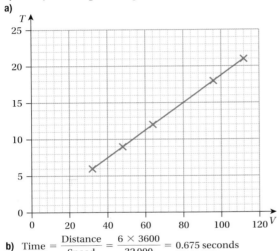

b) Time $= \dfrac{\text{Distance}}{\text{Speed}} = \dfrac{6 \times 3600}{32\,000} = 0.675$ seconds

4 0.006

5 a)

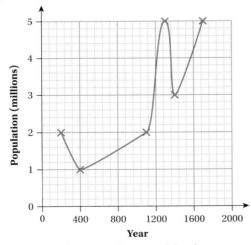

b) The collapse of the Roman Empire and the plague.

c) The estimate is based upon the Domesday Book.

d) You might estimate perhaps 7 million. Advances in medicine and sanitation enabled massive population growth.

6 a)

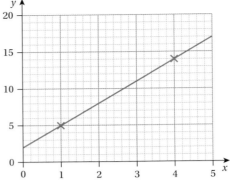

b) 2   **c)** 3   **d)** $y = 3x + 2$

7 a) $y = 3x$ **b)** $y = -x + 7$

    c) $y = 4$ **d)** $y = 2x - 9$

8 a) $l = 48 + 0.33T$

    b) Medical advances may continue to improve life expectancy. However, extrapolation is problematic because there might be wars/epidemics.

9 a) 3 cm

    b) The spring would eventually break/lose its elasticity.

    c) $e = 2T$

10 a) $-\dfrac{1}{16} = -0.0625$

    b), c) (0, 1.4) This extrapolation has no physical significance since p cannot be greater than 1. However, 1.4 is the intercept which occurs in the equation of the line.

(22.4, 0) This predicts that at 22.4 metres and over, the player has zero chance of scoring. Again, this will not be the case. The graph can be expected to curve round and slowly approach zero height as D increases.

    d) $p = 1.4 - \dfrac{D}{16}$ or $p = 1.4 - 0.0625D$

    e) 71%

11 a)

b) Approximately 1 : 50

c)

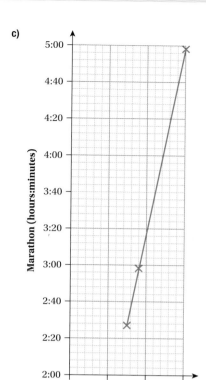

d) Approximately 3:27

Exercise 11B

1 a) Linear   **b)** None   **c)** Linear
   d) Cubic   **e)** None   **f)** None
   g) Linear   **h)** Quadratic   **i)** None

2 a)

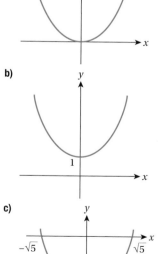

d)

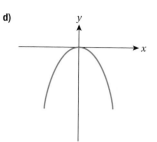

3 a) -20, **0**, **6**, **4**, **0**, **0**, 10

b)

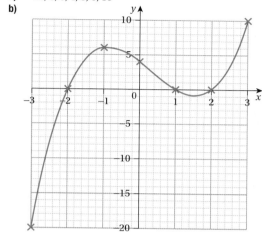

c) -1.7, 0.2, 2.5

4 a)

x	0	1	2	3	4	5	6	7
y	1.7	2.62	3.38	3.98	4.42	4.7	4.82	4.78

x	8	9	10	11	12	13	14	15
y	4.58	4.22	3.7	3.02	2.18	1.18	0.02	-1.3

b)

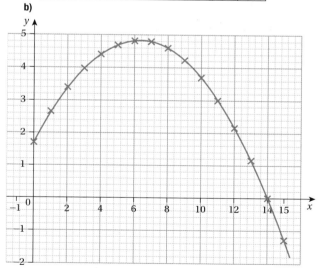

c) The throw is from shoulder height.
d) 14 m
e) 4.8 m

5 a)

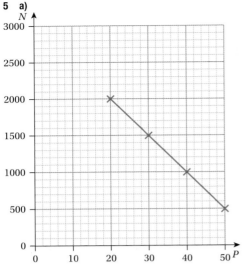

b) Linear

c) $N = 3000 - 50P$

6 a) N items are sold and the profit on each is £$(P - 15)$.

b)

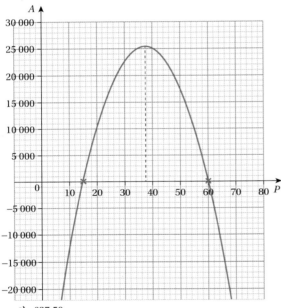

c) £37.50

Exercise 11C

1 a) i 4.8

   ii 3.2 or 3.3

b) $y = 50 - 10x$

c) 2.8

2 a)

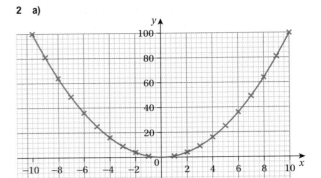

b) i 3 and -8

   ii 3.5 and -5.5

3 a), b)

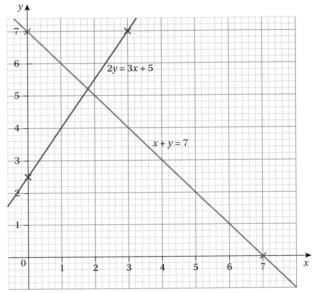

c) $x = 1.8$, $y = 5.2$

4 a) Countries and companies extract more from existing oil fields. New oil fields become viable, for example in the North Sea or the South Atlantic.

b) More use is made of other sources of heating or energy.

c) £70

Consolidation exercise 11

1 Using the photograph, one can estimate the height of the towers (approximately 20 m) and the distance between them (approximately 200 m). For these distances the equation would be $y = 0.002x^2$ where the origin is at the centre of the bridge.

2 a) Anand's rating has stayed constant whilst Carlsen's rose rapidly and is now levelling off.

b) Carlsen was only 13 years old in 2002.

c) Anand: $R = 2800$.
You will be better able to model the curve for Carlsen when you have worked through Chapter 13. However, you could use a quadratic curve with a maximum point at $(0, 2900)$; for example, $R = 2900 - 6y^2$.

d) To analyse the accuracy of your model you could give a table of predicted and actual ratings for several years.

e) The rating difference was about 80 points giving a score of 60%. Carlsen actually scored $6\frac{1}{2}$ out of a possible 11 points, or 59%.

3 a) $S = 3000 - 10P$

b)

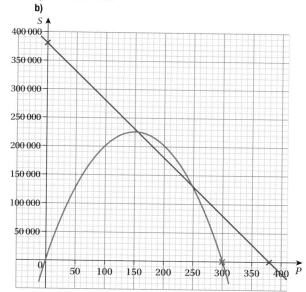

c) The revenue is £P for each of S bicycles and so the total revenue is £$SP = £3000P - 10P^2$.

d) i £155 **ii** £200

4 a) £$(500 + 0.0225P)$

b)

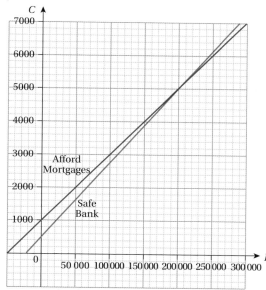

c) For more than £200 000 use Afford Mortgages. Otherwise use Safe Bank.

5 a) $T \approx 1.5$, approximately September 2010

b) $a = 3$ and $b = 3$

c) You cannot use your equation to extrapolate to 2015. If you attempted this you would find that a market share of over 100% would be predicted. In fact, Samsung's share dropped from a peak of over 30% to 20% by mid-2015.

Chapter 12

Exercise 12A

1 a)

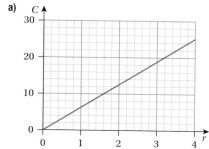

b) 2π. Each 1 cm increase in the radius increases the circumference by 2π cm.

2 a) $\frac{100}{10} = 10$ cm per year. On average, Kim grows 10 cm per year.

b) i Courtney was 8 cm longer at birth but is growing at a rate which is 1 cm per year less.

ii 8

c) It is not possible to answer this question. Growth rates change so much throughout life that extrapolation makes no sense for this context.

3 a) $-46\,°C$

$41\,°F - 140\,°F$

$68\,°F$

$34\,°C$

$66\,°C$

b) 1.8 Each change of 1 °C is equivalent to a change of 1.8 °F.

c) $\frac{1}{1.8} = 0.\dot{5}$

d) 1.8

4 a) The cost for each extra kWh.

b) Tariff A has a standing charge of £5 per month even if no energy is used. However, the charge for each extra unit is higher on Tariff B.

c) This is where the total charge would be the same on either tariff. A family should choose A if they use more energy than this and otherwise should choose B.

d) They simply say the standing charge for Tariff B is zero!

5 a) 12

b) If x changes by 1, y changes by 3. Then z changes by 4 times this amount, i.e. by 12.

6 a) Assume that the Earth is a perfect sphere and that the orbit is circular (x is constant).

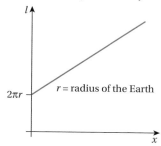

b) 2π this is the same as in question 1.

Exercise 12B

1 a) The gradient is initially large and positive. It gradually decreases to zero at $(0, 0)$. It then increases again and becomes large and positive.

b) Approximately 12

c) Approximately 3

d) 12 3 0 3 12

e) $3x^2$

If you can't see the relationship from the numbers, you could plot a graph of gradient against x to see the relationship.

2 a) 2 °C

b) Approximately $\frac{1}{2}$°C per minute

c) The bottle of milk is warming up by $\frac{1}{2}$°C per minute.

d) 7 °C (from 2 °C to 9 °C)

e) A rate of warming of $\frac{1}{2}$°C per minute would produce a 10 °C rise. However this rate is gradually decreasing.

3 a) 3 mm

b) 1.5 mm per day

c) Graph with point (10, 1.5) marked.

d) 35 days

4 a) 100 in 3 days. Approximately 33 per day.

b)

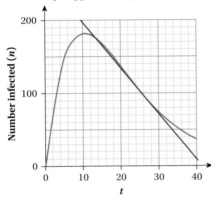

c) −7 per day. The number infected is reducing by 7 per day.

d) Assuming no fresh outbreaks, the gradient will tend to zero.

Exercise 12C

1 a) i 3 ms⁻¹   **ii** 12 ms⁻¹

b) The roller coaster car starts from rest and accelerates for 5 seconds to reach a steady speed of 3 ms⁻¹. The car maintains this speed for 15 seconds until it reaches the top of the first incline. The car then picks up speed rapidly as it starts to descend before slowing down when it comes to the next incline.

2 a) 20 ms⁻¹ is the gradient of the graph at time $t = 0$.

b) 10 ms⁻¹

3 a) The ball is at rest when its velocity is zero. 1.5 s

b) The ball's velocity starts at 15 ms⁻¹ and then steadily decreases.
After 1.5 seconds its velocity becomes negative.

c) Think about how the ball could be thrown and how its velocity would change. The velocity decreases to zero and then becomes negative, which means the ball starts moving in an opposite direction to its initial motion. The ball is thrown vertically upwards and then starts to fall back down.

d) The ball stops moving upwards and its velocity is zero just before it starts to fall downwards. The point where the velocity is zero is the highest point the ball reaches.

4 a) $3.125 \times 120^2 = 45\,000$ m, or 45 km.

b) $\frac{45\,000}{120} = 375$ ms⁻¹

c)

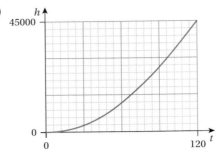

d) 750 ms⁻¹

5 a)

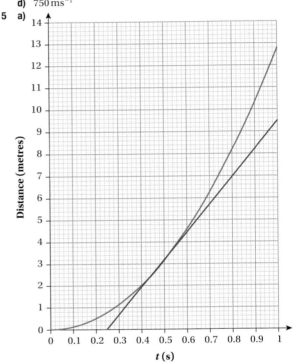

b) Approximately 13 ms⁻¹

c) 0 ms⁻¹. The tangent line is horizontal.

d) The gradient (in ms⁻¹) at $t = 1$.

Consolidation exercise 12

1 a) 30 ms⁻¹

b)

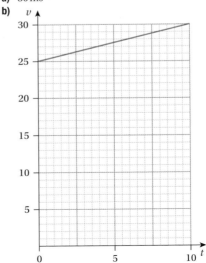

c) 275 metres

d) $\frac{275}{10} = 27.5\,\mathrm{ms}^{-1}$

2 a) 1st January and 1st October

b) +110

c) 45. The stock is increasing at a rate of approximately 45 cars per month.

d) −22. The stock is reducing at a rate of approximately 22 cars per month.

e) Extra cars might have been acquired during this period. All that is known is that the net stock fell by 110.

3 a) 5.4 seconds

b)

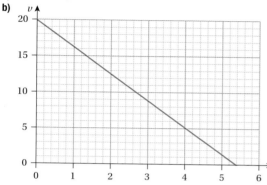

c) $-3.7\,\mathrm{ms}^{-2}$. This is the acceleration due to gravity.

d) Possibly Mars.

4 a) Approximately −100 and −30 (°C per hour).

b) The temperature of the tea is falling at these rates.

c) As the tea cools its rate of cooling rapidly reduces.

d) $t = 0.205$ when tea was discovered. This is approximately 12 minutes after it was poured.

So the time of pouring is about 11:48 am.

Assumptions are that the tea was at 100 °C when poured (it was probably a few degrees less than this) and that the conditions in the flat were the same when the forensic scientist obtained the curve as when the tea was poured.

5 a) 2.4 metres

b) 1.2 seconds

c) $-10\,\mathrm{ms}^{-2}$ or $10\,\mathrm{ms}^{-2}$ downwards.

d) $t = 0.4$ This is when the ball has reached its maximum height and is about to start falling down.

Chapter 13

Exercise 13A

1 a) 2^{-x} Base 2, Exponent $-x$

c) $\left(\frac{1}{2}\right)^{2x}$ Base $\frac{1}{2}$, Exponent $2x$

e) 10^x Base 10, Exponent x

Note that **g)** is *not* called an exponential function because the base is not constant.

2 a) 1.37×10^{11} **b)** 15.6   **c)** 0.0243

3 a) $x = 4$ **b)** $x = 3$ **c)** $x = 0$ **d)** $x = -2$

4 a) $y = 3^x$ **b)** $y = \left(\frac{1}{3}\right)^x$ or 3^{-x}

5 a) £1040   **b)** £1216.65 **c)** £1000(1.04)^x

6 a) $3.6 \times 10^{-8} \times 28\,000 = 0.001\,\mathrm{m} \approx 1\,\mathrm{mm}$

b) 1 cm, 10 cm Each increase in magnitude of 1 unit increases the amplitude by a factor of 10.

7 a) 400 mg   **b)** 204.8 mg   **c)** $t = 6.21$ so time is $18:13$

Exercise 13B

1 a) 20.1   **b)** 0.14   **c)** 26.2

2 a) 0   **b)** −0.18   **c)** 1.6

3 a) 98 °C   **b)** 50 °C   **c)** 22 °C

d)

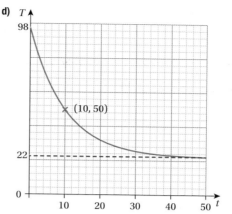

e) i A is the amount by which the initial temperature exceeds room temperature.

ii B is room temperature.

4 a) 389   **b)** 3.7

Exercise 13C

1 a) $e^{-2x} = 0.1$

$x = 1.15$

b) $e^{0.05x} = \frac{1}{7}$

$x = -38.9$

2 1.73 years

3 3.67

4 a) $1600e^{0.078t} = 500\,000$

$t = 73.6$ so 1974 (which is the correct date)

b) For $t < 35$ there were no cane toads so the model is not valid for small values of t.

For large values of t it is also not valid since the area cannot exceed the area of Australia.

5 a) $y = 20 + 7e^{0.220x}$

b) $y = 50 - 20e^{-0.229x}$

Exercise 13D

1 3 million is a reasonable estimate which will be used in answers to questions **2** and **3**.

2 $2^n = 3\,000\,000$

$n = \frac{\ln 3\,000\,000}{\ln 2} \approx 21.5$

On the 22nd square

3 $2^n = 3 \times 10^{12}$

$n = \frac{\ln (3 \times 10^{12})}{\ln 2} \approx 41.4$

On the 42nd square

Consolidation exercise 13

1 a) 243   **b)** 4.66   **c)** 0.18

2 a) 1.5   **b)** $\frac{\ln 5}{\ln 3} = 1.46$ **c)** $\frac{\ln 5.7}{\ln 1.3} = 6.63$

d) $5x = \ln 17 \Rightarrow x = 0.57$

3 a) $e^x = \frac{4}{7} \Rightarrow x = -0.56$

b) $e^{2t} = \frac{1}{2} \Rightarrow t = -0.35$

4 a) For 1801: $n = 0$ and $1.6(1.11)^0 = 1.6$

For 1901: $n = 10$ and $1.6(1.11)^{10} \approx 4.54$

b) For 1851 it is sensible but it is not sensible to extrapolate to 1951. See the answer to part **d)**.

c) The model gives 2.7 million which is 7% out.

d) $1.6(1.11)^{21.5} \approx 15$ million. This is wildly inaccurate since the actual figure is roughly 7.4 million.

5 a) $P_0\,e^{-1.097} \approx \frac{1}{3}P_0$

b) $\frac{1}{2}P_0 = P_0\,e^{-0.000124h}$

$h = \frac{\ln 2}{0.000124} = 5600\,\mathrm{m}$

c) Weather conditions

6 a) When $t = 0$: $\theta = 20 + 80 = 100$, the boiling point of water

b) $60 = 20 + 80e^{-\frac{k}{2}}$
$40 = 80e^{-\frac{k}{2}}$
$k \approx 1.39$

c) $30 = 20 + 80e^{-1.39t}$
$t \approx 1.496$
90 minutes

d) The temperature of the room containing the container. The water temperature will get closer and closer to $20\,°C$.

7 a) m_0 grams is the mass of carbon-14 in the organism when it died.

b) $\frac{1}{2}m_0 = m_0e^{-0.000\,121t}$
$\frac{1}{2} = e^{-0.000\,121t}$
$\ln\left(\frac{1}{2}\right) = -0.000\,121t$
$-0.693 = -0.000\,121t$
$t \approx 5730$ years

c) m_0 is needed. This is estimated by cross reference with other fossils and with modern day carbon-14 take-up.

d) 0.002 36 grams. The smallness of this quantity is why carbon-14 dating tends not to be used for dating items over 50 000 years old.

e) Above ground nuclear testing.

8 a) $N_0 = 26.5$

b) $26.5e^{12k} = 61.4$
$\Rightarrow \quad e^{12k} = \frac{61.4}{26.5} = 2.32$

c) $\frac{\ln 2.32}{12} = 0.07$

d) $26.5\,e^{0.07 \times 6} = 40.3$ which is an accurate estimate.

Practice questions

Paper 1

1 a) Because of rounding to the nearest thousand.

b) i Sample size too small.
Sample not taken in proportion to group size.

ii Use given sample size (at least 20).
Use a stratified sample.
Calculate sample size $\times \frac{19\,054}{30\,513}$ etc.

2 a) Shoe sizes.

b) i One possibility is to design a (very simple) questionnaire to be used with people leaving a local shoe shop.

ii Ask local shop(s) for the data you need or use the internet for the Office for National Statistics, Shoe manufacturers, ...

3

End of year	1	2	3	4	5
Amount in account (£)	1530	1560.06	1091.81	613.65	125.92

Total amount withdrawn = £500 + £500 + £500 + £128.44
$= £1628.44$

4 Assumptions should include:
- typical lengths of golf shots
- length of route that could be taken by someone playing golf shots.
  Some comment should be made that the surroundings will often prevent long drives, for example when passing through towns and villages.

Other possibilities for assumptions might include:
- lost golf balls can be replaced from approximately where they were lost.

Assuming a route of 1600 km and an average length of 100 m per shot, the number of shots would be
$1600 \times \frac{1000}{100} = 16\,000$.

5 a) Pay as you earn.

b) £24 000 − £10 600 = £13 400
£13 400 × 0.2 = £2680
$\frac{£2680}{12} = £223.33$

c) D2 = B2 × D1

6 a) Box and whisker plot for the newspaper readers should have:
Least time 20, Greatest time 900
Median approx. 170
LQ approx. 77, UQ approx. 355
Comments should include:
- range for newspaper readers is smaller
- IQR for newspaper readers is much lower
- newspaper readers spent less time on average (justify using median)
- least time, LQ, UQ and greatest times are all less for the newspaper readers.

b) The two readerships might be very different.
Bias may have been introduced by the way the surveys were carried out.

7 a) $C = 2000$, $t_1 = 1$, $t_2 = 2$, $i = 0.15$

b) $2000 = \frac{A}{1.15} + \frac{A}{1.15^2}$
$2000 \approx 1.626A$
£A ≈ £1230.23

8 a) If it is needed urgently.
If it is much more convenient to (for example) shop locally.
If any difference in cost for such a product is likely to be minimal.
If it costs more to travel to purchase the product than the saving made.

b) Assumptions should include:
- ignoring standing charges that have to be paid irrespective of distance driven
- the cost of petrol
- distance travelled per litre
- size of a petrol tank
- smaller costs for tyres etc.

Other possibilities for assumptions might include:
- including (or not) the cost of one's time.

The main calculation should then have a form such as:

Cost of petrol   130p per litre
Distance per litre   10 miles
Size of a tank   50 litres

For a garage 1 mile away.
Cost of fuel used (2 miles) $= \frac{130}{5} = 26$p
Money saved on tank of fuel = 50p
This gives a rough figure that a 1p per litre saving justifies travelling to a garage 2 miles away. Including other costs (and the cost of one's time) reduces this figure.

Paper 2

1 a) • No label on the vertical or horizontal axis.
- Different intervals mean the bars have misleading relative heights.

b) Assume a maximum salary, for example £120 000.

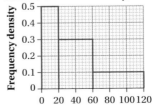

The frequency density for salaries above £60 000 must be in proportion with your assumption for the maximum salary.

2 **a)** He repays £1200 so he is correct to say he pays £200 interest.

   b) The crucial point is that he does not borrow the full £1000 for the two years. So, after one year at 10%, he can consider that he is paying back £500 as well as the interest of £100. However, in the second year he would then be paying interest of £100 on a loan of only £500, that is, 20% interest.

3 **a)** The multiple for English resits is $\frac{100\,239}{20\,544} \approx 4.88$. It is perfectly reasonable to say this 'is five times'. The multiple for maths resits is $\frac{110\,811}{27\,579} \approx 4.02$. This is *not* 'six times'.

   b) The fine would take resources away from schools and could potentially reduce the likelihood of success with the next cohort of students. However the money raised by the fine could improve the chances of the large number of students resitting at FE colleges.

   c) Individually (both for a student and for a school/college), extra well-targeted effort will improve results **relative** to other students and schools/colleges.
However, the results of the entire cohort of students are effectively fixed by this cohort's results at key stage 2.

Paper 2A

4 Sample mean = 390.4
$\frac{4}{\sqrt{5}}$ seen, 1.96 seen

$390.4 \pm 1.96\frac{4}{\sqrt{5}} = [386.9, 393.9]$

The measurements are significantly raised from typical outside measurements.
However, this poses no environmental risk since the level is well within the typical range for indoor levels.

5 **a)** The gradient is 1.35.
For each **extra** mark on Paper I, candidates tend to score 1.35 more marks on Paper II.

   b) (Strong) positive correlation. The two papers measure very similar skills.

   c) The point for Candidate I is furthest from the line of best fit. A total mark of $33 + 50 = 83\%$ could be considered.

6 Features should include:
- bell-shaped histogram
- correctly labelled axes
- centred on 37 °C
- values between roughly 35.8 °C and 38.2 °C
- maximum height at frequency density approximately 1.

Comments could include:
- bell-shaped OR symmetrical
- centred on 37 °C
- comments about proportions lying within $\pm 1, \pm 2, \pm 3$ standard deviations of the mean.

7 **a)**

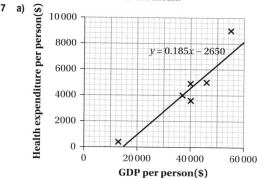

b) $y = 0.185x - 2650$

c) Either from the equation or from the graph, the estimates are $4000 and $-1500.
The estimate for Italy can be expected to be reasonably accurate because it has many similarities with other countries in this list. (In fact it is significantly higher than the actual figure of roughly $3200.)
The estimate for India is obviously impossible. This illustrates the dangers both of extrapolation and of applying figures from (largely) developed countries to other countries.

8 **a)** s.d. = 0.1
$\Phi\left(-\frac{0.01}{0.1}\right) = 1 - \Phi(0.1) = 0.460$

   b) s.d. $= \frac{0.1}{5} = 0.02$
$\Phi\left(-\frac{0.01}{0.02}\right) = 1 - \Phi(0.5) = 0.309$

Paper 2B

4 **a) i** This appears true – there are precisely 500 females and 500 males.
(There are also 60 girls and 60 boys.)

     ii This is also true – there are 83 sufferers out of 1000, or 1 in 12.05.

   b) Conclusions, which must be supported by numerical data, include:
- asthma is more common amongst boys than girls
- asthma is more common amongst women than men
- asthma is more common amongst children than adults.

5 **a)**

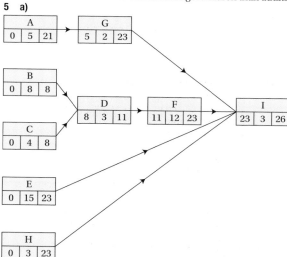

   b) He must start preparing the meal by 6.30 pm.
The only way to achieve this is to prepare the vegetables and spices in the first 12 minutes and then fry them before adding water and cooking the curry.
Laying the table and serving the chutneys and poppadoms can be done whilst the curry is cooking.
The rice can be cooked at any convenient point in this process.

6 **a)** The expected gains in £s are
$\frac{100\,000}{2\,000\,000} + \frac{10\,000}{400\,000} + \frac{250}{10\,000} + \frac{50}{5000} + \frac{20}{250} + \frac{10}{250} = 0.23$
The expected loss is therefore 77p per ticket.

   b) The probability of winning a prize is
$\frac{1}{2\,000\,000} + \frac{1}{400\,000} + \frac{1}{10\,000} + \frac{1}{5000} + \frac{1}{250} + \frac{1}{250} = 0.008303$
This is 1 in 120.4 and so the stated odds are correct.
What is misleading is the emphasis on the top prize that a person has very little chance of winning.

7 a) Undertake an action if, and only if, the expected extra benefit from undertaking the action is greater than the expected extra cost.

b) The extra costs of the proposed expansion in £s per month are

$$5000 + \frac{20 \times 100}{12} \approx 5167$$

The definite extra benefits are

$$50 \times 60 = 3000$$

In addition there is a probability, p say, of a monthly benefit of

$$\frac{500\,000}{12} \times \frac{10}{100} \approx 4167.$$

The break-even point would be if

$$5167 = 3000 + 4167p$$

so $p \approx 0.52$

The chance of winning the contract must be greater than 0.52.

Paper 2C

4 a) There will be fixed costs (for example of cabling) for any capacity of solar panel system.

b) $m = 1.5$, $c = 1500$

c)

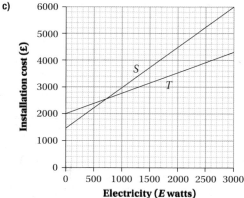

Installation cost (£) vs **Electricity (E watts)**

The 'break-even' point is where the two lines cross. For capacities of less than 667 watts the solar panels are cheaper to install.

5 a) 1680 million (By graph or by algebra)

b) $e^{50x} = \frac{811}{230}$

$50x = \ln\left(\frac{811}{230}\right)$

$x = 0.0252$

$230e^{100x} = 2860$ (3 sf)

2860 million

(The UN forecast is 2300 million. This assumes some reduction in fertility rates.)

6 a) Graph A shows none of the volatility of the price.

b) The scale on the x-axis is not uniform.

The y-axis has a false zero.

c) i The graph is (roughly) symmetrical about a single minimum point.

ii $a = 719$, $b \approx 5$ (for example)

Date	Data (£)	Model (£)	Difference (£)
24th	740	739	−1
25th	722	724	2
26th	719	719	0
27th	726	724	−2
28th	739	739	0

7 a) His speed is slowest over the first 20 m, $\frac{20}{2.93} \approx 6.8\,\text{ms}^{-1}$.

- His speed increases to a maximum of $\frac{20}{1.61} \approx 12.4\,\text{ms}^{-1}$.
- It then reduces slightly to $\frac{20}{1.67} \approx 12.0\,\text{ms}^{-1}$.
- After the first 20 m, his speed changes very little.

(His average speed for the entire race is $\frac{100}{9.63} \approx 10.4\,\text{ms}^{-1}$.)

b) i $0\,\text{ms}^{-1}$

ii $6\,\text{ms}^{-1}$ to $7\,\text{ms}^{-1}$

(Probably obtained from considering a tangent line)

c) METHOD 1

Plot a graph of velocity against time.

Recognise a straight line (at least near $t = 1$).

Find the gradient and give correct units for example $8\,\text{m/s/s}$ or $8\,\text{ms}^{-2}$.

METHOD 2

Use the graph to obtain two velocities for example at $t = 0.5$ and $t = 1.5$.

Use of $\dfrac{\text{Change in velocity}}{\text{Change in time}}$ for example $\frac{10 - 2}{1} = 8\,\text{ms}^{-2}$

(More precise analysis gives a maximum acceleration of $9.5\,\text{ms}^{-2}$).

Statistical tables and formulae sheet

Table 1   Normal distribution function

The table gives the probability, p, that a normally distributed random variable Z, with mean = 0 and variance = 1, is less than or equal to z.

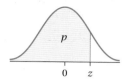

z	0.00	0.01	0.02	0.03	0.04	0.05	0.06	0.07	0.08	0.09	z
0.0	0.50000	0.50399	0.50798	0.51197	0.51595	0.51994	0.52392	0.52790	0.53188	0.53586	0.0
0.1	0.53983	0.54380	0.54776	0.55172	0.55567	0.55962	0.56356	0.56749	0.57142	0.57535	0.1
0.2	0.57926	0.58317	0.58706	0.59095	0.59483	0.59871	0.60257	0.60642	0.61026	0.61409	0.2
0.3	0.61791	0.62172	0.62552	0.62930	0.63307	0.63683	0.64058	0.64431	0.64803	0.65173	0.3
0.4	0.65542	0.65910	0.66276	0.66640	0.67003	0.67364	0.67724	0.68082	0.68439	0.68793	0.4
0.5	0.69146	0.69497	0.69847	0.70194	0.70540	0.70884	0.71226	0.71566	0.71904	0.72240	0.5
0.6	0.72575	0.72907	0.73237	0.73565	0.73891	0.74215	0.74537	0.74857	0.75175	0.75490	0.6
0.7	0.75804	0.76115	0.76424	0.76730	0.77035	0.77337	0.77637	0.77935	0.78230	0.78524	0.7
0.8	0.78814	0.79103	0.79389	0.79673	0.79955	0.80234	0.80511	0.80785	0.81057	0.81327	0.8
0.9	0.81594	0.81859	0.82121	0.82381	0.82639	0.82894	0.83147	0.83398	0.83646	0.83891	0.9
1.0	0.84134	0.84375	0.84614	0.84849	0.85083	0.85314	0.85543	0.85769	0.85993	0.86214	1.0
1.1	0.86433	0.86650	0.86864	0.87076	0.87286	0.87493	0.87698	0.87900	0.88100	0.88298	1.1
1.2	0.88493	0.88686	0.88877	0.89065	0.89251	0.89435	0.89617	0.89796	0.89973	0.90147	1.2
1.3	0.90320	0.90490	0.90658	0.90824	0.90988	0.91149	0.91309	0.91466	0.91621	0.91774	1.3
1.4	0.91924	0.92073	0.92220	0.92364	0.92507	0.92647	0.92785	0.92922	0.93056	0.93189	1.4
1.5	0.93319	0.93448	0.93574	0.93699	0.93822	0.93943	0.94062	0.94179	0.94295	0.94408	1.5
1.6	0.94520	0.94630	0.94738	0.94845	0.94950	0.95053	0.95154	0.95254	0.95352	0.95449	1.6
1.7	0.95543	0.95637	0.95728	0.95818	0.95907	0.95994	0.96080	0.96164	0.96246	0.96327	1.7
1.8	0.96407	0.96485	0.96562	0.96638	0.96712	0.96784	0.96856	0.96926	0.96995	0.97062	1.8
1.9	0.97128	0.97193	0.97257	0.97320	0.97381	0.97441	0.97500	0.97558	0.97615	0.97670	1.9
2.0	0.97725	0.97778	0.97831	0.97882	0.97932	0.97982	0.98030	0.98077	0.98124	0.98169	2.0
2.1	0.98214	0.98257	0.98300	0.98341	0.98382	0.98422	0.98461	0.98500	0.98537	0.98574	2.1
2.2	0.98610	0.98645	0.98679	0.98713	0.98745	0.98778	0.98809	0.98840	0.98870	0.98899	2.2
2.3	0.98928	0.98956	0.98983	0.99010	0.99036	0.99061	0.99086	0.99111	0.99134	0.99158	2.3
2.4	0.99180	0.99202	0.99224	0.99245	0.99266	0.99286	0.99305	0.99324	0.99343	0.99361	2.4
2.5	0.99379	0.99396	0.99413	0.99430	0.99446	0.99461	0.99477	0.99492	0.99506	0.99520	2.5
2.6	0.99534	0.99547	0.99560	0.99573	0.99585	0.99598	0.99609	0.99621	0.99632	0.99643	2.6
2.7	0.99653	0.99664	0.99674	0.99683	0.99693	0.99702	0.99711	0.99720	0.99728	0.99736	2.7
2.8	0.99744	0.99752	0.99760	0.99767	0.99774	0.99781	0.99788	0.99795	0.99801	0.99807	2.8
2.9	0.99813	0.99819	0.99825	0.99831	0.99836	0.99841	0.99846	0.99851	0.99856	0.99861	2.9
3.0	0.99865	0.99869	0.99874	0.99878	0.99882	0.99886	0.99889	0.99893	0.99896	0.99900	3.0
3.1	0.99903	0.99906	0.99910	0.99913	0.99916	0.99918	0.99921	0.99924	0.99926	0.99929	3.1
3.2	0.99931	0.99934	0.99936	0.99938	0.99940	0.99942	0.99944	0.99946	0.99948	0.99950	3.2
3.3	0.99952	0.99953	0.99955	0.99957	0.99958	0.99960	0.99961	0.99962	0.99964	0.99965	3.3
3.4	0.99966	0.99968	0.99969	0.99970	0.99971	0.99972	0.99973	0.99974	0.99975	0.99976	3.4
3.5	0.99977	0.99978	0.99978	0.99979	0.99980	0.99981	0.99981	0.99982	0.99983	0.99983	3.5
3.6	0.99984	0.99985	0.99985	0.99986	0.99986	0.99987	0.99987	0.99988	0.99988	0.99989	3.6
3.7	0.99989	0.99990	0.99990	0.99990	0.99991	0.99991	0.99992	0.99992	0.99992	0.99992	3.7
3.8	0.99993	0.99993	0.99993	0.99994	0.99994	0.99994	0.99994	0.99995	0.99995	0.99995	3.8
3.9	0.99995	0.99995	0.99996	0.99996	0.99996	0.99996	0.99996	0.99996	0.99997	0.99997	3.9

Table 2 Percentage points of the normal distribution

The table gives the values of z satisfying $P(Z \leqslant z) = p$, where Z is the normally distributed random variable with mean $= 0$ and variance $= 1$.

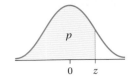

p	0.00	0.01	0.02	0.03	0.04	0.05	0.06	0.07	0.08	0.09	p
0.5	0.0000	0.0251	0.0502	0.0753	0.1004	0.1257	0.1510	0.1764	0.2019	0.2275	0.5
0.6	0.2533	0.2793	0.3055	0.3319	0.3585	0.3853	0.4125	0.4399	0.4677	0.4959	0.6
0.7	0.5244	0.5534	0.5828	0.6128	0.6433	0.6745	0.7063	0.7388	0.7722	0.8064	0.7
0.8	0.8416	0.8779	0.9154	0.9542	0.9945	1.0364	1.0803	1.1264	1.1750	1.2265	0.8
0.9	1.2816	1.3408	1.4051	1.4758	1.5548	1.6449	1.7507	1.8808	2.0537	2.3263	0.9

p	0.000	0.001	0.002	0.003	0.004	0.005	0.006	0.007	0.008	0.009	p
0.95	1.6449	1.6546	1.6646	1.6747	1.6849	1.6954	1.7060	1.7169	1.7279	1.7392	0.95
0.96	1.7507	1.7624	1.7744	1.7866	1.7991	1.8119	1.8250	1.8384	1.8522	1.8663	0.96
0.97	1.8808	1.8957	1.9110	1.9268	1.9431	1.9600	1.9774	1.9954	2.0141	2.0335	0.97
0.98	2.0537	2.0749	2.0969	2.1201	2.1444	2.1701	2.1973	2.2262	2.2571	2.2904	0.98
0.99	2.3263	2.3656	2.4089	2.4573	2.5121	2.5758	2.6521	2.7478	2.8782	3.0902	0.99

Volume and surface area

Shape	Volume	Surface area
Cone	$V = \dfrac{1}{3}\pi r^2 h$	$A = \pi r l + \pi r^2$
Sphere	$V = \dfrac{4}{3}\pi r^3$	$A = 4\pi r^2$
Pyramid	$V = \dfrac{1}{3}\text{base} \times h$	

Financial calculation – AER

The annual equivalent interest rate (AER), r, is given by

$$r = \left(1 + \frac{i}{n}\right)^n - 1$$

where i is the nominal interest rate, and n the number of compounding periods per year.

Note: the values of i and r should be expressed as decimals.

Financial calculation – APR

The annual percentage interest rate (APR) is given by

$$C = \sum_{k=1}^{m}\left(\frac{A_k}{(1 + i)^{t_k}}\right)$$

where £C is the amount of the loan, m is the number of repayments, i is the APR expressed as a decimal, £A_k is the amount of the kth repayment, t_k is the interval in years between the start of the loan and the kth repayment.

It may be assumed that there are no arrangement or exit fees.

Index